Advances in Fossil Energy R&D 2017

Topical Conference at the 2017 AIChE Annual Meeting

Minneapolis, Minnesota, USA
29 October – 3 November 2017

ISBN: 978-1-5108-5778-0

Printed from e-media with permission by:

Curran Associates, Inc.
57 Morehouse Lane
Red Hook, NY 12571

Some format issues inherent in the e-media version may also appear in this print version.

Printed by Curran Associates, Inc. (2018)

For permission requests, please contact AIChE
at the address below.

AIChE
120 Wall Street, FL 23
New York, NY 10005-4020

Phone: (800) 242-4363
Fax: (203) 775-5177

www.aiche.org

Additional copies of this publication are available from:

Curran Associates, Inc.
57 Morehouse Lane
Red Hook, NY 12571 USA
Phone: 845-758-0400
Fax: 845-758-2633
Email: curran@proceedings.com
Web: www.proceedings.com

Advances in Fossil Energy R&D 2017

Topical Conference at the 2017 AIChE Annual Meeting

Minneapolis, Minnesota, USA
29 October – 3 November 2017

AIChE
The Global Home of Chemical Engineers

TABLE OF CONTENTS

Some abstracts in this book refer to images that were not made available by the publisher

499952 Influence of Solvent Molecular Structure on Energy Consumption of Post-Combustion CO_2 Capture Processes

Abstract:

Influence of Solvent Molecular Structure on Energy Consumption of Post-Combustion CO_2 Capture Processes

Kevin G. Joback[1] J.R. Heberle[2] and Abhoyjit S. Bhown[2]

[1]Molecular Knowledge Systems, Bedford, NH 03110-0755 USA

[2]Electric Power Research Institute, Palo Alto, CA 94304, USA

We present an updated physical property-based model for calculating the energy requirement for the post-combustion capture of carbon dioxide by absorption in aqueous amine solutions for any arbitrary amine. The only input the model requires is the amine's molecular structure.

Equations 1 through 6 show the chemical reactions between carbon dioxide, water and an amine:

$$CO2g \rightarrow CO2aq \quad (1)$$

$$2H2Ol \rightarrow H3O++OH- \quad (2)$$

$$CO2aq+2H2Ol->H3O++HCO3- \quad (3)$$

$$H2Ol+HCO3-=H3O++CO3-2 \quad (4)$$

$$RR'R''Naq+H3O+=RR'R''NH++H2Ol \quad (5)$$

$$HCO3-+RR'NH(aq)=H2Ol+RR'NCOO- \quad (6)$$

Primary and secondary amines undergo reactions 5 and 6 while tertiary amines undergo only reaction 6, i.e., tertiary amines do not form carbamates.

Our updated model uses:

- An equation for the Henry's law constant of carbon dioxide in aqueous solvent mixtures (Equation 1). This equation was developed from experimental data on carbon dioxide solubility in water and nitrous oxide solubility in pure amines.
- Equilibrium constants regressed from data on the water dissociation reaction (Equation 2) and the carbon dioxide ionization reactions (Equation 3 and Equation 4).
- A new group contribution technique for estimating a solvent's pKa as a function of temperature (Equation 5).
- A new, general equation for the equilibrium constant of the carbamate formation reaction for primary and secondary amines (Equation 6). This equation requires only knowledge of the amine's molecular structure; no additional experimental data is needed.
- A group contribution methods to estimate the liquid-phase activity coefficients.
- A group contribution method to estimate pure component heat capacities.
- A simple model used to estimate the heat capacity of aqueous amine solutions loaded with carbon dioxide.

Combined, these equations are used to calculate the concentration of chemical species, carbon dioxide partial pressure and the enthalpies of reaction over a range of solvent concentrations, loadings, and system temperatures. Finally, the effective heat capacity is used to calculate the equivalent electrical load (reduction in net electrical output from a power plant) using a simplified process-independent model.

The development of the model was guided by a simplified sensitivity analysis. We varied values for parameters and inputs—e.g., Henry's law constants, pKa as a function of temperature, equilibrium constants, activity coefficients, and heat capacities—to identify those parameters that had the greatest affect on overall model accuracy. Those parameters with the greatest significance became the focus of improvement. The final model developed has good accuracy when compared with experimental data for species concentrations, partial pressures, and enthalpies of reaction for several common aqueous amine solvents. We also show that our new model has accuracy comparable to several current models that have been regressed directly from experimental data.

While the model uses multiple equations and regressed parameters, we reemphasize that the only input needed to evaluate the performance of any amine solvent is the amine's molecular structure. We used our model to computationally evaluate hundreds of candidate solvent structures. For the most promising candidates we calculated the species concentrations, partial pressures, enthalpies of reactions and imposed loads. From these calculations, we have drawn some general conclusions on the influence basic physical properties on the energy consumption of post-combustion carbon dioxide capture processes.

503008 Highly Efficient Warm Gas Carbon Capture System for IGCC Power Plants

Abstract:

TDA is developing a novel sorbent that removes CO_2 via physical adsorption from synthesis gas. The relatively strong affinity of the sorbent to CO_2 enables effective operation at temperatures up to 300°C (well above the dew point of synthesis gas stream generated by commercial gasifiers). However, because the sorbent and the CO_2 do not form a true covalent bond, the energy needed to regenerate TDAâ€™s sorbent (5.4 kcal per mol of CO_2) is much lower than that observed for either chemical absorbents (e.g., 29.9 kcal/mol CO_2 for sodium carbonate) or amine-based solvents (e.g., 14.2 kcal/mol CO_2 for monoethanolamine). TDAâ€™s sorbent can be regenerated isothermally and CO_2 can be recovered at pressure (as high as 150 psia). Thus, the energy needed to regenerate the sorbent and compress the CO_2 for sequestration is significantly lower than that for any other technology reported to date. The high surface area and favorable porosity of the sorbent also provides a unique platform to introduce additional functionality, such as active groups to catalyze the water-gas-shift (WGS) reaction or to remove trace metals (e.g., Hg, As). Hence, TDAâ€™s CO_2 sorbent could be used to do a warm gas capture of CO_2 above the dew point of the synthesis gas in an Integrated Gasification Combined Cycle (IGCC) power plant. TDA's pre-combustion CO_2 capture system achieves a net plant efficiency of 34% on a coal high heating value basis when operated on bituminous coal, which is significantly higher than that can be achieved with the Selexol™ scrubber at 31.6% (state-of-the-art pre-combustion carbon capture technology). The levelized cost of electricity including the transport, storage and monitoring costs for CO_2 is calculated as \$92.9/MWh for TDA's system, in comparison to \$105.2/MWh for Selexol™. In a current DOE sponsored project, TDA is carrying out field tests with the warm gas CO_2 capture unit at 0.1 MWe scale at the National Carbon Capture Center (NCCC) in Wilsonville, AL and has so far completed over 500 hours of field tests achieving greater than 90% carbon capture. In this paper we present the results from the system analyses and the field test with the 0.1 MW_e pilot unit at NCCC

486301 the Development of Data Mining-Molecular Modeling Approach to Screen Physical Solvents for Gas Separation

Abstract:

Integrating the design of both process and solvent for gas separation involves simultaneously optimizing the process structures, operating conditions and solvent molecules. This task is challenging partly due to the large number of non-linear equations for process models, the large number of variables (such as the operating conditions) and the large solvent chemical space to be explored. One way to alleviate this complicated problem is through sub-optimizations by restricting the search space in each step of process design and solvent development. In this work, we focus on solvent design to physically and selectively absorb one gas from the other in the gas mixture. Specifically, a hybrid computational approach was developed to systematically screen solvents to separate CO_2 from a fuel gas mixture which contains CO_2, H_2 and H_2O. The objective of this work is to search for solvents which have desired properties such as hydrophobicity, high CO_2 loading (volume based), high CO_2/H_2 solubility selectivity, non-foaming, low viscosity, low heat capacity, good thermal stability, no environmental and safety issues, and cheap price. The computational approach involves searching the NIST database for pure organic compounds, the in-house computational database development, and atomistic simulations. By using this computational approach, several promising solvents have been identified. For example, we have identified one super-hydrophobic solvent, which has a high boiling point of ~275 °C, large CO_2 loading of 1.06 mol/MPa/L at 25 °C, CO_2 heat of absorption of 10.6 kJ/mol (not high), large CO_2/H_2 solubility selectivity of 50 at 25 °C, low viscous (~4 cP) at ambient condition, heat capacity of 1.64-1.87 J/g°C at 25-100 °C (not high and much less than the heat capacity value for water), non-foaming, and reasonably cheap. The predictions are being confirmed by the experiments at NETL. In addition, we will present the largest CO_2 solubility and CO_2/H_2 solubility selectivity in any organic compound, which would be useful to implement the techno-economic analysis in process modeling to estimate the minimum operating and capital costs.

498520 Analysis of Post-Combustion Inertial CO_2 Extraction System

Abstract:

Analysis of Post-Combustion Inertial CO_2 Extraction System

Adam H. Berger[1] Yuqi Wang[1*] Abhoyjit S. Bhown[1]

Robert Kielb,[2] Anthony Castrogiovanni[2] Vladimir Balepin[3]

[1]Electric Power Research Institute (EPRI), Palo Alto, CA, USA

[2]ACENT Laboratories, Bohemia, NY, USA

[3]Orbital ATK, Ronkonkoma, NY, USA

Abstract

One method for removing CO_2 from flue gas that has been proposed for post-combustion CO_2 capture is cryogenic separation. To cool the flue gas to the necessary temperature to remove CO_2 as a solid, Orbital ATK and ACENT Labs have proposed the Inertial CO_2 Extraction System (ICES) as shown in Figure 1. ICES uses the cooling effect from accelerating flue gas to supersonic speeds in order to promote the formation of solid CO_2 particles. The solids can then agglomerate to form larger particles that grow sufficiently to be inertially separated from the remaining gas flow by an appropriately-curved gas flow duct. At the duct curve, CO_2 solids can be removed using a simple knife edge separator, sent to a cyclone, and pressurized in dense-phase, minimizing the energy of compression. The remaining CO_2-free gas stream that is still at high speed in the duct but at low temperature and pressure can recover its pressure in a diffusor for emission to the atmosphere. This work provides a thermodynamic analysis of this process and the effects of condensing material in supersonic, two phase flow in the duct.

When gas is isentropically accelerated to supersonic speeds through a nozzle, the temperature and pressure of the gas are significantly lowered to a point where CO_2 forms a solid as described by isentropic supersonic flow equations. However, these simple relations are complicated by the introduction of multi-phase flow, heat of sublimation released during phase change, friction losses, and other non-idealities. In addition, to ensure that the solid particles that are formed can grow to the appropriate mass to inertially migrate towards the solid-capture duct, nucleation sites can be introduced in the form of recycled solid CO_2 particles. In order to analyze the performance of the ICES capture system for post-combustion CO_2 capture, the accelerating flow through a converging-diverging nozzle, solid CO_2 capture, and subsequent pressure recovery in a decelerating diffusor were computationally simulated at a range of initial conditions and with different levels of solid CO_2 recycle.

Figure 1. Inertial CO_2 Extraction System.

To simulate the ICES process, an equation of state for CO_2 valid in the solid-gas region was implemented. This was combined with real-gas property data to calculate the phase behaviour and thermodynamic state of the system at the temperatures and pressures required for solid CO_2 formation. The equation of state with phase equilibrium calculations was used to calculate the thermodynamic state of the process at each velocity during acceleration and deceleration. The solids separation and gas deceleration in the diffusor were subsequently calculated as well using similar techniques. The simulation results were verified through forwards and backwards comparison to isentropic flow, single-phase supersonic flow, and Rayleigh flow. This integrated simulation provides an understanding of the ideal flow case and allows analysis of the ICES process.

This paper will present the results of these calculations, including the effect of inlet temperature, inlet pressure, and the presence of a recycled medium to increase nucleation. Several lessons learned from the simulations and parametric tests of the process will be described. Thermodynamic limits to supersonic cryogenic CO_2 capture processes will also be discussed, for post combustion CO_2 capture from coal-derived flue gas as well as for other gas streams of interest.

*Present Address: Department of Mechanical Engineering, University of California, Los Angeles

499031 Process Modeling and Optimization of a Novel Membrane-Assisted Chilled Ammonia Process for CO_2 Capture

Abstract:

Recent focus on climate change and clean energy has led to a large research focus into the area of post-combustion carbon capture. According to the Intergovernmental Panel on Climate Change, fossil fuels are a leading cause on climate change[1]. Monoethanolamine (MEA) has been the standard of solvents used for carbon capture but there has been recent interest in ammonia. Ammonia is an attractive substitute due to its low energy requirements for regeneration, high CO_2 loading capabilities, and its resistance to oxidative and thermal degradation[2]. One drawback of ammonia is the solvent loss due to the high vapor pressure of ammonia. Two methods, implemented separately or concurrently, are currently being employed to reduce absorber slip for the ammonia system; operating the absorber at low temperatures as done in what is known as the chilled ammonia process (CAP) to lower ammonia vapor pressure, and adding a wash system to remove ammonia from the flue gas before it is released to the atmosphere. This work focuses on developing a full process model of the CAP coupled with a water wash section to reduce the ammonia loss. To reduce the energy penalty in the stripper for the water-wash section as well as CO_2 regenerator, a novel membrane-based water separation technique is investigated.

The electrolyte NRTL model is used to calculate liquid properties and PC-SAFT equation of state to calculate vapor properties. A rate-based model of the towers with two-film characterization and true components is developed. Model validation for the CAP process has been done in the literature but is relatively sparse. Due to the extreme non-ideality of these systems, parameter regression can greatly improve the performance of these models. Simultaneous regression of mass transfer coefficients, interfacial area, diffusivity, and reaction kinetics are carried out by using wetted wall column (WWC) and pilot plant data together. This results in a model that is valid over a large range of operating conditions and equipment sizes. Furthermore, a model of a reverse osmosis membrane is developed to be used in the water wash and CO_2 desorber sections. The membrane model is developed for multi-component separation following the solution diffusion mechanism. Finally, the capture system is optimized to minimize the annuitized cost by considering tradeoffs between the capital cost of the membranes and the energy benefits that are achieved by using them.

REFERENCES:

1 IPCC, 2014: Climate Change 2014: Synthesis Report. *Contribution of Working Groups I, II and III to the Fifth Assessment Report of the Intergovernmental Panel on Climate Change [Core Writing Team, R.K. Pachauri and L.A. Meyer (eds.)]*. IPCC, Geneva, Switzerland, 151 pp.

2 Puxty,G., Rowland, R., Attalla, M., Comparison of the rate of CO2 absorption into aqueous ammonia and monoethanolamine. *Chemical Engineering Science*. 2010; 65, 915-922

495282 Design and Operations Optimization of Membrane Separation for Flexible Carbon Capture from Natural Gas Combined Cycle Systems

Abstract:

We explore the concept of flexible carbon capture using membrane separation. Flexible carbon capture has been studied in recent years as a measure to decarbonize fossil-fired power generation in response to electricity market conditions, including varying electricity prices and/or fluctuating electricity supply requirements. The importance of flexible operation of carbon capture systems is highlighted by the increasing penetration of intermittent energy generation from renewable sources such as wind and solar PV.

To accommodate the integration of renewable energy into the grid, fossil-fired power plants would need to generate varying electricity load, producing flue gas with varying characteristics and at varying volumes. Carbon capture systems in a high-renewables grid will be required to operate flexibly to respond to these changes.

Flexible carbon capture has been studied for amine absorption, [1, 2] a leading CCS technology. This work explores the possibility of flexible carbon capture using membrane separation, a promising alternative to amine absorption. In particular, membranes are considered to be very responsive to system changes and require very short start-up times. [3] In addition, membrane separation has advantages such as having a smaller footprint, being more environmentally benign (no corrosive chemicals involved in the separation process), and potentially incurring lower separation energy.

In this work, we perform optimization to determine the optimal process design and time-varying operations of a polymeric membrane system separating CO_2 from a natural gas combined cycle (NGCC) with wind energy integration. The total net present value (NPV) of the gas turbines and membrane system is maximized. Both design and operations of the capture plant are optimized. Design decision variables include membrane process configuration, membrane size, CO_2/N_2 selectivity and CO_2 permeance (two key membrane properties relevant to separation performance), and compressor and vacuum pump sizes. These parameters are determined before a capture unit is built. After the plant is built, operational decision variables include gas flowrates, the pressure ratio across the membrane and permeate-side (low-pressure side) pressure. These are parameters that can be adjusted, within limits, given electricity market conditions for a given time period.

Time-varying electricity output from gas turbines as well as the associated flue gas flowrate and composition will be determined by HyPPO, an in-house software developed at Stanford University for modeling and optimization of flexible and renewable-integrated power systems. HyPPO models beneficial operating strategies for a set of statistically representative days. HyPPO results will be used with membrane separation models for various process configurations as modeled in MATLAB.

[1] Cohen, S.M., G.T. Rochelle, and M.E. Webber. "Optimizing post-combustion CO_2 capture in response to volatile electricity prices." *Int. J. Greenhouse Gas Control* 8 (2012): 180â€“195.

[2] Mac Dowell, N., and N. Shah. "The multi-period optimisation of an amine-based CO_2 capture process integrated with a super-critical coal-fired power station for flexible operation." *Comput. Chem. Eng.* 74 (2015): 169â€“183.

[3] Brunetti, A., F. Scura, G. Barbieri, and E. Drioli. "Membrane technologies for CO_2 separation." *J. Membr. Sci.* 359 (2010): 115â€“125.

494893 Technical and Economic Feasibility Study of Membranes for Carbon Capture from Low CO_2 Concentration Sources

Abstract:

In the past decade, membranes and membrane processes have been studied extensively for carbon capture from various gas sources in the energy sector, such as syngas (~35% CO_2) from an integrated gasification combined cycle, flue gas (11 â€“ 15% CO_2) in coal-fired power plants, and gas turbine flue gas (~4% CO_2) of a natural gas combined cycle. However, there are other more challenging candidates for carbon capture where the CO_2 concentration is low but the total CO_2 emission is large. In this presentation, a first-of-a-kind membrane process is proposed for carbon capture from low CO_2 concentration (~1%) sources. The low CO_2 concentration sources have included the treated flue gas in coal-fired power plants in which 90% CO_2 has been removed by a primary carbon capture unit, e.g., membrane or aqueous amine absorption, as well as coal mine gas streams. The low CO_2 concentration feed gas is passed to a 2-stage enriching membrane cascade where the CO_2 is sequentially enriched to 95% purity. Vacuum is pulled on the permeate sides of both membrane stages to provide the transmembrane driving force. The techno-economic analysis shows that a high CO_2/N_2 selectivity is crucial to achieve the 95% CO_2 purity. If the CO_2 permeance is higher than 1000 GPU (1 GPU = 10^{-6} cm^3 (STP)·cm^{-2}·s^{-1}·$cmHg^{-1}$), a minimum selectivity of 80 is required. A high selectivity of about 200 is preferred from a cost and energy standpoint. Further increasing the selectivity shows little benefit. By using the state-of-the-art amine-containing facilitated transport membrane with a CO_2 permeance of 1100 GPU and a CO_2/N_2 selectivity of 200, a capture cost of $294.7/tonne CO_2 (in the 2011 dollar basis) can be achieved for 90% CO_2 recovery. This corresponds to an increase in the cost of electricity of 21.2%. The membrane process is feasible for CO_2 mitigation with a feed CO_2 concentration spectrum of 0.6 â€“ 2.0%. If the feed CO_2 concentration falls within this range, 90% CO_2 recovery and 95% purity can be achieved with similar total capital and operational expenditures. It is also found that the 90% CO_2 removal may not be the most economic option for low CO_2 concentration sources. Reducing the CO_2 recovery leads to a smaller energy penalty, and a lowest capture cost is reached at 70% recovery. These cost calculations in the 2011 dollar basis have included both membrane module installation cost and 20% process contingency.

492531 Doping Polybenzimidazole (PBI) with Phosphoric Acid to Enhance Membrane H_2/CO_2 Separation Properties

Abstract:

Membranes that permeate H_2 and reject CO_2 at temperatures above 150 °C are of great interest for low cost CO_2 capture in pre-combustion processes. The current leading polymeric material for this application is poly[2,2'-(m-phenylene)-5,5'-bisbenzimidazole] (PBI), which exhibits good thermal stability, strong size-sieving ability and good H_2/CO_2 selectivity. In this study, we demonstrate that H_2/CO_2 selectivity in PBI can be significantly enhanced by doping with phosphoric acid (H_3PO_4). H_3PO_4 can bond strongly with imidazole rings in PBI, and the complexes are thermally stable up to 200 °C under vacuum. We prepare a series of H_3PO_4 doped PBI samples by immersing PBI thin films (12 μm) in solutions containing H_3PO_4 and methanol. The H_3PO_4 concentration can be varied from 0.05 wt.% to 1.0 wt.% to achieve different doping levels in the PBI films (0.3 - 1.0, defined as the molar ratio of H_3PO_4 to PBI repeating unit). Increasing the doping level increases H_2/CO_2 selectivity and decreases H_2 permeability. For example, pure PBI exhibits pure-gas H_2 permeability of 27 Barrers and H_2/CO_2 selectivity of 14 at 150 °C, while the PBI with a doping level of 0.44 exhibits H_2 permeability of 6.1 Barrers and H_2/CO_2 selectivity of 59 at 150 °C. As the doping level increases to 1.0, the PBI shows an impressive H_2/CO_2 selectivity of 136, though the H_2 permeability decreases to 1.3 Barrers at 150 °C. This performance is above the upper bound in the Robeson's plot for H_2/CO_2 separation. This presentation will also discuss the effect of H_3PO_4 doping on structural changes such as free volume, as well as CO_2 sorption and diffusion, and the structure/property relationships in these H_3PO_4 doped PBI thin films.

496585 The U.S. Department of Energy's R&D Program for Carbon Use and Reuse

Abstract:

The U.S. Department of Energy Office of Fossil Energyâ€™s National Energy Technology Laboratory has an active carbon capture and storage (CCS) program. A focus area within the carbon storage program is Carbon Use and Reuse. The objective of this area is to identify and develop a suite of technologies to beneficially use CO_2 to produce useful products excluding enhanced oil recovery technologies. Identification of candidate chemistries and economically feasible approaches that utilize large amounts of CO_2 as a feedstock for high-demand product is very challenging while also ensuring that the overall process has a smaller greenhouse gas footprint that traditional pathways. Current projects in the portfolio include algae, cements/concretes and novel concepts.

490858 Design of New Bimetallic Catalysts for More Selective CO_2 Hydrogenation to Olefins and Methanol

Abstract:

CO_2 recovered from industrial flue gas can be hydrogenated using H_2 produced from H_2O using renewable energy, for manufacturing chemicals (lower olefins such as ethylene and propylene, methanol, and carbonates), and fuels (such as liquid transportation fuels including gasoline, diesel and jet fuels, methanol/DME, or synthetic natural gas). The challenges are to develop highly selective and active catalysts and energy-efficient catalytic processing, apart from the need to lower the cost of H_2 production using renewable energy. This presentation will discuss some new bimetallic catalysts (M1-M2, M1=Fe, Pd, Co, M2=Co, Cu, Ni, Pd) based on our recent research at Penn State and based on our collaboration with researchers in China, Thailand, and Japan. The hydrogenation of CO_2 over novel bimetallic catalysts can selectively produce either C_2-$C_4^=$ olefins (and C5+ liquid hydrocarbons) using Fe-Co catalysts, or methanol over Pd-Cu catalysts. Nano-structured catalysts based on hollow zeolite-encapsulated Ni nanoparticles can create a highly stable and carbon-resistant catalysts for co-conversion of CO_2 and CH_4 in bio-gas and natural gas to syngas for chemicals and ultra-clean fuels. Some recent advance in industrial CO2 conversion will also be briefly discussed.

497138 Direct Carbonation of $Ca(OH)_2$ Using Liquid and Supercritical CO_2

Abstract:

Direct carbonation of Ca(OH)2 using liquid and supercritical CO2

Daniel Klingenberg, Ali Zolhadr and Joseph J. Biernacki

Abstract:

The production of portland cement is responsible for about 7% of the global carbon dioxide (CO2) emissions. Roughly one ton of CO2 is produced for every ton of cement. The direct carbonation of calcium hydroxide (Ca(OH)2) has recently been suggested as a possible CO2-neutral replacement for portland cement in some applications, e.g. prefabricated cement-based construction components. The art of burning limestone ($CaCO_3$), the formation of slaked lime (calcium hydroxide) and subsequent carbonation has been practiced since at least Roman times as a form of cement. The process, however, is generally too slow to be practical for most modern applications. This study examines the use of liquid and supercritical carbon dioxide to accelerate carbonation. Discs of Ca(OH)2 were compacted at various pressures and thicknesses (different mass) to produce samples of different density and size. The primary goal of this study was to determine the practical limits of carbonation Ca(OH)2 and to glean relevant kinetic and morphological (microstructural) information. X-ray diffraction and electron microscopy was used to characterize carbonated samples.

UCLA Engineering
Civil and Environmental Engineering

CARBON UPCYCLING

CO_2 upcycling by mineralization of a carbonate-based cementation material: Processing-property relationships of CO_2NCRETE™

Prepared By: Gaurav N. Sant, Ph.D.
Associate Professor and Henry Samueli Fellow
University of California, Los Angeles (UCLA), CA 90095
Phone: (310) 206-3084, Email: gsant@ucla.edu

Collaborators: G. Falzone, H. Kweon, I. Mehdipour, B. Wang, et al.
Team Carbon Upcycling UCLA: http://www.co2upcycling.com

Developed for: Session 138: CO_2 use and reuse, AIChE Annual Meeting 2017, Minneapolis, MN, October 30, 2017

UCLA Engineering
Civil and Environmental Engineering

CARBON UPCYCLING

Presentation outline

- Making a case for CO_2 utilization
- Critical requirements for ensuring scalable CO_2 utilization and why construction offers a target of opportunity
- CO_2 utilization by the direct carbonation of $Ca(OH)_2$
- Demonstration of CO_2NCRETE^{TM} production by direct CO_2 conversion from flue gas (i.e., dilute CO_2 streams)
- Summary and conclusions

UCLA Engineering

Civil and Environmental Engineering

CARBON UPCYCLING

Background: Why we should utilize CO_2

- Capture costs $ 60-90 per ton *(2016)*
- There are additional costs associated with disposing CO_2 in reservoirs (e.g., saline aquifers, O&G reservoirs ...)
- Questions regarding permanence of sequestration solution, and suitable sites are not always readily accessible
- Unless "sovereign guarantees" are provided CO_2 sequestration is difficult
- There is a need to mitigate industrial CO_2 emissions, and create a waste-to-value or carbon-to-value economy

UCLA Engineering

Civil and Environmental Engineering

CARBON UPCYCLING

Why I count on concrete (and you should too)

- Concrete is truly the foundation of modern infrastructure (bridges, ...)
- The production of OPC (ordinary portland cement) at industrial scale in the late 1800s set off our ability to construct like never before
- Why: cheap, abundant, forgiving, and offers an "open time" for transport and placement (casting)
- We produce > 20 billion tons of concrete annually (global) from a production base of > 4 B of OPC

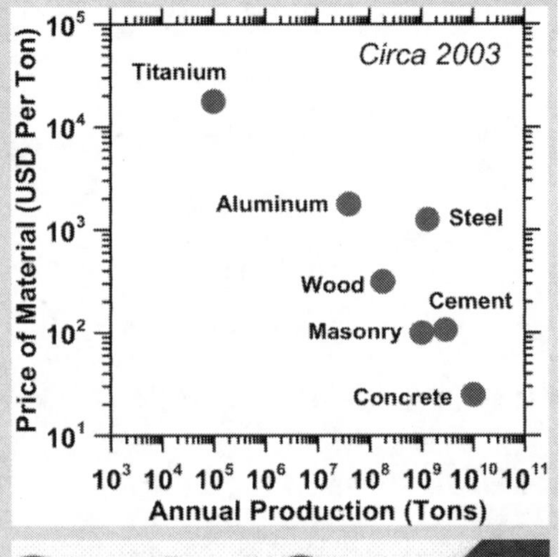

Count on Concrete

UCLA Engineering

Civil and Environmental Engineering

CO2 CARBON UPCYCLING

But CO_2 emissions from OPC production ...

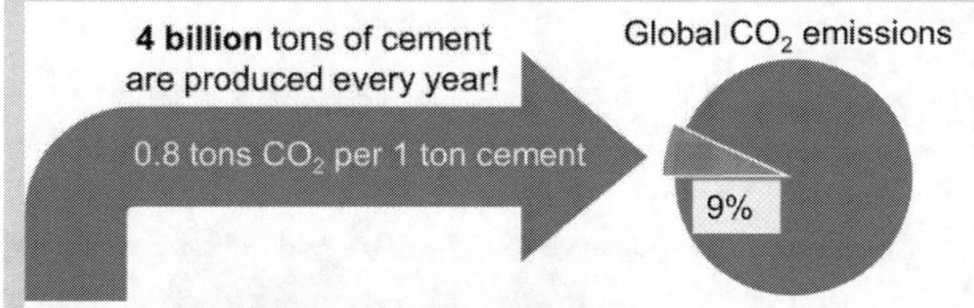

Calcination

$CaCO_3 \rightarrow CaO + CO_2$

- **Industry risk**
- CO_2 penalty
- Impact on OPC pricing
- Impact on construction
- In time, 9% will turn to 25% (2050)

UCLA Engineering

Civil and Environmental Engineering

CO2 CARBON UPCYCLING

What do we do about CO_2 (from cement) ?

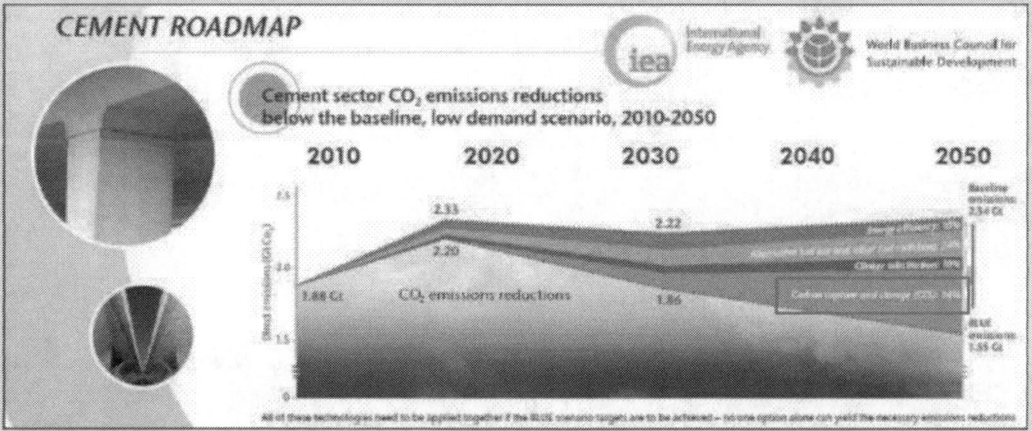

56 % of CO_2 emissions reductions is projected to come from carbon capture and storage (CCS). Clearly, this is unaffordable for a commodity (chemical) which costs $ 100 per ton. This would impede our ability to construct buildings, infrastructure and beyond.

UCLA Engineering
Civil and Environmental Engineering
CARBON UPCYCLING

Why CO_2 mineralization ?

Gaurav Sant | AIChE 2017 Annual Meeting, Minneapolis, MN | Slide 7 of 21

UCLA Engineering
Civil and Environmental Engineering
CARBON UPCYCLING

Mineralizing CO_2 into [Ca, Mg] carbonates

@ Brucite is expected to offer similar carbonation behavior and potential

*** Maintain mineralized CO_2 within closed-loop*

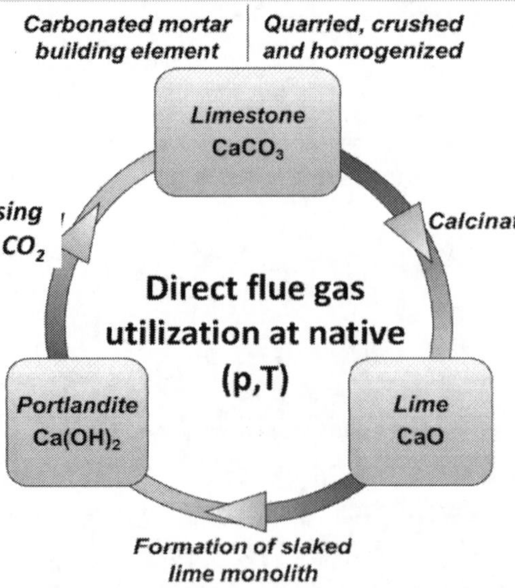

 # Portlandite can take up around 50 mass % CO_2 cheaply ($ 100/t)

CAUTION
Low temperature

** Portlandite can be produced by low-temperature pathways*

Gaurav Sant | AIChE 2017 Annual Meeting, Minneapolis, MN | Slide 8 of 21

Why focus on cement and concrete?

- Gigaton scale CO_2 utilization, e.g., by replacing OPC and concrete aggregates; estimate of 1-to-5 billion tons by 2030
- Low(er) temperature routes for cement production (8% global industry energy)
- Allows use of stack-tap CO_2, i.e., with SO_x and NO_x and no need for capture
- The construction industry offers ready markets and it could accommodate the largest quantity of CO_2 for lowest cost
- Offers a multi-gigaton market without "parasitic energy or financial" losses

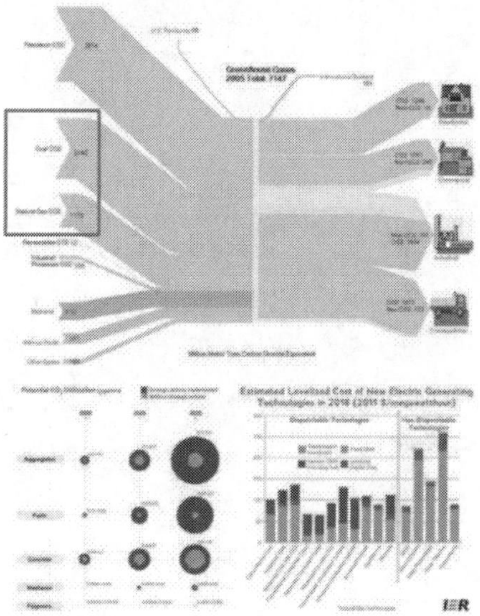

CO_2 mineralization reactions

- The reaction of gas phase $CO_2$with mineral to form products that embed CO_2 within a stable solid
- $Ca(OH)_2$ (and $Mg(OH)_2$) is a very attractive carbonation agent, i.e., far more so than CaO for example
- This approach has been historically used to produce lime mortars, but the rate or carbonation has been far too slow to be practical
- The formation of limestone from through solution reactions is known to offer cementation (sea-shells)

The carbonation product(s)

- When $Ca(OH)_2$ is reacted with CO_2 in gas, liquid, or supercritical states calcite forms as the stable polymorph (of $CaCO_3$)
- Thermogravimetric analysis (TGA) readily quantifies the extent of carbonation reaction (i.e., the quantity of CO_2 that is taken up into the products)
- Water that is released during carbonation is also detected – an important feature in catalyzing portlandite carbonation

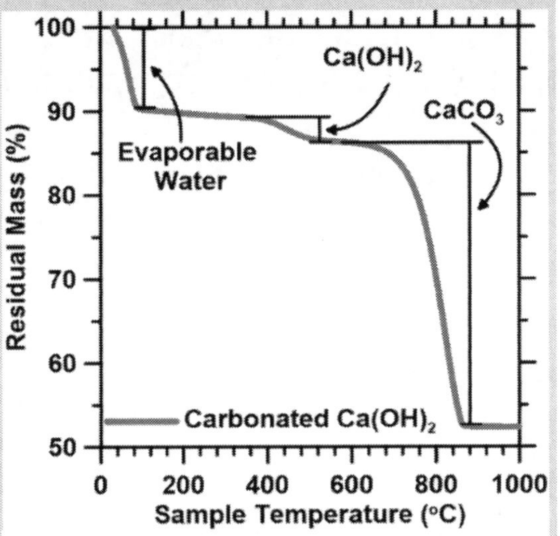

$$Ca(OH)_2 + CO_2 \rightarrow CaCO_3 + H_2O$$

Insights into the carbonation reaction

- Instantaneous surface nucleation
- Solid reactant separated from fluid by boundary layer; process switches to diffusion-control
- Reaction proceeds by OH^--for-CO_3^{2-} exchange (carbonate entry appears to be rate limiting)
- OH^- recombines with H^+ resulting in formation of water – which in turn "catalyzes" the process
- Also confirmed by DFT calculation – 32% reduction in energy (ΔG)

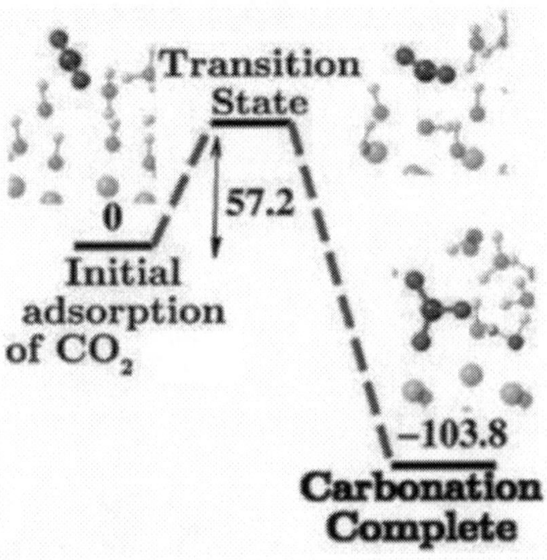

UCLA Engineering
Civil and Environmental Engineering
CO2 CARBON UPCYCLING

Rapid carbonation kinetics

- For $Ca(OH)_2$ particulates that react with pressurized CO_2, carbonation is rapid; reaction achieves 80% completion in 2 h
- Kinetics suggest initial region of surface control (surface area is influential); and a secondary region of diffusion control
- In the secondary step, reaction proceeds by CO_2-for-H_2O exchange (intermediate C and O species; not really "CO_2")

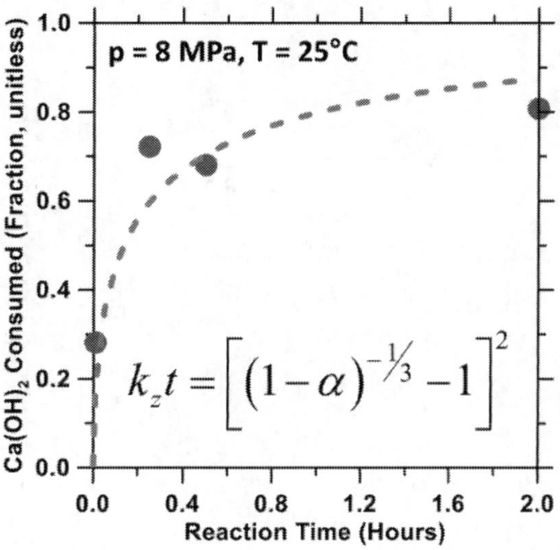

UCLA Engineering
Civil and Environmental Engineering
CO2 CARBON UPCYCLING

Role of (p,T) on portlandite carbonation

- $Ca(OH)_2$ powders, d_{50} = 10 µm
- Little dependence on 'p', slightly more so on 'T'. This is a critical reason for use for $Ca(OH)_2$
- Arrhenius analysis reveals the activation volume (ΔV_R = -27.3 cm^3/mole) and activation energy to be: (E_a = 7.5 kJ/mole)
- Carbonation proceeds rapidly for RH > 20%; so some water (even adsorbed) is required for progress of reaction

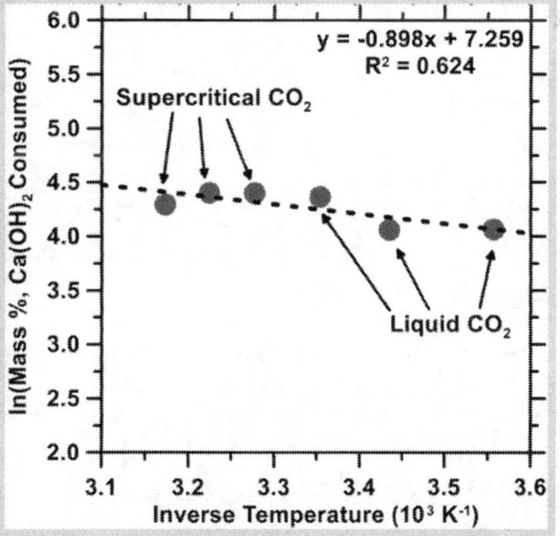

Progress of carbonation at p = 0.1 kPa, T ≥ 45°C

- Examined carbonation progress of bicomponent $Ca(OH)_2$ + OPC slurries formed into cylindrical samples: d x l [100 mm x 150 mm] in 17% CO_2 (v/v) over 24 h
- Previously, slurries were dried at 75°C for 12 hours to react OPC, and create open pore pathways
- In general, progress of reactions is rapid although reaction rate is maximized around 60°C (related to exothermic nature of reaction and Le Chatelier's principle)

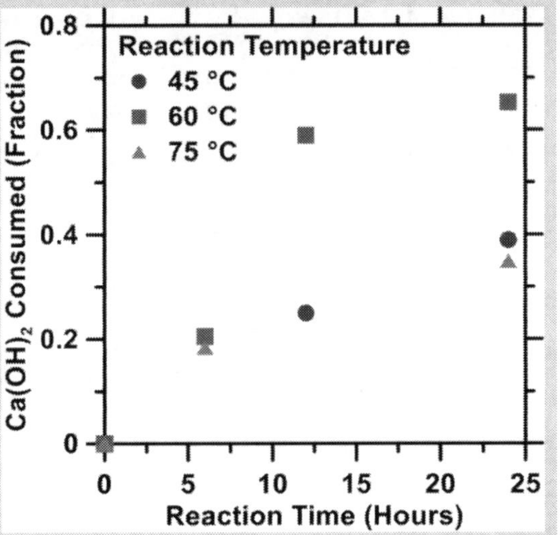

Cementation due to CO_2 mineralization?

- $CaCO_3$ products are expected to bridge and 'cement' sand grains in the style OPC does in concrete
- This is seen in SEM observations
- Based on our data, the gradient in carbonation is small, so limited impact on section properties
- Moreover, carbonation results in a 10% increase in solid volume
- This increase in volume should in-fill the porosity, ensure strength gain with increasing carbonation

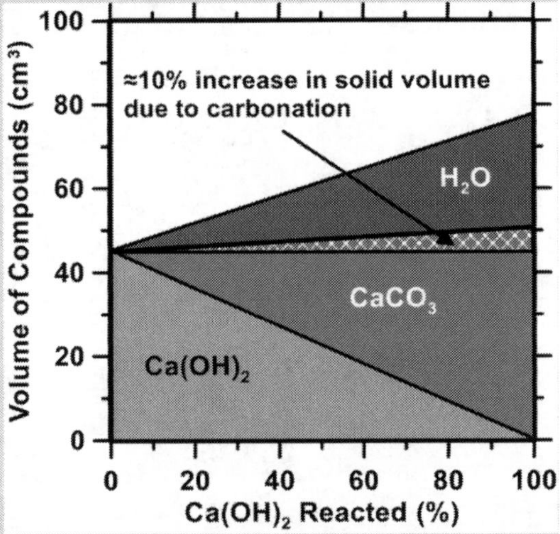

UCLA Engineering

Civil and Environmental Engineering

CO_2 CARBON UPCYCLING

Mechanical strength gain with carbonation

- Created a range of formulations which contain portlandite, fly ash (filler), OPC (to offer green strength), sand, stone and water *a la* concrete
- In general, carbonation results in around 35-to-50% strength increase based on formulation in binders that may contain ≤ 30 mass % OPC, or are OPC-free
- Expect to achieve 35 MPa after 7 days of aging and processing

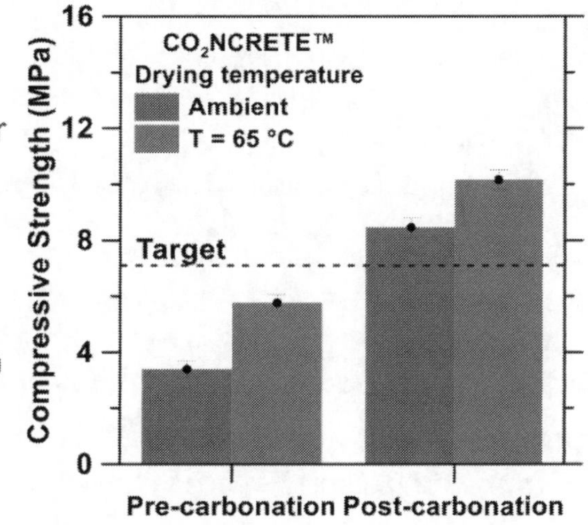

UCLA Engineering

Civil and Environmental Engineering

CO_2 CARBON UPCYCLING

CO_2NCRETE™ pilot-scale demonstration

- Designed, developed and validated system of 6 reactors to process 200 kg of CO_2 in 24 hours
- Input as a N_2-CO_2 mixture for CO_2 concentration range of 5-17% (v/v)
- Seeks to replicate conditions of fossil fuel power plant with heat secured from flue gas; T < 100°C
- System has achieved > 65% CO_2 capture (mineralization) efficiency per 24 hours carbonation period
- Embeds 0.2 g of CO_2 / g of reactant

$CO_2NCRETE^{TM}$ workflow

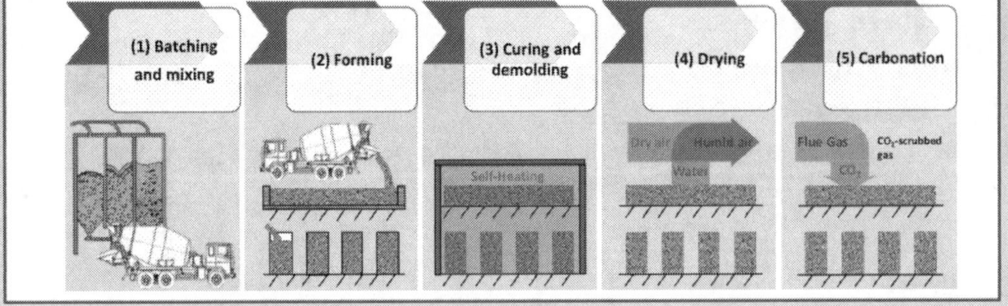

- $CO_2NCRETE$ is batched/mixed using traditional concrete production equipment, and then formed into the desired geometry (e.g., slabs)
- The forms are removed after $CO_2NCRETE$ gains green strength (i.e., ≤ 3 MPa) to maximize surface area that is available for CO_2 contact
- Batch production in parallel reactors allows continuous CO_2 utilization

Ensuring CO_2-neutral cementation ...

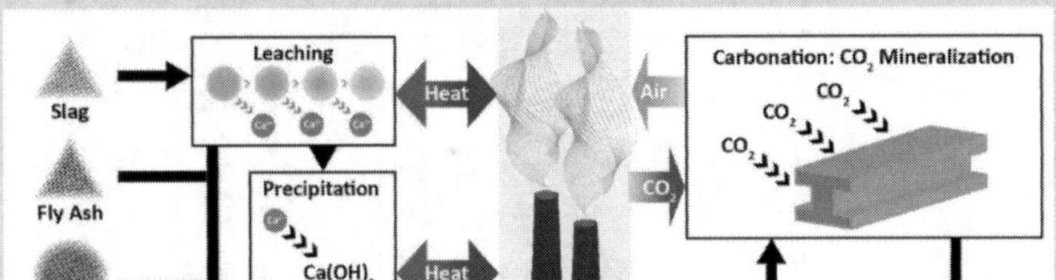

- *Upcycling* low-value industrial wastes into a CO_2-negative concrete
- The process utilizes Ca (and Mg) -extracted from wastes as precursors to produce $Ca(OH)_2$ at low temperatures for CO_2 mineralization

UCLA Engineering

Civil and Environmental Engineering

CO_2 CARBON UPCYCLING

Summary and conclusions

- Direct carbonation of $Ca(OH)_2$ using dilute CO_2 streams is a viable route towards large-scale CO_2 utilization
- CO_2 mineralization occurs rapidly, and results in the rapid development of strength; indicative of cementation
- The process has been scaled up to demonstrate a pilot-scale process 200 kg of CO_2 in a day; while producing a range of element geometries including cylinders and slabs
- The success of this approach provokes new pathways for CO_2 mineralization that could achieve CO_2-neutral (and perhaps CO_2-negative) actions

Gaurav Sant | AIChE 2017 Annual Meeting, Minneapolis, MN | Slide 21 of 21

498101 CO_2 Conversion to Fuels and Chemicals through Microbial Electrolysis Cells

Abstract:

With a fully oxidized carbon atom, CO_2 is notoriously stable under most environmental conditions. Thus, the conversion of CO_2 to fuels and chemicals has been pursued through chemical, photochemical, electrochemical, and biological means, in addition to reforming. Through these different routes, CO_2 can be converted to chemicals such as formic acid, acetic acid, and urea; alcohols such as ethanol and methanol; and gases such as methane, syngas (CO), and acetylene (C_2H_2). The majority of these reactions require a reducing agent, most commonly hydrogen (H_2). However, hydrogen is quite expensive, so the resulting products would not be cost-competitive; thus, the need for alternative approaches is obvious.

To reduce CO_2 to fuels and chemicals, our laboratory has enriched a microbial community that has demonstrated repeatable high rate production of methane and organic acids from CO_2. Two unique features make this process extremely interesting for further exploration, and profoundly attractive for potential technological development. First, CO_2 conversion to value-added products takes place in a one-chamber membrane-less Microbial Electrolysis Cell (MEC). The methane production rate is 10-56x higher than typically reported for methane release from CO_2 in two-chamber MEC units. Second, the enriched microbial community thrives on a defined medium without the presence of any organic carbon and nitrogen sources.

In this talk, we will present our recent data on this project. Results, such as the microbial community composition, the enrichment process, electrode modification and fundamentals related to the single chamber MECs will be discussed. All of these will assist in future research and development in making MEC a promising platform for converting CO_2 to useful commodities.

489037 High-Selectivity, Gas Fermentation of CO_2 to Ethanol

Abstract:

LanzaTech has developed a fully integrated, platform for commercial-scale production of fuels and chemicals from sustainable biomass and waste feedstocks. The platform comprises of a highly-optimized, gas-utilizing host organism, an advanced strain engineering and process optimization pipeline, validated metabolic and process models, and scalable reactor designs with demonstrated performance.

LanzaTechâ€™s innovative gas fermentation platform can capture and utilize CO_2, H_2 and CO for the production of fuels and chemicals. The ability of the fermentation platform to utilize gas mixtures of varying compositions enables the use of a diverse range of feedstocks, including syngas generated from agricultural residues or municipal solid waste and waste gases from industrial sources (e.g. refineries and steel mills).

Refinery waste gases which are primarily composed of H_2 and COÂ
products directly from low value gas streams while reducing CO_2 and air emissions at the refinery site. Furthermore, the ethanol gas fermentation pathway exhibits a high selectivity which significantly exceeds the selectivity of catalytic pathways for conversion of CO_2, H_2 and CO. This ability of the LanzaTech gas fermentation platform to selectively produce ethanol from dilute refinery waste gases provides the next step in the drive to increase efficiency and reduce waste in the refinery setting.

Production of over 30 molecules has been demonstrated, adding value to carbon emissions and highlighting the opportunities that exist through carbon capture and utilization technologies.

Proprietary, scalable reactor designs and optimized process chemistry, ensure efficient, continuous, single-pass gas conversion with a high selectivity to the product of interest. The process has been successfully scaled up from the laboratory bench through in-lab and in-field pilot plants to fully integrated 100,000-gallon/year pre-commercial demonstration plants with a total of over 55,000 hours on stream. First full commercial units are in construction in US, Europe and Asia.

Carbon utilization technologies such as LanzaTechâ€™s offer both feedstock holders a path to add value, and chemical users a path to off take large volumes of recycled carbon products.

493824 Role of Nitrogen Addition to Reduced Graphene Oxide-TiO_2 Nanocompostites in Enhancing CO_2 Photoreduction

Abstract:

Carbon dioxide (CO_2) capture and utilization is a promising potential technology to address global climate change. Photocatalytic conversion of CO_2 into hydrocarbon fuels has attracted abundant research attention in recent years. However, there have been few reports about CO_2 capture and conversion on the same material so far. Addition of nitrogen has been reported to be a feasible approach to enhance materialsâ€™ CO_2 capture performance. In this work, we demonstrated two ways to fabricate the nitrogen contained reduced graphene oxide (NrGO)-TiO_2 nanocomposites and discussed the function of nitrogen surface modification, as well as the nitrogen structure doping. We found that both nitrogen surface modification and structure doping could enhance the CO_2 photoreduction of rGO-TiO_2 materials significantly.

It was demonstrated that NrGO (N surface modified rGO or N structure doped rGO) with an appropriate N quantity and N-bonding configuration acted as a dual-functional promoter, simultaneously enhancing CO_2 adsorption on the catalyst surface and facilitating electron-hole separation, while eventually boosted the photocatalytic performance. Experimental results in this work provide a better understanding of the critical roles of nitrogen addition in the synthesized composites and also inspire the ongoing interest in a better design of other N-contained graphene based materials for photoreduction of CO_2.

Reference:

Lin, L. Y., Nie, Y., Kavadiya, S., Soundappan, T., & Biswas, P. (2017). N-doped reduced graphene oxide promoted nano TiO_2 as a bifunctional adsorbent/photocatalyst for CO_2 photoreduction: Effect of N species. *Chemical Engineering Journal*, *316*, 449-460.

498868 High Performance CO_2 Electrolyzers

Abstract:

This paper will consider the combination of CO_2 electrolysis and GTL processing as a way to convert waste CO_2 into useful fuels and chemicals. First electrolysis is used to convert the CO_2 to CO and O_2

The main emphasis of the talk will be on the CO_2 electrolysis. A cell with optimized cathode and anion exchange membrane has run for over 3500 hrs at 200 mA/cm^2 and 3V at room temperature with stable high selectivity of ~95%. Currents rise to 600 mA/cm^2 at 50°C. A new flow-field design has been created that allows large cells to be easily manufactured Solid. Initial results on coupling the devices to fuel production systems will also be discussed. These results pave the way for the commercialization of CO_2 conversion to useful fuels.

500835 CO Conversion to Selected Chemicals Using Electrochemically Generated CO from CO_2

Abstract:

Dioxide Materials has developed new electrochemical cell technology capable of generating CO from the electrochemical reduction of CO_2 at high current density and Faradic efficiency. The electrochemical cell operates in a current density range of 200 to 600 mA/cm^2 at voltages in theÂ 2.9 â€" 3.2 V range. The conversion selectivity of CO_2 to CO is in the range of 95% to 98%.

The electrochemical cell utilizes a two-compartment design, consisting of an anode and cathode compartment that are separated by a special anion exchange membrane. The anode consists of a GDE utilizing a nanoparticle IrO_2 catalyst deposited onto a carbon fiber paper support and using a potassium bicarbonate anolyte solution that is recirculated in anode flow field. The cathode utilizes a nanoparticle Ag-based catalyst in combination with a Sustainion^{TM} ionomer that is deposited on a GDE carbon paper cathode to efficiently reduce the CO_2feed to CO at a low cathodic overpotential.

The special membrane utilized in the cell design is a highly conductive Sustainionâ„¢ X-37 anion exchange membrane, developed by Dioxide Materials, which allows the transport of OH^- and HCO_3^-ions from the cathode to the anode.

The surprisingly high conversion performance and current density of this electrochemical CO cell at low voltages allows us to consider the use of this technology, when scaled-up, in potentially producing selected renewable chemicals when operating on renewable energy sources such as solar, wind, and nuclear.

In this presentation, we will discuss the details of the new electrochemical cell technology in addition to identifying the production of potential renewable chemicals such as methanol and acetic acid using the cell CO product.

482789 Nano-Engineered Catalyst for the Utilization of CO_2 in Dry Reforming to Produce Syngas

Abstract:

The objective of this study was to develop a novel catalytic reactor containing nano-engineered catalyst for the utilization of CO_2 (captured from coal-fired power plants) in dry reforming of methane (DRM) (CO_2 + CH_4 â†' 2 H_2 + 2 CO) to produce synthesis gas (syngas). The technology aims to reduce CO_2 emissions by developing beneficial uses for CO_2 from coal-fired power plants. It also offers an alternative to mitigate CO_2 emissions in areas where geologic storage may not be an optimal solution and/or utilization could significantly offset the costs of carbon capture and sequestration.Â

The nano-engineered Ni-based catalyst was prepared by atomic layer deposition (ALD). The Ni particles were as small as ~2-4 nm. The nano-engineered catalyst supported on porous Î³-Al_2O_3 particles showed extremely high methane reforming rates at different temperatures (1840 $Lh^{-1}g_{Ni}^{-1}$ at 850 °C, 1740 $Lh^{-1}g_{Ni}^{-1}$ at 800 °C, 1320 $Lh^{-1}g_{Ni}^{-1}$ at 750 °C). These are the highest DRM reaction rates reported to date as compared to data in the literature. The Ni-based ALD catalyst also showed good stability in DRM reaction during a 300-hour continuous operation at temperatures that ranged from 700°C to 850°C. The good stability is due to a strong bonding between the nanoparticles and substrates since the particles were chemically bonded to the substrate during the ALD process. The high thermal stability maintains high dispersion of Ni nanoparticles, which can inhibit coke formation, because their step edges are small enough to limit carbon nucleation and growth.

498262 Ethylene Production from Shale and Coal Derived Flue Gases

Abstract:

As an inexpensive and abundant source of energy, coal, provides ~41%[1] of the worldâ€™s electricity needs. Yet, coal based power plants are under serious scrutiny due to their damaging impact on the environment. In 2015, although coal contributed to 33%[2] of the total electricity production in the U.S., it was responsible for 71%[3] of the total CO_2 emission from the U.S. electric power sector for the same year. One of the most attractive approaches to combat high GHG emission from coal-fired plants is CO_2 capture and conversion to higher value products.

Contemporaneously, thanks to the shale gas growth in the U.S., an excess supply for pure (extracted) ethane and other light hydrocarbons is available currently at low cost. As a result, an increasing number of steam cracker plants are using ethane as raw material to produce ethylene, the highest volume petrochemical intermediate produced in the world with a current global production capacity of 334 billion lb/year. Ethylene production from steam cracking is also an energy intensive process. A typical ethylene steam cracker plant consumes ~17-21 GJ/ton ethylene energy and emits ~1.0-1.2 ton CO_2/ton of ethylene[4].

Southern Research (SR), in partnership with DOE/NETL is developing a low temperature ethylene production process using CO_2, notably from coal-fired flue gas streams, and readily available low cost ethane as raw materials. The proposed process proceeds through a less energy intensive oxidative dehydrogenation pathway by using SRâ€™s recently developed low-cost mixed metal oxide catalyst which balances the functionalities between ethane activation and ethylene selectivity with a goal to ensure yields (>60%) higher than reported in literature. The process is also anticipated to benefit from the presence of several components of the flue gas from coal-fired plants and ensure lower downstream separation costs.Â IT is anticipated that this new process would (i) reduce the reaction temperature for ethylene production by at least 200°C from the conventional steam cracking and (ii) reduce overall GHG emission by ~35% for the combined coal-fired power plant and ethylene production process. With an existing SR catalyst formulation, high selectivity (99%) and high turn-over rates have already been achieved. This paper will discuss the catalyst screening and key performance results for further optimizing catalysts. The studied parameters include reaction temperature, ratio of ethane to oxidant and oxidizing agents (CO_2/O_2), catalyst deactivation, and impact on separations as a result of product selectivity and yield. References

[1] http://data.worldbank.org/indicator/EG.ELC.COAL.ZS, [2] Electric power monthly, Jan 2017, U.S. Energy Information Administration, [3] Monthly energy review. August 2016. U.S. Energy Information Administration. www.eia.gov/mer. [4] Ren, Tao, Martin Patel, and Kornelis Blok. *Energy* 31.4 (2006): 425-451.

503439 CO_2 Utilization in the Production of Ethylene Oxide

Abstract:

A conventional hydrocarbon feedstock that is of particular interest for CO_2 utilization is ethylene, which is used to make ethylene oxide. Ethylene oxide is an important feedstock for the chemical industry and is used to make many useful products such as polyurethanes, polyols, glycols, nitriles, alcoholamines, and ethers. Ethylene oxide is manufactured by several closely related industrial processes by major chemical makers worldwide (e.g., DOW Chemical, Japan Catalytic Company, Shell International Chemicals, Sumitomo Chemical, BASF, Scientific Design).

The chemical methodology by which ethylene oxide is currently produced is the partial oxidation of ethylene using oxygen (this process is called epoxidation). A silver-based catalyst is used, and either air or preferably oxygen is the oxidant. Ethylene epoxidation is performed at lower temperatures (220 to 280°C) compared with other selective oxidation processes and at high pressure (10 to 30 bar).[1]

The Captured CO_2 Catalyst for the Production of Ethylene oxide (C3-PEO), a technology being developed at RTI International, aims at producing ethylene oxideâ€"a high-value chemicalâ€"while consuming CO_2â€"a greenhouse gas. This technology is based on novel catalysts utilizing the following key discoveries:

1. Abstraction of oxygen from CO_2 using the reduced mixed-metal oxide catalysts
2. Transfer of the abstracted oxygen from CO_2 to react with hydrocarbons to form desired products
3. Operation at temperatures which make these catalysts commercially practical to produce ethylene oxide

Our proposed process uses a cofeed reactor as shown in Figure 1. The proposed ethylene oxide process will have a concentrated CO byproduct stream, which could be used for the manufacture of many products such as methanol, dimethyl ether, acetic acid, acetic anhydride, vinyl acetate, styrene, terephthalic acid, formic acid, *n*-butanol, 2-methylpropanal, acrylic acids, neopentylacids, propanoic acid, dimethyl formamide, and Fischer-Tropsch hydrocarbons.[2] Therefore, the two marketable product streams from the proposed process are ethylene oxide and CO, which are both valuable intermediates for the petrochemical industry.

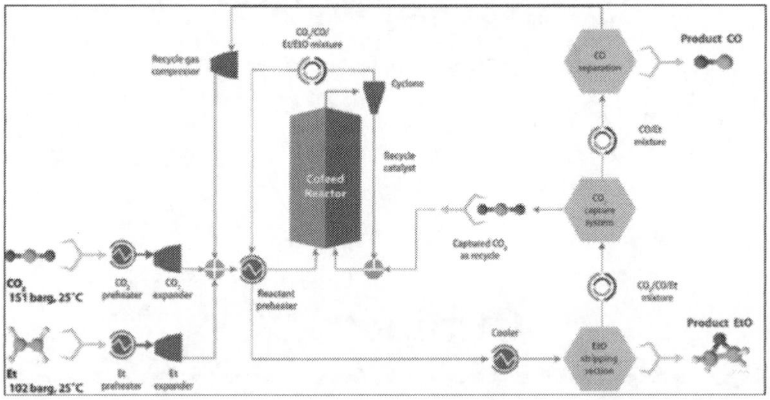

Figure 1. Conceptual schematic of the transport reactor process used in this process.

RTI has improved on its previous catalyst formulations for CO_2 utilization[3] and developed families of catalysts which can both remove oxygen from CO_2 and transfer the oxygen to ethylene to make ethylene oxide. The catalyst families are based on metal oxide phases which were found to be similar to iron in terms of reacting with CO_2, but are more selective than iron for ethylene epoxidation. Improvements on the production of ethylene oxide have been made by the use of promoters, probing the catalyst support to identify a correlation with support acidity, and observing the impact of surface area on dispersion. Ethylene oxide has been produced at small scale in the fixed-bed catalyst test reactor in both a continuous cofeed and a transport mode of operation. However, a cofeed system was found to be optimal and studied for scale-up of the technology.

Life cycle analysis has shown that the process avoids more than 2.8 kg of CO_2 per kg of ethylene oxide produced. Techno-economic analysis results show that the process can produce an IRR above 15%, and sensitivity analyses reveal the most important factors regarding the economic viability of the technology. We have compared the RTI CO_2 oxidation process with the state of art partial oxidation process and found that the new process can be economically favorable depending on market conditions.

(1) Chongterdtoonskul, A.; Schwank, J. W.; Chavadej, S. *Journal of Molecular Catalysis A: Chemical* **2013**, *372*.

(2) Kolb, K. E.; Kolb, D. *Journal of Chemical Education* **1983**, *60*, 57.

(3) Shen, J. P.; Mobley, P. D.; Douglas, L. M.; Peters, J. E.; Lail, M.; Norman, J. S.; Turk, B. *RSC Advances* **2014**, *4*, 45198.

Keywords: CO2 Utilization, Ethylene Epoxidation

FARADAY
TECHNOLOGY, INC.

Pulsed Electrodeposition of High-Activity Gas-Diffusion Electrocatalysts for CO_2 Reduction

2017 AIChE Annual Meeting

Minneapolis, MN

30 Oct 2017

Brian Skinn[a], Sujat Sen[b], Rajeswaran Radhakrishnan[a], McLain Leonard[b], Dan Wang[a], Tim Hall[a], Stephen Snyder[a], Fikile R. Brushett[b], EJ Taylor[a]

[a] Faraday Technology, Inc., Englewood, OH 45315

[b] Dept. of Chemical Engineering, Massachusetts Institute of Technology, Cambridge, MA 02139

BrianSkinn@FaradayTechnology.com

17AIChE
Annual Meeting, Minneapolis, MN

Outline

- Electrochemical Processing
  - Overview
  - Advantages of Pulse-Processing
- Pulsed FARADAYIC® ElectroCatalyzation of GDEs
  - Rationale and Electrodeposition Approach
  - Catalyst Material Selection
  - Prepared GDE Electrocatalysts
  - Electrocatalyst Performance Testing

Electrochemical Processing

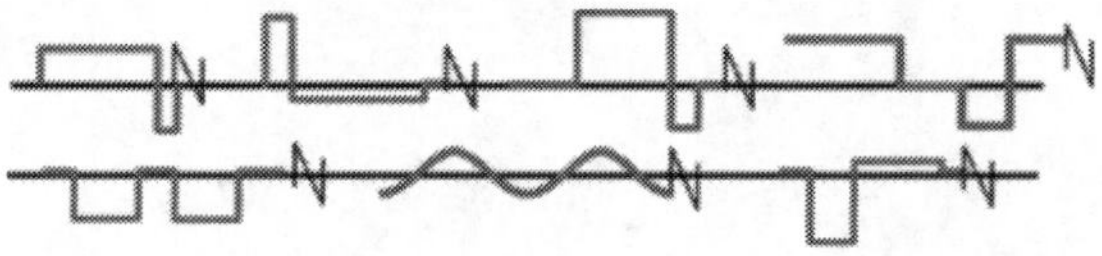

Conventional DC vs Pulse/Pulse-Reverse

Applied Electrochemistry

$$Red \leftrightharpoons n\, e^{-} + Ox$$

Electronic Current

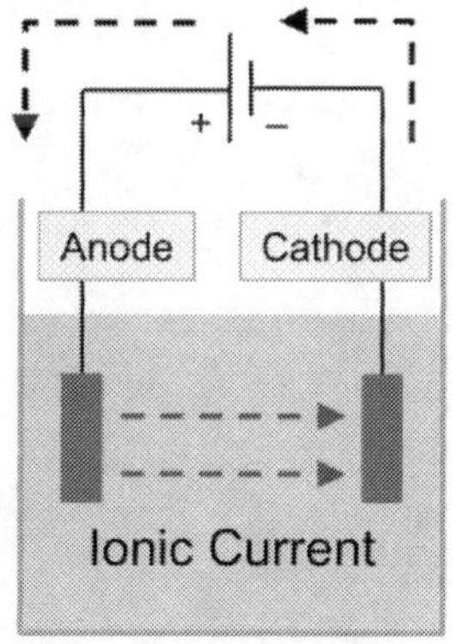

Current is net flow of
POSITIVE CHARGE

Applied Electrochemistry

$Red \leftrightharpoons n\,e^- + Ox$

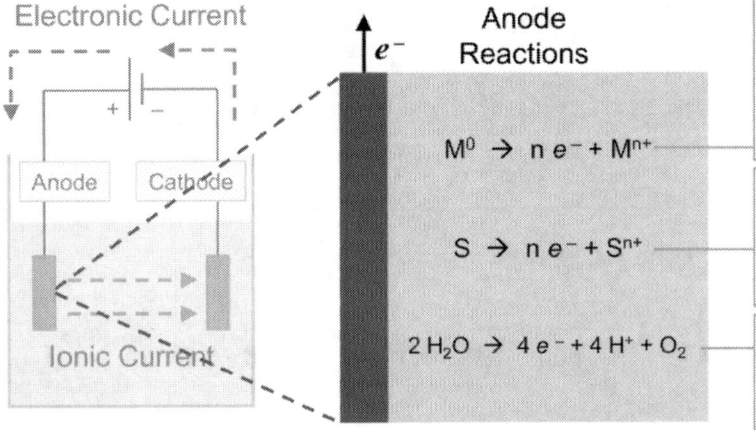

Material Processing
Anodic (subtractive)
Electropolishing

Cathodic (additive)
Electroplating

Chemical Modification
Anodic / Cathodic
Chlor-alkali
CO_2 conversion

Water Electrolysis
Anodic / Cathodic
H_2 / O_2 production
Bubble generation

Current is net flow of
POSITIVE CHARGE

Pulsed Waveforms vs DC

Know: *Linearized Concentration Profile*

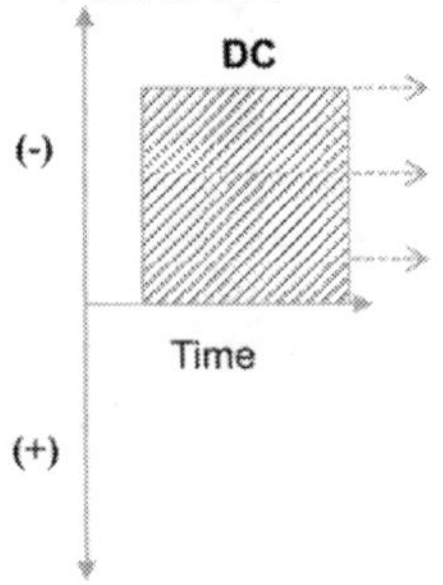

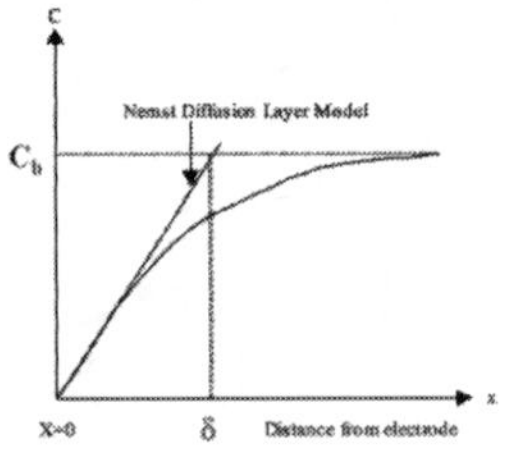

Pulsed Waveforms vs DC

Know: *Linearized Concentration Profile*

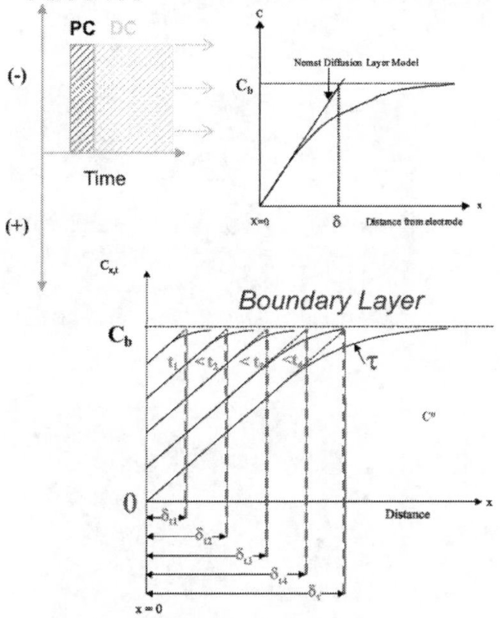

FARADAY TECHNOLOGY, INC.

Pulsed Waveforms vs DC

Know: *Linearized Concentration Profile*

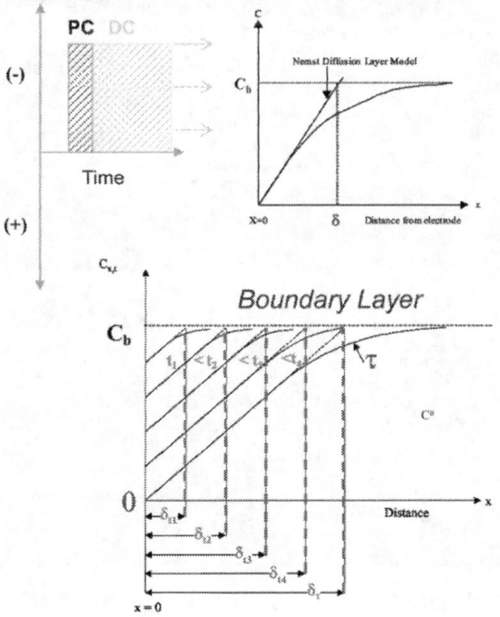

Assume:

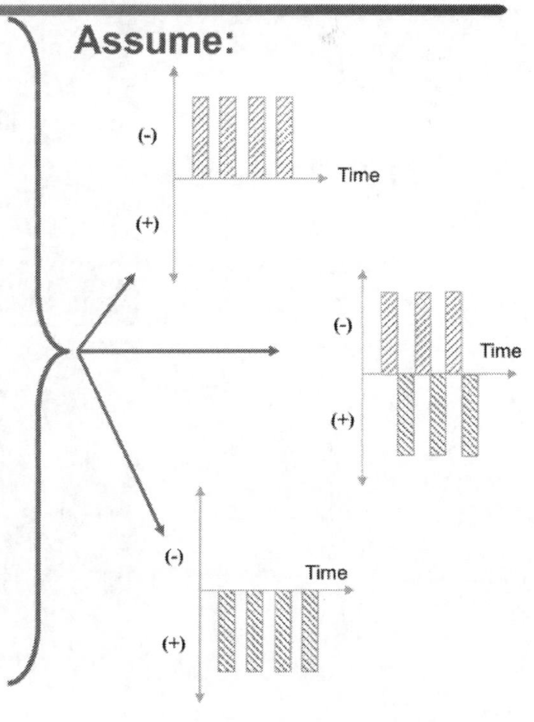

FARADAY TECHNOLOGY, INC.

Prior Work: Copper Electrodeposition for PCBs

Standard DC Recipe

$CuSO_4$
H_2SO_4
"Accelerator"
"Suppressor"
"Leveler"
Other Additives

Pulsed Waveform Recipe

$CuSO_4$
H_2SO_4
Trace Cl^-
Trace PEG

Z-Interconnects Cu-Plated by Pulse-Reverse FARADAYIC® ElectroDeposition

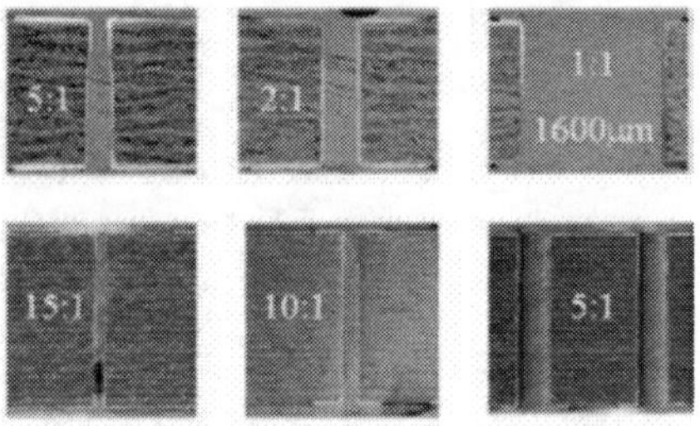

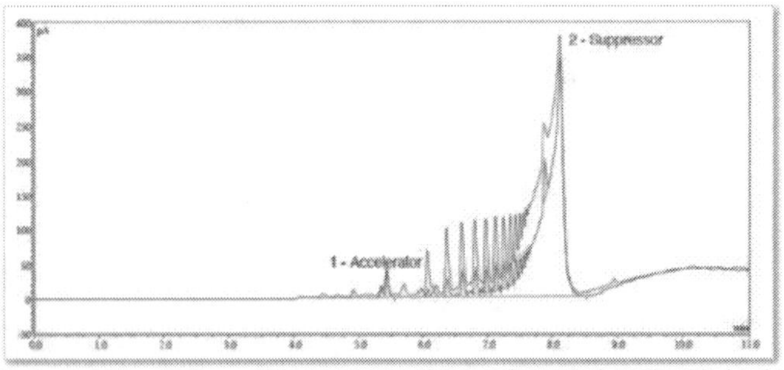

Plante, et al. "Quantization and Characterization of Copper Plating Bath Additives by Liquid Chromatography with Charged Aerosol Detection." Thermo Scientific. http://www.dionex.com/en-us/webdocs/113469-Pittcon12_1820-6_MPlante_PN70008_PlatingBaths.pdf

Prior Work: Lead-Free Tin

Rotating Disk Plating Tool

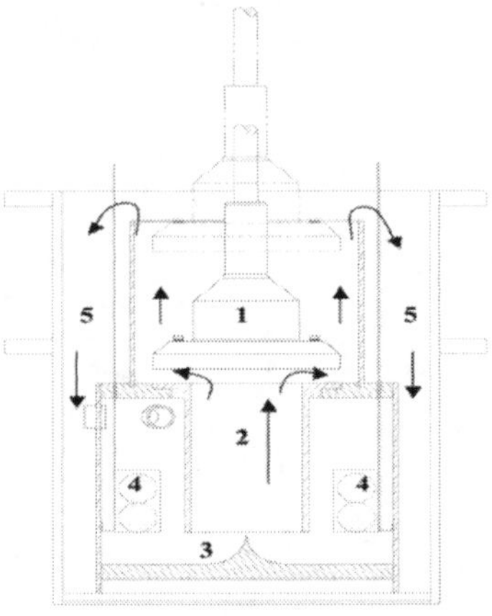

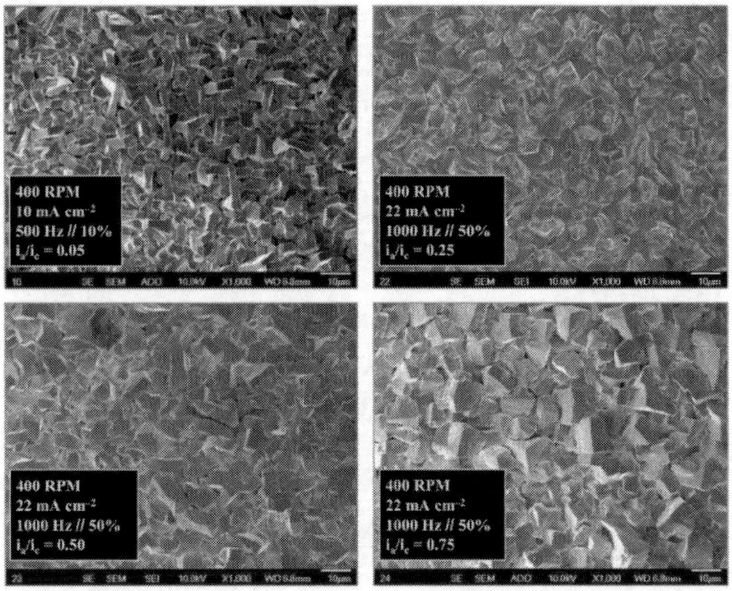

Sn Deposits on Cu Substrates — 1000× Magnification

U.S. Patent 8,603,315 Granted

FARADAY TECHNOLOGY, INC.

FARADAYIC® ElectroCatalyzation

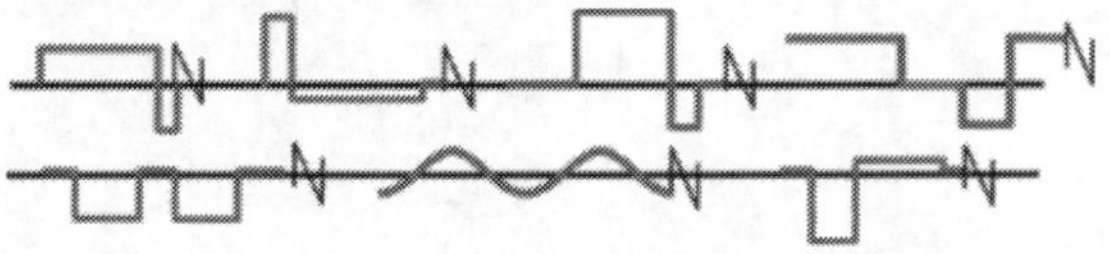

Gas Diffusion Electrode (GDE) Electrocatalysts

FARADAYIC® GDE ElectroCatalyzation

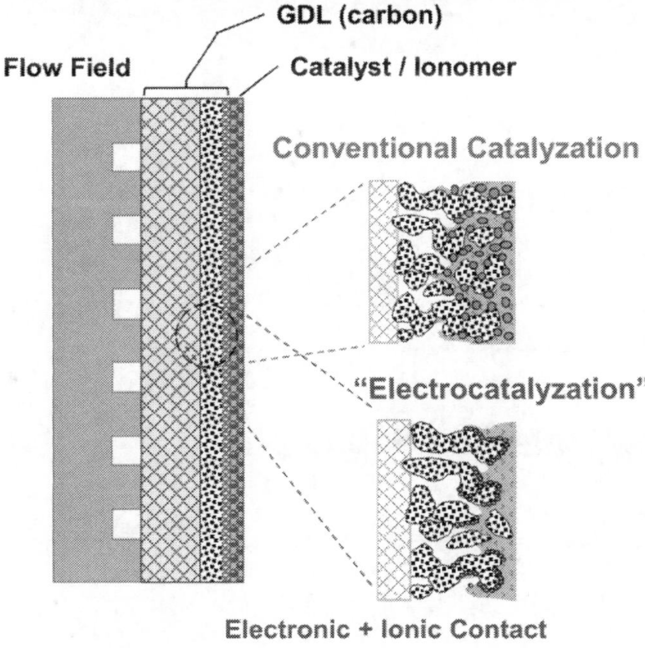

TEM of Pt on Carbon

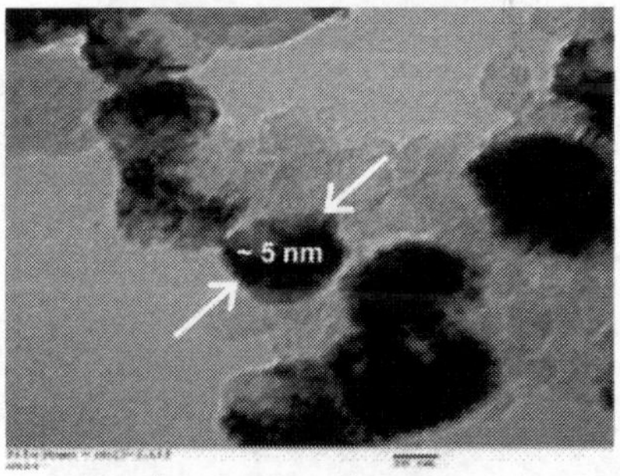

Performance Comparison

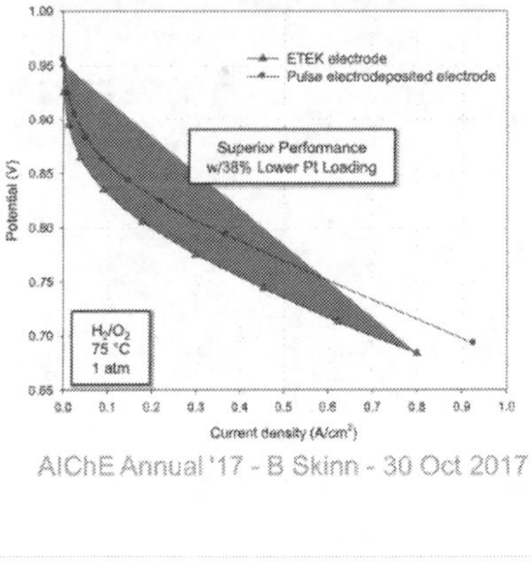

FARADAYIC® GDE ElectroCatalyzation

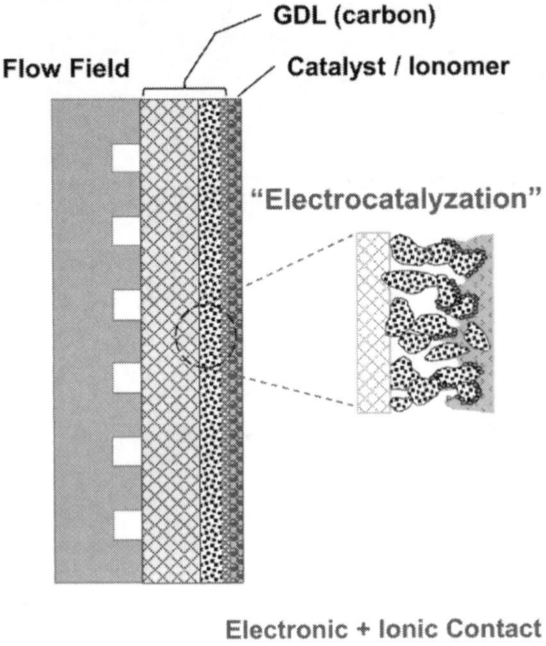

FARADAYIC® GDE ElectroCatalyzation

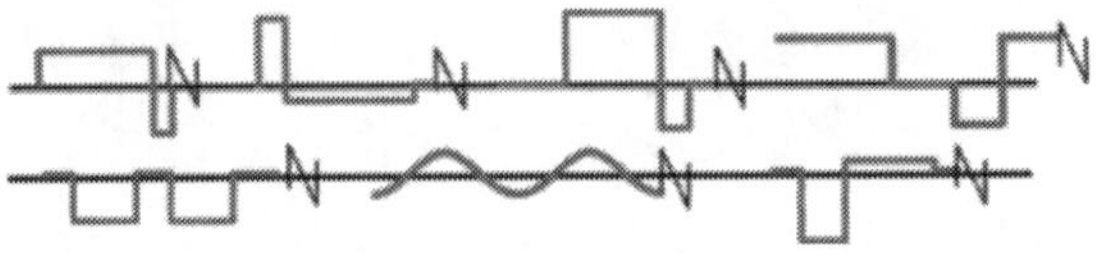

Electrocatalyst Materials Selection

CO_2 Electrocatalyst Selection

Product distribution over single-metal electrocatalysts as determined by Hori et al.
Electrochimica Acta 39(11-12): 1833 (1994).

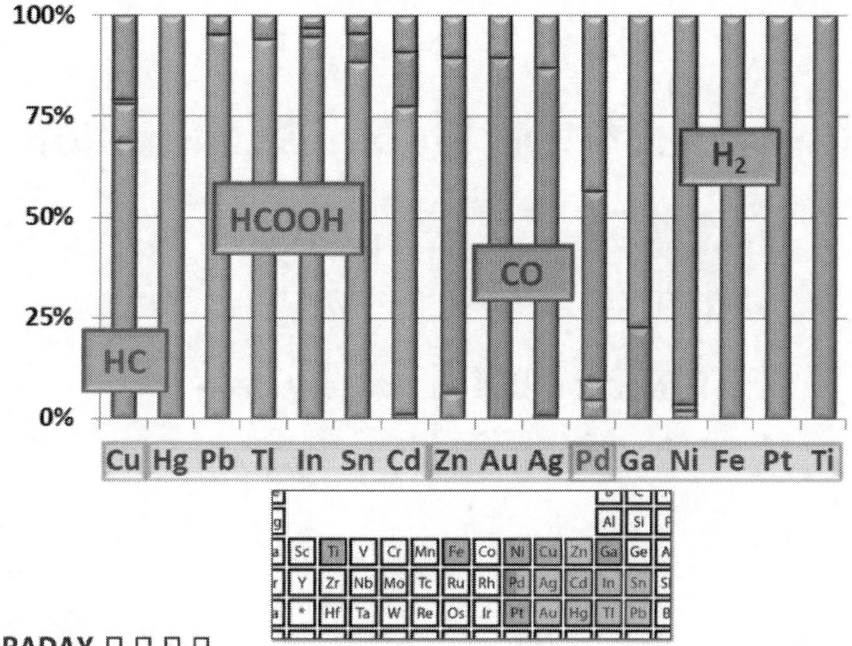

FARADAY TECHNOLOGY, INC.

CO_2 Electrocatalyst Selection

Product distribution over single-metal electrocatalysts as determined by Hori et al.
Electrochimica Acta 39(11-12): 1833 (1994).

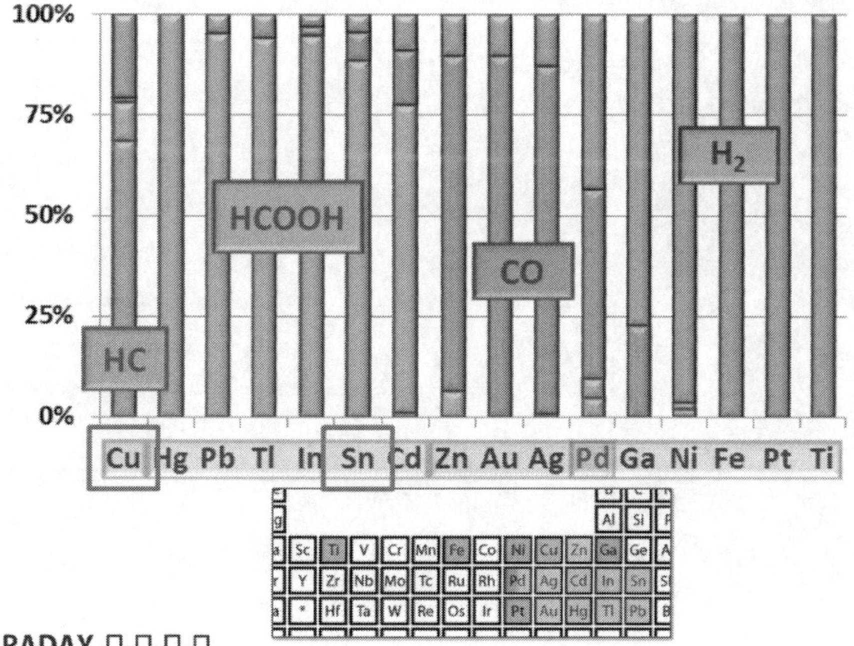

FARADAY TECHNOLOGY, INC.

FARADAY
TECHNOLOGY, INC.

FARADAYIC® GDE ElectroCatalyzation

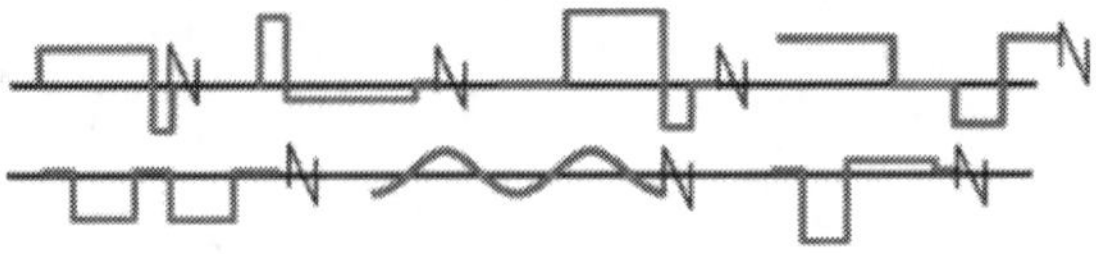

Prepared Electrocatalysts

GDE Catalysts – Cu

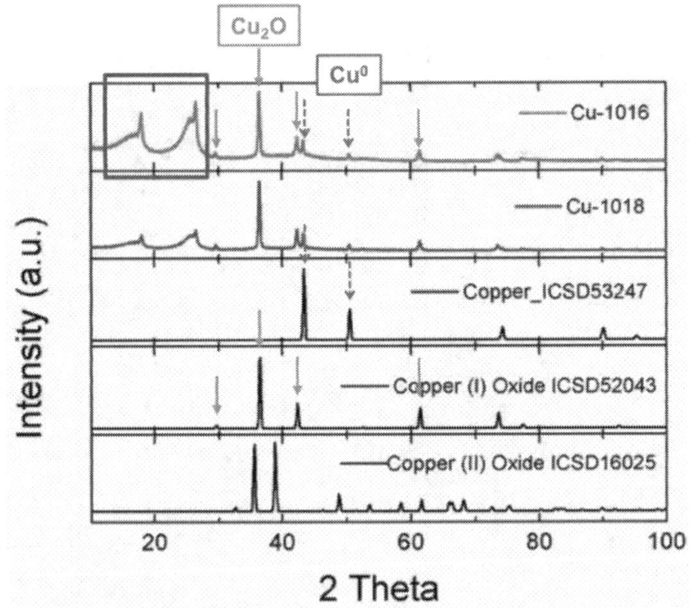

GDE Catalysts – Sn

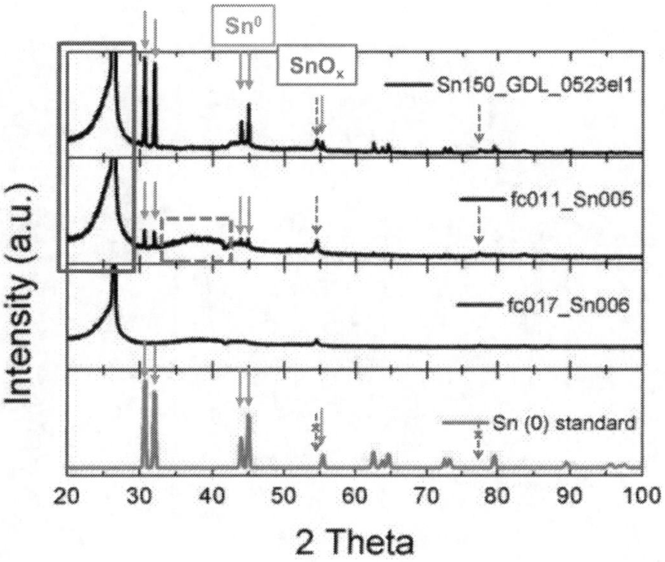

FARADAY TECHNOLOGY, INC.

FARADAY TECHNOLOGY, INC.

FARADAYIC® ElectroCatalyzation

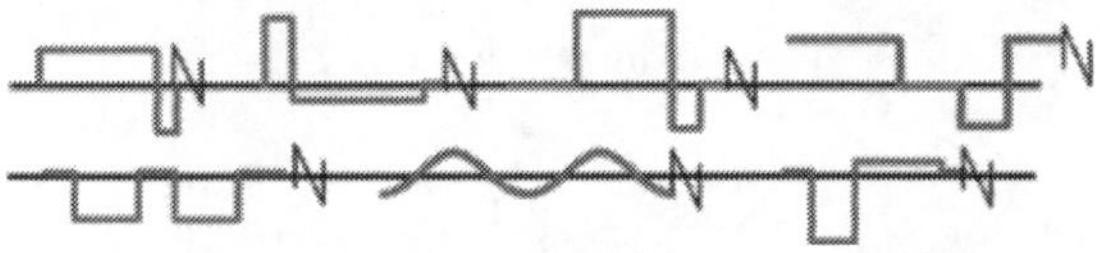

Electrocatalyst Testing

W-Cell Apparatus

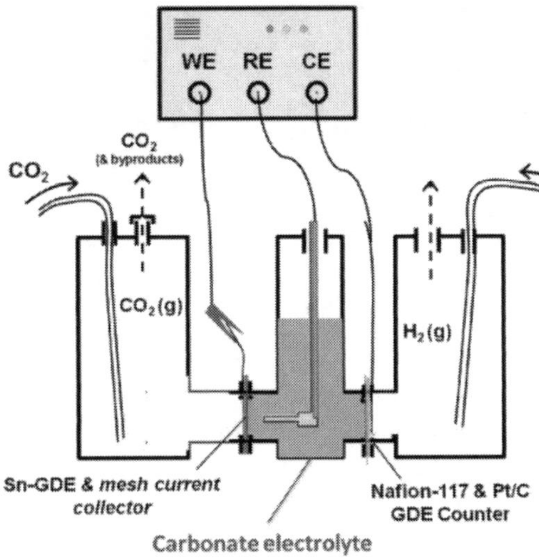

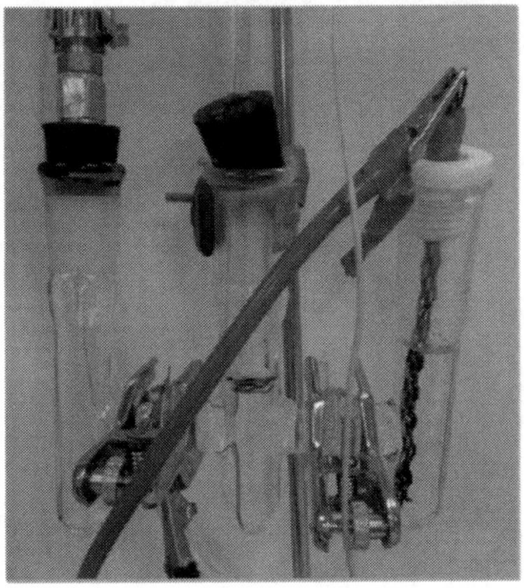

FARADAY TECHNOLOGY, INC.

Flow Cell Apparatus

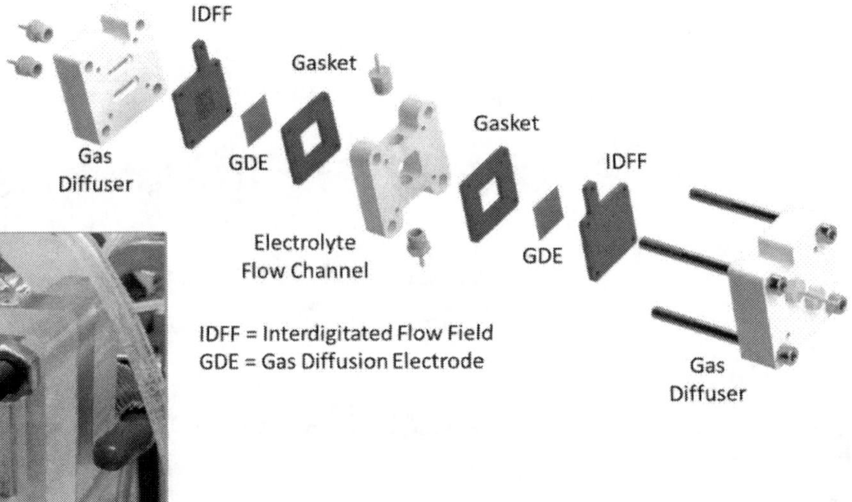

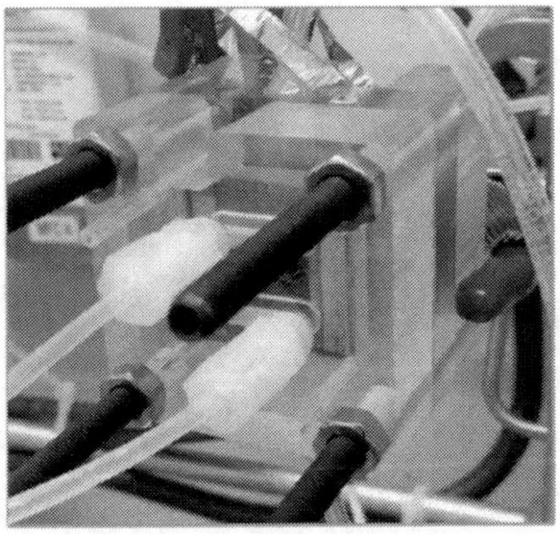

FARADAY TECHNOLOGY, INC.

Sn GDE – W-Cell Electrocatalysis

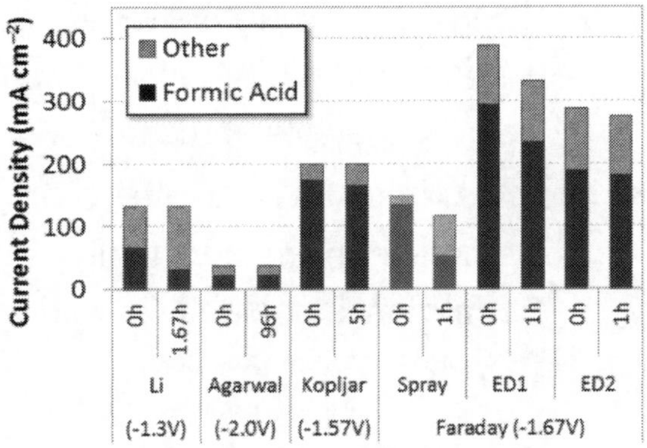

(Half-cell potentials vs SHE)

WE: $CO_2 + 2H^+ + 2e^- \rightarrow HCOOH$

CE: $H_2 \rightarrow 2H^+ + 2e^-$

Cell: $CO_2 + H_2 \rightarrow HCOOH$

Li and Oloman. *J Appl Electrochem* 35: 955, 2005.

Agarwal et al. *ChemSusChem* 4: 1301, 2011.

Kopljgar et al. *J Appl Electrochem* 44: 1107, 2014.

FARADAY TECHNOLOGY, INC. 23 AIChE Annual '17 - B Skinn - 30 Oct 2017

Cu GDE – Flow Cell Electrocatalysis

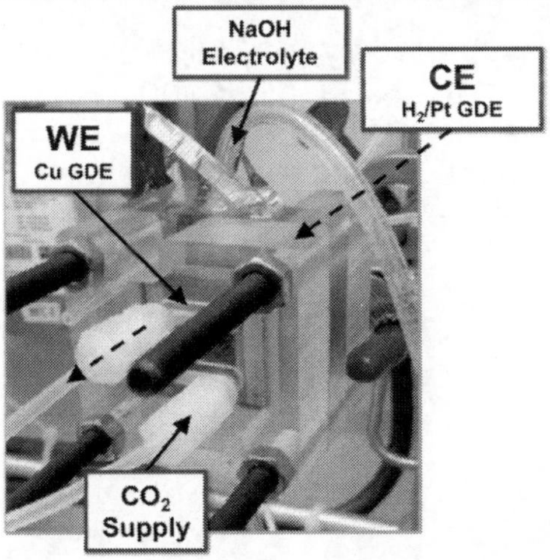

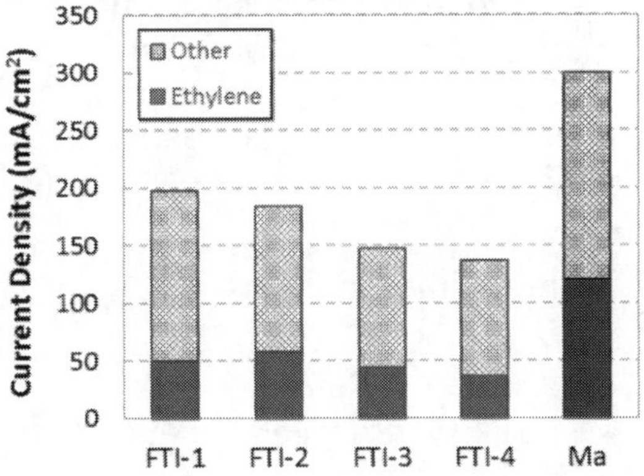

Ma et al. *J Power Sources* 301: 219, 2016.

WE: $2\ CO_2 + 8\ H_2O + 12\ e^- \rightarrow C_2H_4 + 12\ OH^-$

CE: $12\ OH^- \rightarrow 6\ H_2O + 3\ O_2 + 12\ e^-$

Cell: $2\ CO_2 + 2\ H_2O \rightarrow C_2H_4 + 3\ O_2$

FARADAY TECHNOLOGY, INC. 24 AIChE Annual '17 - B Skinn - 30 Oct 2017

Summary

- Wide variety of active GDE catalyst materials can be deposited via pulsed FARADAYIC® ElectroCatalyzation
  - Readily scalable process
- High-activity CO_2 conversion catalysts
  - Demonstrated in industrially-relevant, scalable form factor
  - Performance competitive with or superior to literature
    - Total activities (mA/cm^2)
    - Formate/ethylene selectivity
  - Durability appears promising; further testing is pending

Acknowledgment

- This presentation is based upon work supported by the United States Department of Energy (contracts #DE-SC0015173 and #DE-SC0017199). Any opinions, findings, conclusions and recommendations expressed in this material are those of the authors and do not necessarily reflect the views of the US DOE.

- Contact Information:
  Brian T. Skinn
  Ph: +1 937 836 7749
  Email: BrianSkinn@FaradayTechnology.com

Thank You!

490489 Process Intensification for Electrochemical Utilization of CO_2

Abstract:

Conversion of anthropogenic CO_2 emissions to useful chemical fuels or feedstocks is a desirable strategy for reducing atmospheric CO_2 concentration. Formic acid is a diverse product currently used in the textile and agricultural sectors, as well as a component in cleaning products and preservatives. Formic acid is also valuable as chemical fuel, a potential liquid H_2 carrier in the "centralized de-carbon economy". Production of formic acid with intermittent renewable energy sources would allow for excess energy produced during low demand load to be stored and used during peak load times, while also reducing CO_2 emissions. Progress toward the development of an electrochemical "one-step" CO_2 utilization process to produce formic acid will be discussed. The proposed CAER-ECUS process combines three previously developed components, a protective redox hydrogel, a CO_2 reducing catalyst, and the CAER proprietary Xerogel® electrode, to synthesize formic acid directly from CO_2 in a flexible process that can be modified based on changing economics and market conditions. These three key components work in concert to make the CAER-ECUS process a potentially disruptive technology by: 1) directly producing a viable and flexible chemical feedstock that can be sold or used for energy storage, 2) protecting valuable catalyst against deactivation by overpotentials and faradaic inefficiencies from the electrode or oxidation from flue gas contaminants, and 3) preforming the reaction on a large surface area (> 200 m^2/g) electrode with large pore structures reduce mass transport issues. We show that the redox-active hydrogel increases electrochemically-active surface area, and that the effective charge carrier has a redox potential within the desirable range for CO_2 reduction. Hydrogel properties, catalyst selection, process design and integration, as well as future directions for the CAER-ECUS process will we discussed.

498959 Implications of High Accuracy Thermochemical Kinetics for H + CH_3 (+M) â‡„ CH_4 (+M) on Combustion Models

Abstract:

CH_3, the simplest alkyl radical, is an extremely stable intermediate that persists in large concentrations even in high temperature chemically reacting systems such as flames. Radical-radical recombination with H-atoms (R1: H + CH_3 â‡„ CH_4) is the dominant process for CH_3 removal in high temperature combustion and flame processes. As such, the net result of this reaction is to terminate reactive H-atoms and CH_3 radicals in flames. Numerous studeies have been conducted to directly measure k_1, however, k_1 has only been measured at low temperatures (< 700 K) while all data at high temperatures (> 1300 K) relevant to the pre-flame regime have been measured in the reverse direction. Hence, reconciling k_1 and k_{-1} requires accurate equilibrium constants spanning an extended range of temperatures. In the present work, we include all relevant anharmonic corrections to calculate an accurate Active Thermochemical Tables (ATcT) partition function for CH_3. The resulting nonrigid-rotator-anharmonic-oscillator partition function is used to determine thermochemical parameters (heat capacity, entropy, and enthalpy increment values) for CH_3 and K_{eq} values for reaction R1 over a wide temperature range (200-6000 K). With the newly evaluated equilibrium constant, literature experiments and theory (based on two-dimensional master equation calculations with a first principles energy and angular momentum transfer kernel) are used to obtain an accurate representation of the kinetics for the title reaction. In particular, the present work provides fits to $k_1(T,P)$ for N_2, Ar, and He as colliders for direct use in combustion modeling. The results of the new rate constant evaluation for R1 has a dramatic impact on combustion modeling. In particular, we assess the impact of incorporating anharmonic thermochemistry for CH_3 and the updated fits for k_1 (and k_{-1}) in current widely used literature models for simulations of CH_4/air and n-heptane/air laminar flame speeds and find that flame speed predictions can be affected by as much as 50%.

This work was supported by the Office of Basic Sciences, Division of Chemical Sciences, Geosciences, and Biosciences, U. S. Department of Energy, under Contract No. DE-AC02-06CH11357.

496829 Estimation of Thermodynamic Properties of Polycyclic Molecules By a Linear Regression Model

Abstract:

Though recent advances in *ab initio* methods have opened the door toward accurate calculations of molecular thermochemistry, large-scale theoretical studies are still often limited, at least initially, to the use of empirical methods to rapidly screen out unimportant species, so only the important species are the subject of CPU-time intensive quantum chemistry calculations. Among the empirical methods developed in the past decades, the Benson group additivity scheme is one of the quickest and most convenient methods to determine thermodynamic properties of molecules without needing the 3D molecular structures. It has achieved great success in accurate prediction of thermochemistry of small molecules and has been widely adopted in modeling software for on the fly prediction of thermodynamic parameters. However, because this additivity scheme simply uses the properties of individual chemical groups independently to calculate the entire property for a molecule, the contribution of the overall molecular structure to the property is usually not taken into account. This problem manifests itself for strained structures and can cause significant errors for polycyclic molecules with fused rings. Therefore, the application of the additivity method is often restricted to simple chemical systems without the presence of polycyclic species.

To address this issue, we have developed an improved model to predict the thermochemistry of polycyclic species.Â A regularized linear model was chosen to resemble the simplicity and interpretability of the additivity method. However, instead of following the definitions of the chemical groups of Bensonâ€™s scheme, we generated a comprehensive list of identifiers containing local (atoms, bonds, and angles) and/or nonlocal (rings) structural information and used these identifiers to compose feature vectors for molecules.Â The identifiers carrying nonessential information were eliminated by L1 regularization during model training so that those remained in the final model represent an optimum set of chemical units for the calculation of the thermodynamic property of interest with the contribution of the cyclic structures. These chemical units are human interpretable and conceptually equivalent to the chemical groups defined in Bensonâ€™s scheme but are selected by the model objectively without human intervention.

For a training set of 25,716 cyclic and polycyclic organic molecules made up of C, H, and O atoms, 408 identifiers were found to be needed for the calculation of formation enthalpies. The transferability of the trained model was validated on an independent test set of 2,858 molecules with a mean absolute error of 1.88 kcal/mol.

490830 Hydrogen Binding Versus Dissociation during Soot Formation: Insights from Reactive Molecular Dynamics

Abstract:

It has been proposed that two reactions are prominent in the earliest stages of soot nanoparticle formation; hydrogen abstraction via collision between a growing PAH (polyaromatic hydrocarbon) cluster and a hydrogen atom (yielding H_2 and a dehydrogenated cluster) and carbon addition via the binding of an alkane/alkene/alkyne chain to the cluster. We examine the first part of this process via reactive molecular dynamics using REAXFF C-H potentials. In simulations, cluster structures determined using the previously developed Atomistic Model for Particle Inception (AMPI) code of Violi and coworkers are equilibrated (at a predefined temperature) and a hydrogen atom with a specific velocity and impact parameter is driven towards the cluster. Depending upon the velocity, impact parameter, and collision orientation, we find that pure hydrogen abstraction is rare and instead at low velocities hydrogen binds to clusters. At increasing velocities, hydrogens rebound from the cluster surface and can drive hydrogen facilitated collision induced dissociation (CID) of the cluster. CID events yield smaller alkane/aklene/alkyne chains, which may serve to bind other clusters. Integration over all impact parameters and velocities for collisions (in accordance with the Maxwell Boltzmann distribution) yields the hydrogen binding rate coefficient and hydrogen induced dissociation rate coefficient for a given cluster structure. We discuss how these rates vary with temperature and cluster size.

499595 Effect of Composition Variation of Syngas Mixtures on NO_x Formation at High Pressure

Abstract:

Increasing interest in syngas combustion in gas turbines has led to highlighting the environmental feasibility, in particular the control of nitrogen oxide (NO_x) emissions. Trace amounts of NO_x species may be present during combustion and lean premixed (LPM) combustion method is one of the effective NO_x control technologies for gas turbines. Despite the proven feasibility of syngas, these fuels face variation in their composition due to different feedstock and production methods. Syngas can contain small amounts of hydrocarbons, such as methane, ethane and ethylene, and the presence of hydrocarbon species in the fuel can affect the NO_x emissions in the post-flame region. A detailed investigation of high-pressure syngas combustion is required to provide both practical information regarding NO_x emissions from gas turbine systems and fundamental understanding of reaction pathways to verify and develop chemical kinetic models.

In this study, a high-pressure burner facility capable of operating at pressures up to 15 bar, was designed and fabricated to simulate syngas combustion in a gas turbine. Syngas consisting mainly of H_2 and CO was blended with components such as nitric oxide (NO), nitrogen dioxide (NO_2) and trace amount of hydrocarbons prior to combustion at high pressure. Hot combustion products were supplied to the high pressure vessel by a flat flame burner operating with lean, premixed syngas-air mixtures with various H_2/CO ratios ranging between 0.25-1.0 and equivalence ratios between 0.5-1.0. Detailed NO_x speciation measurements in the post-flame region were conducted for different fuel compositions at various pressures up to 10 bar using Fourier transform infrared (FTIR) spectrometer. The experimental data generated from this project can be used to validate/develop kinetic models to predict NO_x emissions.

Numerical modelling of industrial burners for reduction of NO_x emissions using flamelet methods in combination with a newly developed postprocessor for fast and accurate emission prediction

DDI Werner Pollhammer, K1-MET GmbH, Linz, Austria
DI Dr. Christoph Spijker, Montanuniversitaet, Leoben, Austria
DI Michael Koller, Ebner Industrieofenbau GmbH, Leonding, Austria
Univ.-Prof. DI Dr. Harald Raupenstrauch, Montanuniversitaet, Leoben, Austria

Abstract

Current trends of research on combustion systems show increased interests on prediction of pollution. To meet the demands of modern combustion systems such as industrial burners detailed studies of their behaviour is needed. One of these pollutants which are commonly of interest is NO_x. The aim of the work is to develop a numerical toolbox that is able to predict NO_x emissions of industrial burners by numerical simulation using Computational Fluid Dynamics (CFD).

A set of numerical equations is solved for flow, temperature, pressure and species. Using CFD it is possible to model a combustion system using different approaches like flamelet methods or detailed chemical reaction systems. To predict the emission of NO_x from a flame a detailed chemical reaction system is necessary such as the GRI 3.0. mechanism [1]. The problem of using this mechanism is that it can be applied only for small geometries to avoid long calculation time or processing costs. Using a detailed chemical reaction system for every species a few chemical reactions have to be solved. Using the GRI 3.0. mechanism 325 reactions for 53 species have to be solved [2]. Also with modern computers it is not affordable to model industrial burners using this approach which would calculate NO_x emissions well.

The second main possibility for simulation of burners is to use flamelet methods [3][4]. Choosing this approach the combustion system is pre-calculated and reaction data will be stored in so called lookup tables. During simulation this reaction data will be accessed by reading them from the lookup tables. This approach allows fast calculation of temperature and flow field of gas burners as well as fast chemical reactions like formation of CO_2, H_2O or CO. This is the standard procedure for simulation of industrial burners for analysis of flame shape and temperature fields. The problem is that the described approach is not applicable for slow chemical reactions such as formation of NO_x. Therefor a detailed chemical reaction system is necessary. The problem is that on the one hand there is the detailed chemical reaction system which would predict NO_x emissions very well but takes too long for calculation. On the other hand there is the flamelet method which is fast but delivers only for flow fields, temperature fields as well as some species fields right results but models the formation of NO_x not very well. The idea now is to combine the benefits of both approaches and to postprocess the results out of the flamelet calculation using a detailed chemical reaction system. This means temperature and flow field are calculated by the fast flamelet calculation. After convergence of this calculation the solution for the flow field and temperature field as well as for some species fields are known. Subsequently for slow chemical reactions such as NO_x production, which cannot be predicted by the flamelet method, the postprocessor calculates the reactions using the already existing temperature and flow field from the flamelet calculation. This combination enables a much faster calculation of NO_x formation in the flame and combines the speed of flamelet method with the accuracy of using detailed chemical reaction systems with reasonable calculation costs [5].

Model description

For a more precise description of radiation effects it was necessary to implement the Discrete Ordinate Model (DOM) into the flamelet solver of OpenFOAM [6]. The standard radiation model is only valid for free stream flames with a constant surrounding wall temperature. The radiation model was used in combination with a Weight Sum of Gray Gases Model (WSGGM) which calculates the optical density of the gas by applying a mixture of temperature dependent absorption coefficients for calculation of the radiation beams [7].

For flamelet calculation a steady laminar flamelet for non-adiabatic combustion was used out of literature [3][4]. By defining a range for enthalpy defects in flamelet generation heat transfer in the system and out of the system can be taken into account. Flamelets were generated for different scalar dissipation rates which takes mixing time scales into account and for different enthalpy defects.

For modelling of turbulent effects the k-ε-model was used with a slight change according to literature data. For a more precise prediction of flame acceleration the constant $C_{2\varepsilon}$ was modified to 1.8. [8]. Out of data from Sandia Flame D boundary conditions for turbulence were generated for turbulent kinetic energy k and dissipation of it named ε [1].

The NO_x postprocessor uses a detailed chemical reaction approach the so called partially stirred reactor model implemented in OpenFOAM for calculation of chemical reaction [9]. It takes chemical reaction kinetics, for instance out of GRI 3.0., as well as turbulent mixing effects into account [2]. The solver uses a conservation equation for each species and chemical reaction is implemented as a source term in every species equation. This means for chemical reaction in every mesh cell the whole reaction mechanism consisting of hundreds of Arrhenius equations has to be solved.

To reduce the consumption of calculation power a special strategy of solving chemistry was implemented. As a matter of fact that the system seeks for a steady state solution that means after a certain time species composition will not chance any more in the cells. While post processing there is no change of temperature and turbulence because these fields are calculated by the previous flamelet calculation and they are constant during post processing. The reaction rates of chemistry then only depend on species concentration. If steady state is nearly reached in a cell and the species concentration varies just a bit there is no big difference between reaction rates of the previous and actual time step. An algorithm which analyses these changes was implemented. Using this algorithm the chemical reaction source terms were tabled from time step to time step. If the change in species is lower than 2 percent of the relative concentration of every species in a cell then the source terms of the previous time step is used. Using this algorithm about 95 percent of chemistry can be accessed by tabling and only 5 percent of the cells have to be calculated. This leads to huge saving of computational power and to a big speed up in simulation.

To avoid unphysical states there is also a check if one of the species concentrations goes under zero which is physically incorrect. If this occurs then chemistry will be calculated in the next time step. To avoid pseudo steady state solutions every 200 time steps the whole domain will be calculated.

The model validation was done by means of a comparison with measurements on the Sandia Flame D [1]. Therefor a two dimensional mesh consisting of 200 000 hexahedra was generated using blockMesh and analysed in terms of mesh independency. The benefit of validation on Sandia Flame D is that temperature, velocity, turbulence and species decomposition is also known at certain points in the flame. This means it is feasible to analyse the whole flame shape and its position dependent characteristics and not only the properties at the outlet. For reliable results it is important to be able to predict the temperature maximum very well. Comparison of species decomposition shows very good agreement in terms of oxidizer consumption and CO_2 formation [1]. Using the newly developed postprocessor a very accurate description of NO_x formation is possible which fits well to literature data [1].

Model validation on Sandia Flame D and results

First the flamelet model was used to describe the temperature, velocity and turbulence field of the flame. The following Figure 1 shows the solution of the Sandia Flame D case and its temperature field.

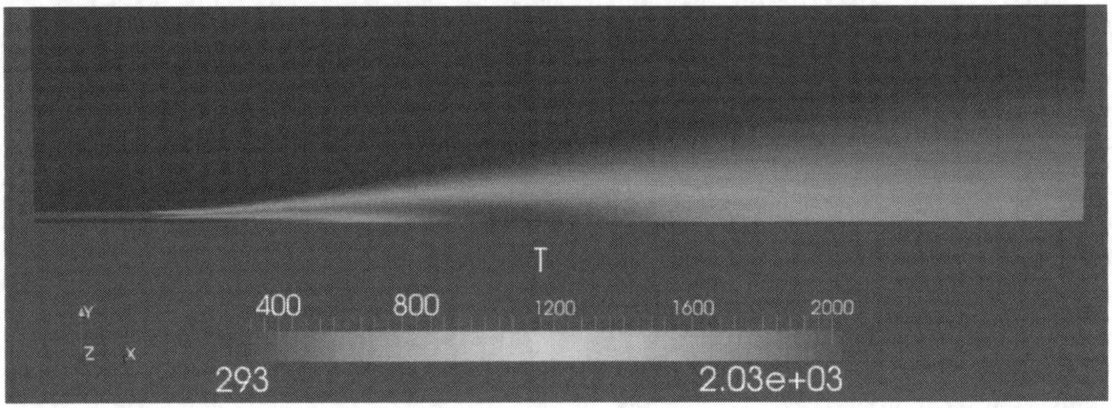

Figure 1: Temperature distribution in Sandia Flame D calculated using flamelet model

This simulation results were then compared to literature data [1] which is displayed in Figure 2. As it can be seen is that the temperature along the rotational axis fits very good to measurements. This is very important for an accurate prediction of nitrogen oxide formation.

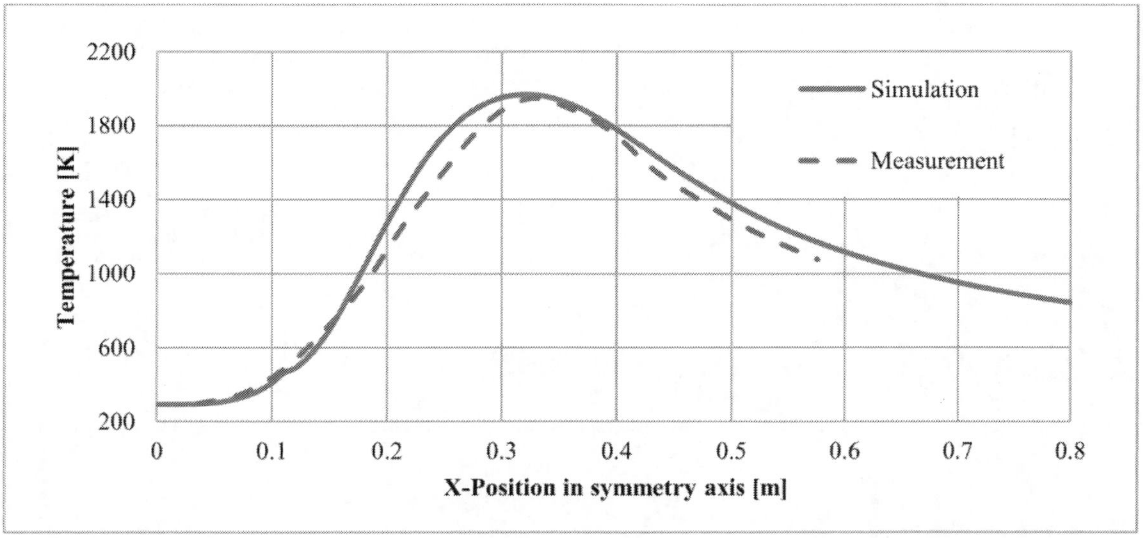

Figure 2: Plot of flame temperature comparing simulation and measurement data from Sandia Flame D [1]

Using this data for temperature, velocity, turbulence and species which can be predicted by flamelet the post processing was performed to calculate species like NO, N_2O, NO_2 which cannot be predicted accurate enough by flamelet. For this species a detailed chemical reaction mechanism was used. In this case the simulations were performed using GRI 3.0. [2]. The following Figure 3 shows the distribution of NO in the Sandia Flame D with three sections through the flame. At these positions a comparison with measurement data was made which is displayed in Figure 4.

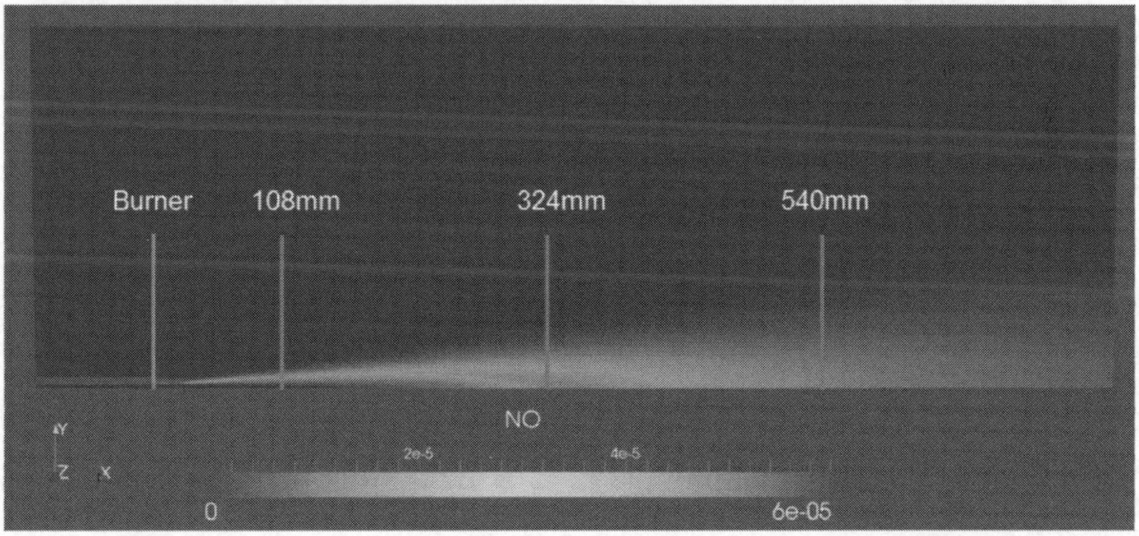

Figure 3: NO field of Sandia Flame D after convergence of postprocessor

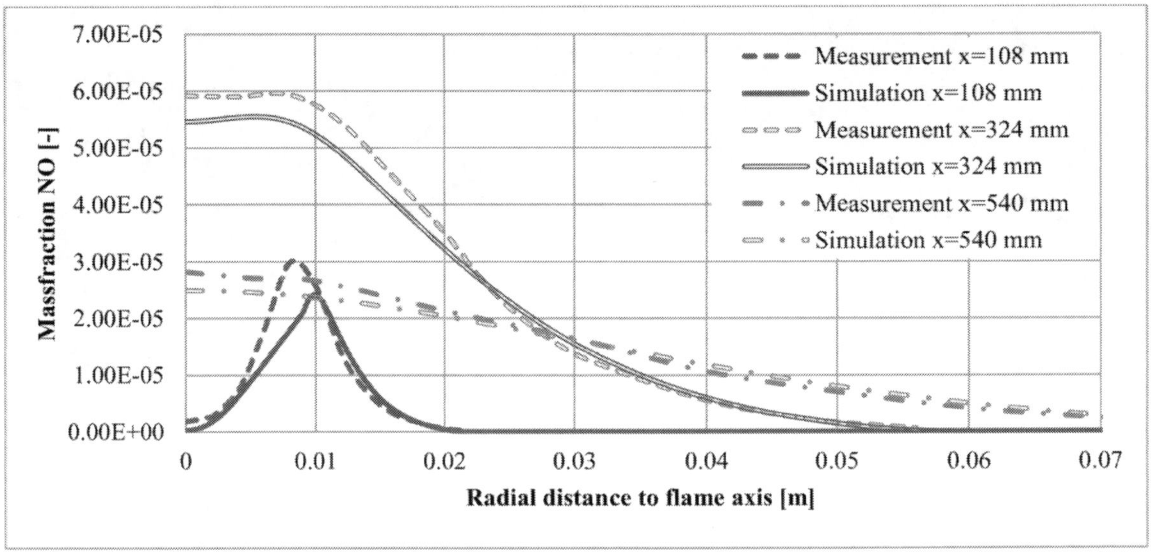

Figure 4: Comparison between simulation and measurement at three cuts through Sandia Flame D as shown in Figure 3 [1]

The result of the first validation is that the generated postprocessor is able to calculate nitrogen oxide formation in Sandia Flame D very well. It predicts slightly lower concentration but the accordance is good in the whole flame. So the developed model can be used for detailed investigations on nitrogen oxide formation in industrial burners.

Model validation on test chamber and results

A second model validation was done by means of an industrial test where a burner was operated in a test chamber. Measurements of wall temperatures by thermocouples and a detailed exhaust gas analysis were performed. Out of the measurements of 20 thermocouples at different positions on the furnace wall a temperature profile was generated to implement it as boundary condition for the flamelet calculation. This lab test was modelled using the developed simulation toolbox. A mesh of the whole test geometry was generated using cfMesh [10]. Out of mass flow and temperature measurements boundary conditions for simulation were calculated. The first validation for the model was a comparison between exhaust gas temperature in the lab test and simulation. These data were then used to start the NO_x post processing. Therefor the developed transient postprocessor was used to calculate the formation of NO_x and various other species which are not predicted well by the flamelet method. At that point it is only necessary to calculate the chemical reaction source terms without solving flow and temperature fields which are now constant in this calculation and already known from the flamelet calculation. This way the simulation is much faster and calculation of NO_x formation now is possible in affordable time using the detailed chemical reaction mechanism with the benefit of a much more accurate description of NO_x formation. For the whole geometry a mesh independency study was performed on the one hand for the flamelet model and on the other hand for the postprocessing. The mesh was generated with a cell size which delivers mesh independent results. For the test chamber with a height of 2.3, a width of 1.7 and a depth 2.5 meters the mesh consists of 5.26 million polyhedral cells.

Figure 5 shows the result of the post processing of the test chamber simulation.

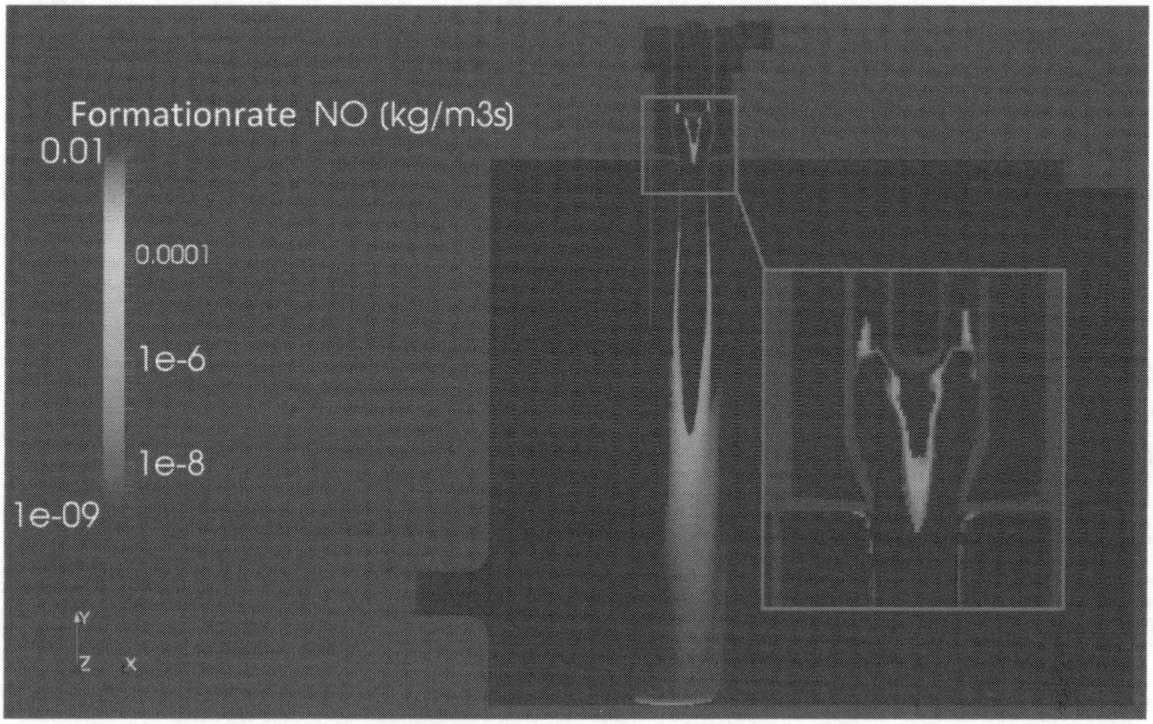

Figure 5: Areas of NO formation in test chamber, calculated using developed postprocessor

This test was performed twice. One test was made with air at ambient temperature and one with preheated air with 199°C.

Table 1: Comparison between simulation results and measurements in test chamber

Formation of NO_x	Simulation [ppm]	Measurement [ppm]
Cold burner air with 28°C	63	65
Preheated burner air with 199°C	78	92

As it can be seen in Table 1 there is also a good accordance for the postprocessor with measurements in the big test chamber. The case with air at 28°C shows the best accordance. The second lab test using preheated air shows the right trend by increasing NO_x formation in terms of the cold air case but shows lower NO_x emissions like the measurement at the test chamber. This effect of lower NO_x concentrations can also be seen in Figure 4 that the model predicts lower nitrogen oxide emissions than the measurements. Important is that the model predicts relative improvements and trends. So now the model was used to perform a geometry study and to test different improvements of the burner geometry by changing the geometry and using the postprocessor for analyzation.

Outlook

Now the model is used to analyse the formation zones of nitrogen oxides in the burner. The next steps are further improvements of the code efficiency and to find more strategies to reduce the calculation time. After some improvement steps of simulation a prototype of the new developed burner will be build a new chamber test will be performed to check these improvements also in a real test.

Acknowledgements

The authors gratefully acknowledge the funding support of K1-MET GmbH, metallurgical competence center. The research program of the competence center K1-MET is supported by COMET (Competence Center for Excellent Technologies), the Austrian program for competence centers. COMET is funded by the Federal Ministry for Transport, Innovation and Technology, the Federal Ministry for Science, Research and Economy, the province of Upper Austria, Tyrol, and Styria and the Styrian Business Promotion Agency.

References

[1] Barlow R, Frank J. *Piloted CH4/Air Flames C, D, E, and F – Release 2.1*. Available at: http://www.sandia.gov/TNF/DataArch/FlameD/SandiaPilotDoc21.pdf; 2007 [accessed 07.04.2017].

[2] Smith GP, Golden DM, Frenklach M, Moriarty N, Eiteneer B, Goldenberg M, Bowman T, Hanson RK, Song S, Gardiner WC, Lissianski, Jr., Vitali V., Qin Z. *GRI Mech 3.0*. Available at: http://www.me.berkeley.edu/gri_mech/; 1999 [accessed 11.04.2017].

[3] Cuoci A, Frassoldati A, Faravelli T, Ranzi E. Formation of soot and nitrogen oxides in unsteady counterflow diffusion flames. *Combustion and Flame* 2009;156:2010–22, doi:10.1016/j.combustflame.2009.06.023.

[4] Cuoci A, Frassoldati A, Faravelli T, Ranzi E. Kinetic Modeling of Soot Formation in Turbulent Nonpremixed Flames. *Environmental Engineering Science* 2008;25:1407–22, doi:10.1089/ees.2007.0193.

[5] Spijker C. *Unsteady Laminar Flamelet Modellierung zur Beschreibung von Mündungsmischbrennern*. Masterthesis, Montanuniversitaet. Leoben; 2010.

[6] Thynell ST. Discrete-ordinates method in radiative heat transfer. *International Journal of Engineering Science* 1998;36:1651–75, doi:10.1016/S0020-7225(98)00052-4.

[7] Barlow RS, Karpetis AN, Frank JH, Chen J-Y. Scalar profiles and NO formation in laminar opposed-flow partially premixed methane/air flames. *Combustion and Flame* 2001;127:2102–18, doi:10.1016/S0010-2180(01)00313-3.

[8] Pfeiler C, Spijker CJ, Raupenstrauch H. CFD als Werkzeug in der Industrieofentechnik. *Berg Huettenmaenn Monatsh* 2011;156:347–52, doi:10.1007/s00501-011-0018-z.

[9] OpenFOAM. *OpenFOAM 2.4.0*. Available at: https://openfoam.org/; 2017.

[10] cfMesh. *Professional Meshing Solutions - cfmesh.com*. Available at: http://cfmesh.com/; 2017 [accessed 11.04.2017].

491775 A CFD Model-Based Optimization of a Low NO_x Burner in Integrated Process System

Abstract:

Although process burners are frequently used in many industries such as refeinring, petrochemical and chemical processes, there is a serious concern about emission control of these equipment. Recently, the deveopment of new numerical approaches such as computational fluid dynamics (CFD) has led to significant improvements in emission control of industrial processes. The proper analysis and testing leads to more efficient combustion, less physical testing and lower cost for full scale equipment burners.

In this study, we used CFD simulations to design, develop and optimize a low NOx burner which produces heat for methane steam reforming. The CFD method helped to design a burner with angled air holes arranged in a swirl pattern. This optimized design increased the mixing near the burner plate that led to lower peak flame temperature and therefore lower NOx emissions. The designed burner was built and experimentally tested and showed a major NOx reduction in steam reforming hydrogen generation process. In addition, reformer performance improved significantly due to the better temperature distribution by the new CFD optimized burner.

496078 Mapping the Stable Region of Operation of Homogenously Charged Compression Ignition (HCCI) Engine for Methane and DMF Fuel Blend

Abstract:

Abstract:

The emission legislations are transmuting to be rigid in accordance to reduce the emissions from the IC engines. Research has been focused to ameliorate both Spark Ignition and Compression Ignition engine in order to make them more efficient. Over the previous decade, major research has been focused on Homogenous Charge Compression Ignition (HCCI) engines. This engine has the potential to consummate the eminent feature of Spark Ignition engine, with no emission of NOx and Soot leading to cleaner combustion, the best of CI engine with higher efficiency and low fuel combustion along with lower carbon dioxide emissions[1].

The focus of this work is to map the stable regions of operation of HCCI engine using Methane and Dimethyl Furan blend as fuel. DMF is comparable to gasoline since it has an energy density of 40% higher than ethanol. Also, Burning DMF requires 33% less air compared to the same amount of gasoline since its stoichiometric air to fuel ratio is 10.72. The Simulations of HCCI engine were carried out in CHEMKIN, where the model used was a closed transient CSTR[2]. The main objective of the study was to depict the stable regions of operation of HCCI engine with and without Exhaust gas recirculation. EGR refers to a part of the exhaust stream recycled back to the engine inlet along with the fuel mixture[3].

Parameters such as EGR Temperature, EGR percentage and equivalence ratio were varied to study the effect on the sustainable emissions at the exhaust for CO, CO_2 and NOx. Detailed Kinetics of DMF obtained from NCBI was used. Initial understanding of the working of HCCI engine was done using Methane as source fuel, keeping EGR temperature and HCCI engine temperatures as a constant at 600K and 447K and the equivalence ratio were varied from 0.5 to 0.7 and the EGR percentage were varied from 0 to 30%. The in-cylinder temperature, pressure and species profile were obtained as a function of Crank Displacement angle (CAD).

The Figure (1) shows the effect of equivalence ratio on the in-cylinder temperature, pressure and species profile for a engine temperature of 447K and EGR temperature of 600K for 30% EGR. Increase in equivalence ratio states the presence of higher amount of fuel where as smaller ratios represent deficit of fuel in the mixture. It could be observed that Peak temperature and pressure increases with increase in equivalence ratio. The mole fraction of CO increase with increase in equivalence ratio, this is because of the lack of oxygen available for complete combustion. In case of Thermal NO production rate, it increases with increase in equivalence ratio due to increase in temperature which leads to increased production of NO at higher temperatures. It was observed from the initial study using Methane as a fuel that, for Lesser exhaust emissions, the fuel must be lean with a higher percentage of EGR.

Fig 1: Variation of In-cylinder temperature, pressure, mole fraction of CO, CO_2 and Thermal NO production rate with Equivalence ratio

The study was further extended to DMF as an additive to Methane and the parameters such as EGR percentage, Equivalence ratio and EGR temperature were varied to find the stable region of operation of HCCI engine for Methane and DMF blend. The effect of Equivalence ratio on the in-cylinder temperature, pressure and species properties for 0% EGR for an engine temperature of 447K and EGR temperature of 600K were studied initially. The figure (2) shows the effect of Equivalence ratio on Peak CO_2 mole fraction in the cylinder for different percentage of Methane and DMF blend. It could be observed that the peak mole fraction of CO_2 increases with increase in Equivalence ratio and also increases with increase in blend percentage. The latter is due to the increase in the carbon availability in the cylinder with increase in fuel.

Fig 2: The Variation of Peak Mole Fraction of CO_2 with percentage of the DMF blend for different Equivalence ratios

As a next step, it is intended to study the effect of EGR percentage and EGR mixing temperature on Methane and DMF blend as shown in the Table 1 and analyse the Peak in-cylinder temperature, pressure and species properties for the different parameters combinations and map the stable region of operation of HCCI engine for Methane and DMF fuel blend.

Table 1: Parameters to be estimated for the Blend

Blend	Parameters
Methane & DMF blend	EGR Percentage – 0%, 10%, 20% & 30% EGR temp – 400K, 500K & 600K Equivalence ratio – 0.5, 0.6 & 0.7 Blend Ratio – 0%, 50% & 90%

Keywords:

Exhaust Gas Recirculation, DMF, Crank Rotation angle.

References:

1. Harrison R. M. ed. (1996) Pollution, Causes, Effects and Control. 3rd Ed. Cambridge, The Royal Society of Chemistry.

2. Xu H., Liu, M., Gharahbaghi, S., Richardson, R., Wyszynski, M., Megaritis, T., "Modelling of HCCI Engines: Comparison of Single-zone, Multi-zone and Test Data", SAE 2005-01-2123

3. Heywood JB, Internal Combustion Engine Fundamentals, Mc Graw Hill; 1988

497085 Mechanistic Insight for Oxidative Coupling of Methane on Mg6MnO8-Based Redox Catalysts in a Chemical Looping System

Abstract:

Methane activation has been the topic of interest for many decades for its upgradation to value added chemicals. With the high availability of shale gas due to advancements in fracking technology, understanding this C-H bond activation becomes important as it is a key reaction step in the direct synthesis of chemicals from methane. Traditionally, methane is first converted to syngas, which can then be processed and used for producing chemicals and fuels. This indirect method is both process and energy intensive and thus incentivizes direct conversion of methane to chemicals. One such method for direct conversion is oxidative coupling of methane (OCM), which provides a single step path towards synthesizing higher hydrocarbons from methane, especially ethylene. A popular way for methane oxidation in OCM is through co-feeding molecular oxygen with methane over a catalyst. Conversely, the chemical looping method utilizes the lattice oxygen in a catalytic oxygen carrier (COC) for methane oxidation. A Mg_6MnO_8-based redox system was studied under the chemical looping method for understanding the mechanism of OCM. The atomistic thermodynamics methods and density functional theory (DFT) calculations were carried out to elucidate the lattice oxygen diffusion and vacancy formation mechanism during the oxidation and reduction process.

Additionally, experiments in a fixed bed reactor were performed to validate the computational findings, by running cyclic redox reactions on a fixed amount of COC. Fixed bed experiments were performed with methane as the reducing gas and air as the oxidizing gas. Several cofeed experiments were also performed in addition to the redox experiments to get a holistic view of the process. These results were coupled with thermogravimetric analysis (TGA) to investigate the effect of temperature on the OCM reaction. A dopant screening was also investigated computationally, to potentially increase the selectivity towards the desired hydrocarbon products. Several analysis techniques such as X-ray diffraction (XRD), X-ray photoelectron spectroscopy (XPS) were used to characterize the COC in its oxidized and reduced form. It was observed that the advent of carbon monoxide in conjunction to the desired hydrocarbons as the OCM reaction progresses gives an insight into the importance of oxygen vacancy formation on the COC surface.

496130 Gold-Palladium Particle Size Effect on Methane Oxidation Activity

Abstract:

Effect of Gold-Palladium Particle Size on Methane Oxidation Activity

Christopher Williams*[a], James Carter[a], Nicholas F. Dummer[a], Robert Armstrong[a], Sara Yacob[b],David J. Willock[a], Randall J. Meyer[b], Stuart H. Taylor[a], Graham J. Hutchings[a]

[a]Cardiff Catalysis Institute, School of Chemistry, Cardiff University, CF10 3AT

[b] ExxonMobil Research and Engineering Company, Corporate Strategic Research, Annandale, NJ 08801, USA

*Email:WilliamsCP1@cf.ac.uk

Natural gas, of which methane is a major component (approx. 70-90%), is a naturally abundant gas with reserves of ca. 190 trillion cubic meters.[1] Both affordable and readily available, natural gas may provide an alternative feedstock from the traditional petroleum reserves in addition to providing a cleaner source of fossil energy. Directly upgrading methane to oxygenates presents a global challenge complicated by the chemistry of methane. Current utilization involves indirect routes via the formation of synthesis gas (CO and H_2), requiring costly and energy intensive processes. A direct methane-to-methanol process at low temperatures provides an attractive and efficient method of producing an energy-dense fuel and important chemical feedstock.

In addition to requiring the use of preformed reactive oxygen species, current systems such as Periana[2] or Palkovits[3] catalysts have required highly acidic reaction media for the oxidation of methane. Alternatively, Hutchings et al. have demonstrated the use of H_2O_2 as an efficient oxidant for the low-temperature oxidation of methane, with noticeable productivities with the application of zeolite catalysts and supported gold-palladium nano-particulate catalysts.[4,5] Gold-palladium catalysts have also been shown to activate C-H bonds present in toluene[6] and in benzyl alcohol[7], in addition to the direct synthesis of H_2O_2 from H_2 and O_2.[8] The in situ generation of H_2O_2 provides a catalytic system for the oxidation of methane.

Figure SEQ Figure * ARABIC 1 : Particle size distribution of sol immobilised AuPd catalysts: Dried (blue) vs. Calcined at 800 °C (red).

It has previously been reported that gold-palladium catalysts, prepared by incipient wetness, produce materials with a bimodal distribution of particles ranging from 5-20 nm and 100-200 nm.[6,8] These 1- wt% AuPd/TiO_2 catalysts gave turnover frequencies of 6.5 h^{-1} and methanol and oxygenate selectivities of 12.1% and 85.4%, respectively.

Furthermore, methanol selectivity was improved with increasing metal weight loading to 5 wt%, whilst retaining high oxygenate selectivity. The bimodal particle size observed with incipient wetness is thought to inhibit reactivity due to (i) small particles that are highly activation.

In this work, a particle size effect is observed for methane oxidation using added H_2O_2 at 50 °C. 1 wt% AuPd/TiO_2 catalysts prepared by sol immobilization initially show relatively low rates of methane activation, forming CO_2 only and decomposing all H_2O_2. The AuPd catalysts are composed of a uniform size distribution ranging 4-8 nm, with mean particle size of 4.9 nm. Heat treatment of the dried catalyst produced a catalyst 3.6 times more active for the oxidation of methane, with 80% oxygenate selectivity and 50% of added H_2O_2 remaining after 0.5 h. Analysis by TEM showed that heat treatment at 800 °C yielded materials with increased AuPd particle sizes, increasing mean particle size from 4.9 nm to 20.3 nm (Fig.1). This particle size increase correlated with an increase in TOF (1.3 h^{-1} to 4.8 h^{-1}) and oxygenate selectivity (0% to 80%). These catalysts demonstrated high selectivity towards methyl hydroperoxide (64%), a key intermediate towards the formation of methanol.

The effect of particle size is an important consideration for the efficient utilization of H_2O_2. Increased particle size results in reduced H_2O_2 decomposition rates which facilitate the selective oxidation of methane to valuable oxygenates. Further work to optimize the synthesis of gold-palladium particles of larger size will be investigated to further improve the oxygenate selectivity.

References/ Bibliography

(1) BP Statistical Review of World Energy June 2016, bp.com/statisticalreview

(2) Periana, R. A.; Taube, D. J.; Gamble, S.; Taube, H.; Satoh, T.; Fujii, H. *Science (80-.).* 1998, *280* (5363), 560–564.

(3) Palkovits, R.; von Malotki, C.; Baumgarten, M.; Müllen, K.; Baltes, C.; Antonietti, M.; Kuhn, P.; Weber, J.; Thomas, A.; Schüth, F. *ChemSusChem* 2010, *3* (2), 277–282.

(4) Hammond, C.; Forde, M. M.; Ab Rahim, M. H.; Thetford, A.; He, Q.; Jenkins, R. L.; Dimitratos, N.; Lopez-Sanchez, J. A.; Dummer, N. F.; Murphy, D. M.; Carley, A. F.; Taylor, S. H.; Willock, D. J.; Stangland, E. E.; Kang, J.; Hagen, H.; Kiely, C. J.; Hutchings, G. J. *Angew. Chemie - Int. Ed.* 2012.

(5) Ab Rahim, M. H.; Forde, M. M.; Jenkins, R. L.; Hammond, C.; He, Q.; Dimitratos, N.; Lopez-Sanchez, J. A.; Carley, A. F.; Taylor, S. H.; Willock, D. J.; Murphy, D. M.; Kiely, C. J.; Hutchings, G. J. *Angew. Chemie - Int. Ed.* 2013.

(6) Kesavan, L.; Tiruvalam, R.; Rahim, M. H. A.; bin Saiman, M. I.; Enache, D. I.; Jenkins, R. L.; Dimitratos, N.; Lopez-Sanchez, J. A.; Taylor, S. H.; Knight, D. W.; Kiely, C. J.; Hutchings, G. J. *Science (80-.).* 2011, *331* (6014), 195-199.

(7) Pritchard, J.; Kesavan, L.; Piccinini, M.; He, Q.; Tiruvalam, R.; Dimitratos, N.; Lopez-Sanchez, J. A.; Carley, A. F.; Edwards, J. K.; Kiely, C. J.; Hutchings, G. J. *Langmuir* 2010, *26* (21), 16568–16577.

(8) Edwards, J. K.; Solsona, B. E.; Landon, P.; Carley, A. F.; Herzing, A.; Kiely, C. J.; Hutchings, G. J. *J. Catal.* 2005, *236* (1), 69–79.

490965 Printable Bioreactors for Bioconversion of Methane to Value-Added Chemicals

Abstract:

An industrial process for the selective oxidation of methane under mild conditions would be highly valuable for controlling emissions to the environment and for utilizing vast new sources of natural gas. Methanotrophic bacteria utilize methane under nearly ambient conditions, suggesting a biochemical process for methane conversion could be more efficient, less capital intensive, and more scalable compared to current industrial practice. However, the industrial use of bioconversion for large-scale applications, especially those involving gas-to-liquid reactions, has been limited by poor mass transfer within stirred-tank reactors. We have embedded the enzyme particulate methane monooxygenase (pMMO) within a 3D printed hydrogel to create a mechanically robust and gas permeable material that overcomes mass transfer limitations to convert methane to methanol. Building up those findings, we also explored encapsulation of whole cells within our material to utilize the entire cell proteome to further oxidize methanol into additional chemicals and expand our production capabilities.

Methods: Biocatalysts including enzymes and/or dehydrated cells were immobilized in poly(ethylene glycol) (PEG) hydrogel during crosslinking by UV-initiated photopolymerization. The utilization of methane by the encapsulated biocatalysts was measured by an infrared-based methane detector over time and high pressure liquid chromatography was used to detect production of lactate and methanol. The stability of the encapsulated biocatalysts was investigated for at least 72 hours.

Results and Discussion: We demonstrated the ability to encapsulate biocatalysts within a gas-permeable hydrogel material while retaining catalytic activity. Encapsulation within our material allowed an increased surface area to volume ratio, thereby overcoming mass transfer limitations and improving methane utilization. We investigated the stability of the encapsulated biocatalysts over time as well as the effect of loading, hydrogel weight fraction, and surface area to volume ratio on biocatalyst activity. We look forward to improving the activity of the encapsulated biocatalyst, and eventually curing the hydrogel material within a poly(dimethyl siloxane) scaffold as part of a flow-through reactor, shown in Figure 1, with which we have demonstrated continuous methane conversion using encapsulated pMMO.

Figure 1: Schematic of enzyme immobilization within hydrogel in a structured polymer scaffold to create a gas-permeable biocatalytic material in a flow-through bioreactor.

This work was performed under the auspices of the U.S. Department of Energy by Lawrence Livermore National Laboratory under Contract DE-AC52-07NA27344.

493072 Influence of Platinum on the Activity of Ga/HZSM-5 Catalyst in Ethane Dehydroaromatization Reaction

Abstract:

Gallium-platinum promoted HZSM-5 is found to be a promising catalyst for ethane dehydroaromatization reaction. In the present work, the influence of Pt as a promoter on the activity of Ga/HZSM-5 catalyst for ethane dehydroaromatization has been investigated. Also comparative study is performed between bimetallic Ga-Pt based (GaPt/HZSM-5) and Mo based HZSM-5, where the GaPt/HZSM-5 showed better aromatic and hydrogen selectivity. Pt promoted Ga/HZSM-5 catalyst exhibited higher activity compared to pure Ga/HZSM-5 catalyst. The presence of platinum in the gallium zeolite considerably accelerates dehydrogenation step in ethane aromatization. In addition, GaPt/HSZM-5 deactivates significantly slower than well-known dehydroaromatization catalyst such as Mo/HZSM-5 and Ga/HZSM-5. TPO study of spent catalysts [Figure 1] revealed that carbonaceous deposit on GaPt/HZSM-5 catalyst were burnt off at lower temperature compared to pure Ga/HZSM-5 catalyst. This happened because the presence of Pt metal on the catalyst facilitates hydrogen spillover resulting in the hydrogenolysis of coke and hence it can moderate the coke deposition. This property of Pt improves the stability of the GaPt/HZSM-5 catalyst. The reaction mechanism associated with aromatic formation is postulated based on the correlation between catalytic performance and surface characterization.

Figure 1: TPO profiles of spent catalysts after running the reaction for 5 h.

496112 Effect of Catalyst Composition and Reaction Mechanism Study on Non-Oxidative Methane Conversion into Higher Hydrocarbons

Abstract:

Direct non-oxidative methane conversion to higher hydrocarbons is a potential approach to convert natural gas into valuable hydrocarbons.[1] Zeolite supported molybdenum (Mo) catalyst is the most effective catalyst reported for direct non-oxidative methane conversion into higher hydrocarbons in which HZSM-5 & HMCM-22 are the two well-known zeolites as support.[2] Mo/HMCM-22 catalyst was reported for effective catalytic performance due to unique pore structure and framework of HMCM-22 zeolite support.[3] Different factors such as reaction parameters (T, P and GHSV), catalyst formulation, such as Mo loading over HMCM-22 zeolite support, and SiO_2/Al_2O_3 ratio (SAR) of HMCM-22 zeolite support significantly affects the reaction. In this regard we have studied here the effect of metal (Mo) loading over HMCM-22 (SAR-55) support and space velocity (GHSV) for non-oxidative methane conversion reaction. Firstly, effect of different Mo loading (2, 5 and 10 wt%) was analyzed and observed that 5 wt% Mo loading is suitable for benzene formation, major product of the reaction as shown in Figure 1(a). Effect of GHSV for the reaction has been shown in Figure 1 (b) for 5wt% Mo loaded HMCM-22 (SAR-55) catalyst, and it was observed that lower space velocity (720 ml/gm-hr) is effective for aromatics (benzene) formation compared to higher space velocity (1200 ml/gm-hr). Higher space velocity gives more ethylene compared to benzene due to lower residence time. Theoretical calculations were performed using density functional theory (DFT) for the Mo/Zeolite catalyst. Molybdenum carbide formed during the reaction has been proposed to be the active phase for C-H activation of methane, C_2 coupling and aromatic formation reactions.[4] Mo_xC_y[5] type nanocluster were supposed to be suitable form of molybdenum carbide species for methane activation and C_2 coupling. In this reference, Mo_4C_2 nanocluster grafted over Al and Si site of zeolite channel was reported for C-H bond activation and a 112 kJ/mol energy barrier was observed when grafted over Al site.[6] In the study we have investigated the C-H bond activation and ethylene (primary intermediate of benzene) formation via C_2 coupling over Mo_4C_2 nanocluster. Schematic reaction paths for methane activation, ethane and ethylene formation over the Mo_4C_2 cluster and also the potential energy diagram for all these steps have been shown in Figure 2(a and b). Methane molecule gets activated over the Mo_4C_2 nanocluster with an activation barrier 112.8 kJ/mol to form CH_3 and H adsorbed on the cluster. In the next step the second CH_4 molecule dissociated over the same site with 115.5 kJ/mol activation barrier to form two CH_3 groups attached to the same Mo atom, which then undergo C-C coupling reaction with an activation barrier of 150 kJ/mole which is comparatively high to form ethane. The energy barrier for C-H bond activation of ethane (89 and 94 kJ/mol) is relatively low and forms ethylene (C_2H_4), which is the major primary unit of benzene formation, over the catalyst as also observed experimentally, in Figure 1(a) and (b).

Figure 1. (a) Product distribution for different Mo loading over HMCM-22 (SAR-55) catalyst (b) Product distribution for 5wt%Mo/HMCM-22 (SAR-55) catalyst at two different GHSV (720 & 1200 ml/gm-hr)

Figure 2. (a) Schematic paths for methane activation and C-C coupling over Mo_4C_2 nanocluster (b) Potential energy diagram for C-H activation and C_2 coupling over Mo_4C_2 nanocluster

References

1. S. Majhi, P. Mohanty, H. Wang, K. K. Pant, *Journal of Energy Chemistry*, 2013, 22, 543-554.
2. S. Ma, X. Guo, L. Zhao, S. Scott, X. Bao, *Journal of Energy Chemistry*, 2013, 22, 1-20.
3. Y. Shu, D. Ma, L. Xu, Y. Xu, X. Bao, *Catalysis Letters*, 2000, 70, 67-73.
4. D. Zhou, S. Zuo, S. Xing, *J. Phys. Chem.* C 2012, 116, 4060-4070.
5. J. Gao, Y. Zheng, G. B. Fitzgerald, J. D. Joannis, Y. Tang, I. E. Wachs, S. G. Podkolzin, *J. Phys. Chem.* C 2014, 118, 4670-4679.
6. J. Gao, Y. Zheng, J. M. Jehng, Y. Tang, I. E. Wachs, S. G. Podkolzin, *Science*, 2015, 348, 62-35.

Acknowledgement

We would like to acknowledge Gas Authority of India Limited (GAIL) for financial support

490990 CO_2 Activation By Methane in a Dual-Bed Configuration Via Methane Cracking and Iron Oxide Lattice Oxygen Transport – Concept and Materials Development

Abstract:

In this work, we explore a novel process configuration for CO_2 activation into CO by methane in a dual-bed reactor configuration by circulating supported iron oxide/iron particles with deposited carbon between two reactors. Iron was chosen as the active phase due to its ability to effectively crack methane at relevant temperatures, its favorable properties for lattice oxygen transport between the two reactors, and its low cost. In reactor 1 Fe_3O_4 is reduced to metallic iron by H_2 that is produced by catalytic cracking of methane fed to the bottom of the reactor. The reduced iron particles and the carbon deposited on them from methane cracking are then introduced to reactor 2, to which CO_2 is fed. There, the carbon is gasified by the reverse Boudouard reaction and the iron is reoxidized back to Fe_3O_4, thereby activating the CO_2 into CO. The merit of cracking the methane in reactor 1 instead of fully combusting it lies in the concentration of the entire carbon fed to the system in the product gas stream from reactor 2. This further allows oxidation of some hydrogen in reactor 1 with air to provide heat for this otherwise highly endothermic process without generating any CO_2 that is diluted with nitrogen from air.

To demonstrate the feasibility of this concept, we pursued a two-fold approach. First, the theoretical potential of the process was evaluated by thermodynamic equilibrium calculations. For both reactors, moving bed configurations with counter-current gas-solid flow and well-mixed fluidized bed configurations were investigated. Moving beds were approximated with a multi-stage equilibrium model, whereas fluidized beds were modeled as CSTRs, with two separate stages for reactor 1 (a cracking and an iron oxide reducing stage, respectively). Results indicated that moving beds offer significant advantages over well-mixed fluidized beds for both reactors. With both reactors realized as counter-current moving bed reactors, about 2.7 mol CO_2 per mol of CH_4 could be potentially activated while providing a product stream of about 82 mol-% CO, with the remainder being CO_2, at a moderate reactor temperature of 850°C.

Second, suitable materials for this process were developed. To prevent sintering of the iron/iron oxide and to facilitate fast methane cracking, reduction and oxidation kinetics, it needs to be combined with a suitable support material. Several potential support materials were investigated based on their electronic, ionic and protonic conduction properties, and cermet materials with 30 wt-% Fe_2O_3 and 70 wt-% support were produced. The reduction kinetics with H_2, the kinetics for methane cracking and the reoxidation kinetics with CO_2 were then evaluated in a TGA in a cyclic manner. Materials were further characterized by X-ray diffraction and SEM. Iron oxide reduction kinetics were fastest with $BaZr_{0.9}Y_{0.1}O_{3-\delta}$ and $SrFe_{0.5}Ti_{0.5}O_{3-\delta}$ as support material. $BaZr_{0.9}Y_{0.1}O_{3-\delta}$ support also exhibited favorable properties for methane cracking, while $SrFe_{0.5}Ti_{0.5}O_{3-\delta}$ proved to be an ineffective support for methane cracking.

In conclusion, our results indicate the high potential of this process for CO_2 activation as evidenced by the high achievable ratio of CO_2 activated per mol of CH_4 in a dual moving-bed configuration. Iron/Iron oxide supported on Zr-based ceramics, in particular $BaZr_{0.9}Y_{0.1}O_{3-\delta}$, exhibited highly promising properties for this application.

504321 Methanol to Propylene Process: An Integration of Catalyst Development and Reactor Engineering

Abstract:

Methanol, which can be produced from coal or natural gas, is one of the most important C1 chemicals. The utilization of methanol to make light olefins including ethylene and propylene has been recently commercialized. However, in many applications, it is desired to maximize the yield of propylene, and thus methanol to propylene (MTP) process becomes increasingly interesting. It is normally hard to achieve high selectivity to propylene in methanol conversion. In this work, we have developed a multi-functional catalyst, over which three reactions, i.e. methanol conversion, ethylene alkylation, and C4+ cracking, can be catalyzed. By integrating these three reactions, we attempted to develop a MTP process with the selectivity to propylene higher than 75%. Evaluation of this catalyst has been carried out in laboratory-scale micro-fluidized bed reactor. It shows that the selectivity to propylene of 50% has been achieved in the methanol conversion reaction. We also studied the ethylene alkylation reaction and C4+ cracking reaction, and identified the optimal operation conditions. Based on the results obtained in the micro-fluidized bed reactor, we proposed an integrated process to maximize the selectivity of propylene. In achieving this goal, a turbulent fluidized bed reactor, in which the optimal operation conditions suitable for both methanol conversion and ethylene alkylation have been considered, has been designed and constructed in our pilot plant. The reactor can be operated with 300t/a methanol feed rate. Pilot plant experiments have been conducted for validating the process concept and scaling up the reactor. It shows the selectivity to propylene of 75%, in addition to a selectivity of ethylene of 10%, can be achieved in the pilot plant. This MTP process can be seen as a successful example of the integration of catalyst development and reactor engineering.

504320 Direct Production of Value-Added Chemicals Via Fischer-Tropsch Synthesis

Abstract:

Fischer-Tropsch synthesis (FTS) is a heterogeneous catalytic process for sustainable production of fuel and value-added chemicals via syngas derived from coal, biomass and natural gas. It is considered as an attractive alternative non-petroleum-based production route. FTS is a very complicated catalytic reaction process with co-production of paraffin, olefin, alcohol and other products. The direct synthesis of high value-added olefins and higher alcohols from syngas has drawn particular attention due to due to its process simplicity, low energy consumption and clean utilization of carbon resource. For the purpose of high selectivity to specific olefins or higher alcohols, it is very desirable to regulate the reaction network for certain reactions to take place preferentially. Obviously, it is necessary to control CO activation and then both carbon chain growth and termination for high selectivity to target product by development new catalyst systems and optimization of reaction conditions.

Reaction network during syngas conversion was firstly investigated by probe molecules techniques [1]. Alcohol, aldehyde, olefin as probe molecules were in-situ introduced into the reaction system. Several parallel reactions and second reactions were clarified. The existence of surface acyl species indicated that the insertion of CO rather than hydroxyl was the underlying mechanism for higher alcohols synthesis over modified FT catalysts along with the support from surface simulation. In addition, the dual active sites for higher alcohols formation was further investigated [2]. Several kinds of dual active sites (Cu/FeC_x, CuCo, and Co/Co_2C) were controllably prepared for higher alcohol synthesis. Both Cu/FeC_x and Co/Co_2C could afford high selectivity for long chain alcohols, while only short chain alcohols were produced for CuCo. The difference of the detailed surface structure and the corresponding catalytic performance was discussed in detail.

Fischer-Tropsch to olefins (FTO) by using new active phase was also investigated. Co_2C nanoprisms with exposed facet of {101} and {020} were prepared and tested for syngas conversion [3]. It is widely accepted the formation of Co_2C is considered to be one of the main reasons for deactivation of the Co-based FT reaction because it possesses very low activity for CO hydrogenation, with methane and CO_2 as the main products. However, due to the strong facet effect of Co_2C, Co_2C nanoprisms afforded very high selectivity for the production of lower olefins (60.8 C%), while little methane (about 5.0 C%) was generated with the ratio of olefin/paraffin amongst the C_{2-4} products being as high 30. The product distribution deviates markedly from the classical ASF distribution with the highest selectivity to propylene. The effect of several promoters and the model for modified ASF distribution were discussed [4]. In addition, the nanoeffects of Co_2C were studied in details to reveal the structure-performance relationship.

References

[1] *Catalysis Science & Technology*ï¼Œ2015, 5, 4224

[2] *Catalysis Science & Technology*ï¼Œ2013, 3, 1591

[3] *Nature*, 2016, 538, 84..

[4] *ACS Catalysis*, 2017, 7, 3622.

504319 An Investigation into the Effect of Li and Mn Promotions on the Activity and Selecivities to Olefins and Alcohols of Co@Co_2C/Activated Carbon (AC) for Fischer-Tropsch Reaction

Abstract:

Abstract:

Utilization of non-precious transition metals for high alcohols and olefins synthesis is of a great importance in heterogeneous catalysis. We synthesized successfully Li and Mn promoted AC supported cobalt metal-carbide (Co@Co_2C/AC) catalysts, which present remarkable activity of more than 200g/kg-cat.h and selectivities of ca. 65 C% both for alpha-alcohols and alpha-olefins via the Fischer-Tropsch reaction at the mild conditions. The formation of the stable cobalt carbide and the Co@Co_2C interface are found essential for the observed reactivity. The results of characterization from In-situ EXAFS, in-situ XRD, HRTEM, XSP, and chemo-adsorption of H_2 and CO experiments showed that Mn promotes the formation of Co_2C species during Fischer-Tropsch reaction and Li increased the CO uptake, while both promoters have little impact on the amount of H_2 uptakes. Density functional theory calculations show that Co_2C is highly efficient for CO non-dissociative adsorption, behaving as noble-metal-like, whereas the Co metal is highly active for CO dissociative adsorption and the subsequent carbon-chain growth. The interface between the cobalt metal and its carbide phase as well as the dual sites available at the interface for facile CO insertion to C_nH_m-groups formed on the Co sites. The remarkably high alpha-alcohol and alpha-olefins selectivities from the Li and Mn promotion effects, namely the formation of more interfacial active sites between Co and Co_2C and enhancement of the ratio for chemo-adsorption CO to H_2 uptake on the surface of Co@Co_2C/AC catalysts, paves a new way for the design of the new catalysts for synthesizing various high value-products in syngas applications.

504154 Direct Production of Gasoline-Range Hydrocarbons from Carbon Dioxide over Iron-Based Multifunctional Catalysts

Abstract:

Converting CO_2 from a detrimental greenhouse gas into value-added liquid fuels not only contributes to mitigating CO_2 emissions, but also reduces dependence on petrochemicals. Unfortunately, the activation of CO_2 and its hydrogenation to hydrocarbons or other alcohols are challenging tasks because CO_2 is a fully oxidized, thermodynamically stable and chemically inert molecule. Most research to date, not surprisingly, is focusing on selective hydrogenation of CO_2 to short-chain products, while few studies on long-chain hydrocarbons, such as gasoline-range (C_5–C_{11}) hydrocarbons. The key to this process is to search for a high efficient catalyst.

Herein, we have succeeded in preparing high efficient, stable, and multifunctional Na–Fe_3O_4/Zeolite catalysts for the direct production of gasoline-range (C_5–C_{11}) hydrocarbons from CO_2 hydrogenation. This catalyst displayed record selectivity towards C_5–C_{11} hydrocarbons (78%) as well as low CH_4 and CO selectivity under industrial relevant conditions, and gasoline fraction are mainly isoparaffins and aromatics thus favouring the octane number. Moreover, the composition of C_5–C_{11} can be tuned by the choice of zeolite type and the integration manner of multifunctional catalyst. When Na–Fe_3O_4 nanocatalyst was matched with HZSM-5 zeolite and integrated by granule-mixing manner, up to 63% of aromatics in gasoline fraction were produced. While Na–Fe_3O_4 was matched with HMCM-22 zeolite, up to 60% of isoparaffins in gasoline fraction were obtained under the dual-bed configuration. Not only that, the multifunctional catalyst exhibited a remarkable stability for 1,000 h on stream, which definitely has the potential to be a promising industrial catalyst for CO_2 utilization to liquid fuels.

In-depth characterizations indicate that, during CO_2 hydrogenation reaction, this catalyst enables RWGS over Fe_3O_4 sites, olefin synthesis over Fe_5C_2 sites, and oligomerization/aromatization/isomerization over zeolite acid sites. The concerted action of the active sites calls for precise control of their structures and proximity. This study paves a new path for the synthesis of liquid fuels by utilizing CO_2 and H_2. Furthermore, it provides an important approach for dealing with the intermittency of renewable sources (sun, wind and so on) by storing energy in liquid fuels.

References

1. G. A. Olah, A. Goeppert and G. S. Prakash, *J. Org. Chem.*, 2008, **74**, 487-498.

2. R. W. Dorner, D. R. Hardy, et al., *Energy Environ. Sci.*, 2010, **3**, 884-890.

3. W. Wang, S. Wang, X. Ma and J. Gong, *Chem. Soc. Rev.*, 2011, **40**, 3703-3727.

4. F. Jiao, J. Li, X. Pan, et al., *Science*, 2016, **351**, 1065-1068.

5. P. Zhai, C. Xu, R. Gao, et al., *Angew. Chem. Int. Ed.*, 2016, **55**, 9902-9907.

6. J. Wei, Q. Ge, R.Yao, et al., *Nat. Commun.*, 2017, **8**, 15174.

Innovative Syngas Production Catalyst for Utilization of CO_2 (CT-CO_2AR™)

Takenori Kanda, Osamu Hirohata, Tomoyuki Mikuriya and Fuyuki Yagi
Chiyoda Corporation, Kanagawa, Japan

1. Introduction

Production of synthesis gas from natural gas and conversion of synthesis gas to fuels and chemicals have become increasingly attractive from the viewpoint of energy resource diversification. The technology of synthesis gas production by the reforming of natural gas with CO_2 (hereafter referred to as "CO_2-reforming") is an effective solution for the issues of efficient energy resource utilization and environmental preservation.

Chiyoda has developed and commercialized a new noble metal based reforming catalyst named CT-CO_2AR™ which has a higher resistance to carbon formation and produces synthesis gas more effectively compared with commercial reforming catalyst.

In this paper, we will elaborate on the characteristics and the competitiveness of the CT-CO_2AR™ catalyst and the results of the operation of the commercial plant.

2. Characteristics of CT-CO_2AR™ catalyst

(1) The characteristics of CO_2/steam reforming

Synthesis gas is used for hydrogen production and as the feed stock of chemicals, and the H_2/CO ratio of synthesis gas depends on the purpose.

For example, hydrogen production requires an H_2/CO ratio of more than 3.0. For ethylene glycol production, the required H_2/CO ratio is 2.0. For chemical production through oxo synthesis, the required H_2/CO ratio is 1.0. For the feedstock of acetic acid or MMA requires only CO or as low as possible H_2/CO ratio.

When hydrocarbons such as natural gas or naphtha and steam are used as reformer feedstocks, the steam reforming reaction shown in Eq. 1 is normal and the H_2/CO ratio of produced synthesis gas becomes more than 3.0.

$$CH_4 + H_2O \Leftrightarrow 3\ H_2 + CO \qquad (1)$$

However, when CO_2 and steam are used as reforming agents, the CO_2 reforming reaction shown in Eq. 2 occurs.

$$CH_4 + CO_2 \Leftrightarrow 2\ H_2 + 2\ CO \qquad (2)$$

Therefore, by changing the ratio of feed CO_2 and steam, various H_2/CO ratios of synthesis gas can be produced.

However, during the reforming reactions, the carbon formation reaction shown in Eq. 3 occurs on the catalyst surface. Deposited carbon causes the deactivation of the catalyst and the occlusion of the catalyst tube[1] leading to an unstable operation.

$$CH_4 \Leftrightarrow C(s) + 2H_2 \qquad (3)$$

To prevent carbon formation for a stable operation, commercial Ni based reforming catalysts require steam and CO_2 greater than the stoichiometric ratio.

Fig. 1 shows the relation between the operation condition and risk of carbon formation[2]. The bold line in the center represents carbon limit line at 600 ℃. In the black area to the left of this line, the process gas has enough high thermodynamic potential for carbon formation at 600 ℃. Operation around this line cannot be carried out with commercial Ni based reforming catalysts. They are usually operated within the excess steam carbon ratio (S/C) zone (blue area).

On the other hand, our new commercialized noble metal based reforming catalyst (CT-CO_2AR™) enables a stable operation under a lower S/C condition (red area) because it has a higher resistance to carbon formation.

In next section, the advantages of a lower S/C operation will be shown.

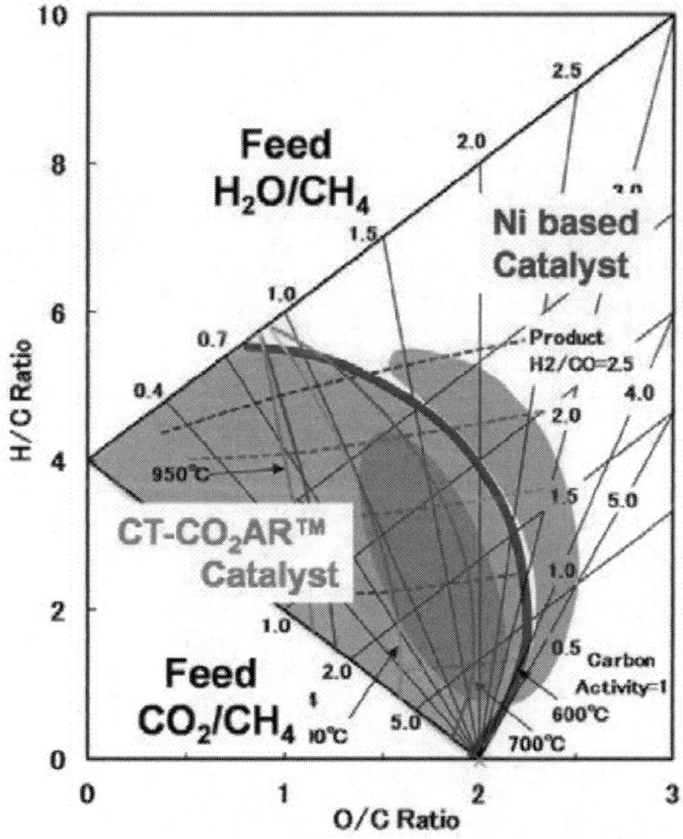

Fig. 1. Carbon Limit Diagram

(2) The advantages of a lower S/C ratio operation

In the case of the synthesis gas production process shown in Fig. 2, the advantages of a lower S/C operation are as follows:

1. Since the amount of steam and CO_2 is reduced, the heat duty to heat steam and CO_2 can be reduced. Therefore, the fuel consumption of the reformer can be reduced.
2. Fig. 3 shows the comparison of synthesis gas concentration at the reformer outlet between a high S/C (=2.2) operation and a low S/C (=1.0) operation in the case of $H_2/CO=1.0$
   In the case of a low S/C operation, the amount of CO_2 can be reduced and as a result, the system volume of CO_2 removal is downsized.
   In addition, because of lower CO_2/C, the capacity of the CO_2 compressor is downsized.
   Therefore, both the CAPEX and the OPEX of CO_2 removal and CO_2 compressor are reduced.
3. Fig. 3 shows that a lower S/C operation can reduce the amount of H_2O at the reformer outlet and it is possible to reduce the amount of process condensate.
   In the case of a low S/C operation, the total amount of produced steam is less than that of a high S/C operation. However, the amount of steam uses as the reforming agent in a low S/C operation is significantly less than that in a high S/C operation.
   As a result, a low S/C operation can increase the amount of export steam.

From these reasons, the CT-CO_2AR™ catalyst is preferable to a commercial Ni based catalyst.

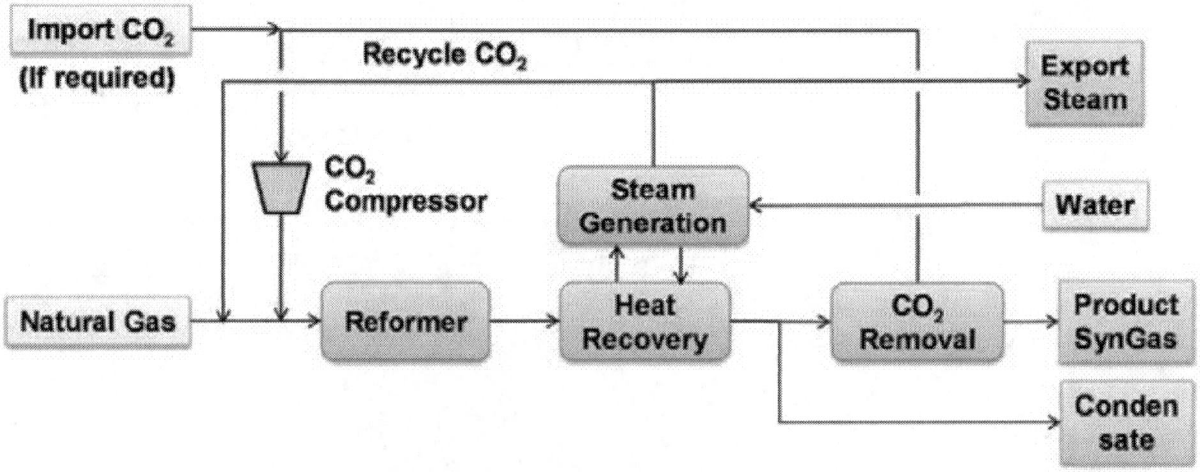

Fig. 2. Block Flow Diagram of Syngas production

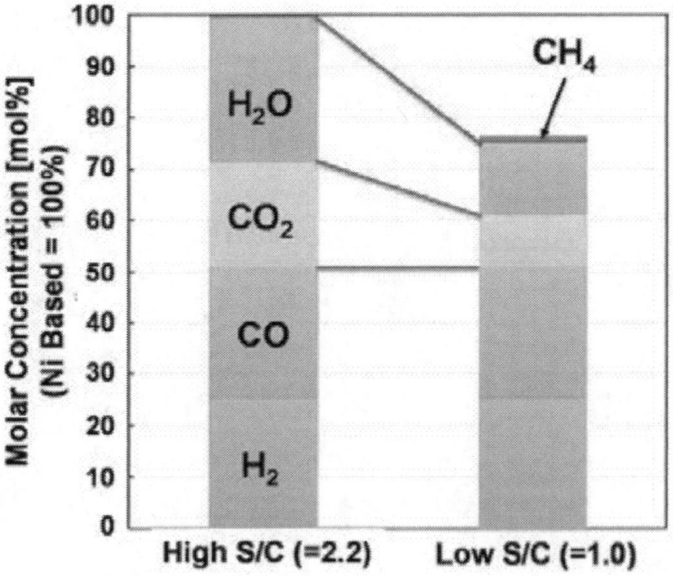

Fig. 3. Typical Example of synthesis gas composition in H_2/CO=1.0

(3) The characteristics of the CT-CO_2AR™ catalyst

Fig. 4 shows the result of EPMA (Electron Probe Micro Analyzer) of the CT-CO_2AR™ catalyst. Generally, the surface reaction on the reforming catalyst is the rate-determining step. Therefore, to reduce the amount of active metal, the active metal is supported on only the outer surface of the carrier like egg-shell.

Fig. 5 shows the accelerated experiment result. The CT-CO_2AR™ catalyst keeps its activity due to its high resistance to carbon formation.

We built the reaction kinetics model shown in Fig. 6 to estimate the behavior of the CT-CO_2AR™ catalyst.

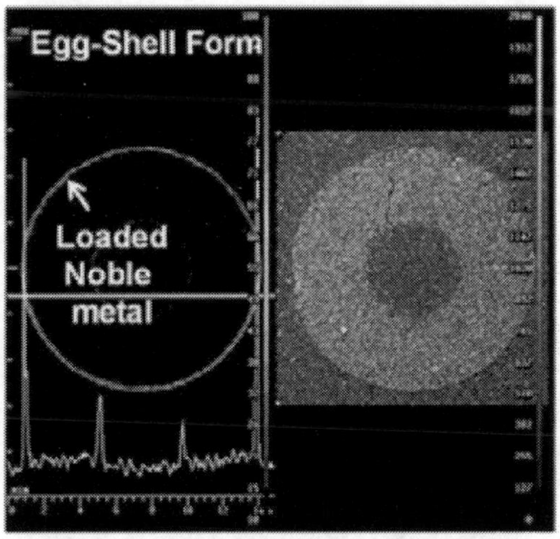

Fig. 4. EPMA result of CT-CO_2AR™ Catalyst

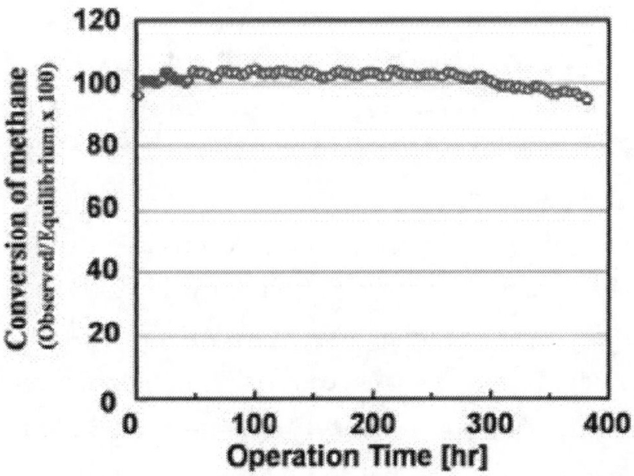

Fig. 5. Result of accelerated experiment

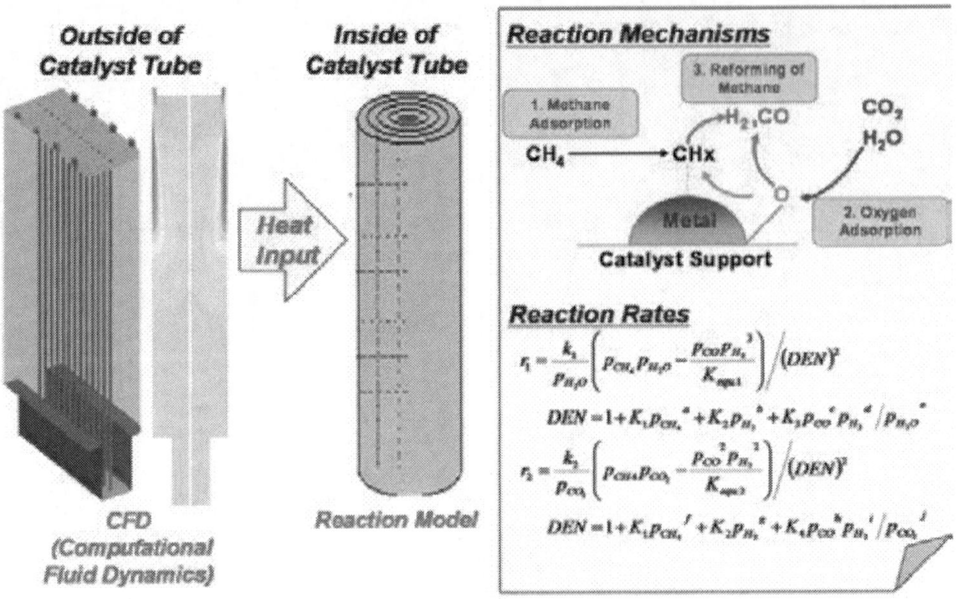

Fig. 6. Reaction model of CT-CO_2AR™ catalyst

3. Advantages of CT-CO_2AR™ catalyst

(1) Study Basis

Regarding the competitiveness of the CT-CO_2AR™ catalyst compared to a commercial Ni based reforming catalyst, we have compared the CT-CO_2AR™ catalyst and Ni based catalysts from the following three points of view based on the block flow shown in Fig. 2.

1. Total energy input (feed + fuel hydrocarbon + utility consumption)
2. Manufacturing costs of synthesis gas
3. CO_2 Emission per final products

[CO_2 Emission] = [CO_2 in Flue and Utility] - [Import CO_2] + [CO_2 in downstream process to produce the final products]

In this study we considered cases with H_2/CO ratios of 1.0 and 0.5. The conditions and estimated final product of each case are listed in Table 1.

(2) Study Result

From Fig.7-9 show the results of the process study. In the case of H_2/CO=1.0, compared to Ni based catalysts, the CT-CO_2AR™ catalyst can cut total energy input by 11% (Fig. 7) manufacturing cost by 12% (Fig. 8).Regarding CO_2 emission, the CT-CO_2AR™ catalyst can achieve a reduction of 22% compared with a Ni based catalyst (Fig. 9). In the case of H_2/CO=0.5, CT-CO_2AR™ catalyst is more competitive. Approximately 30% of both total energy input and manufacturing cost can be reduced and 76% of the amount of CO_2 emission can be reduced. From this study, the CT-CO_2AR™ catalyst is competitive compared with Ni based catalysts both in terms of CAPEX and OPEX.

Table 1. Process study conditions

H_2/CO Ratio	Catalyst	Feed S/C	Amount of (H_2+CO)	Final Products	Capacity
[-]	[-]	[-]	[Nm3/h]	[-]	[ton/hr]
CO (0.5)	Ni Based	3.3	37,500 (25,000 as CO)	Acetic Acid	500,000
	CT-CO_2AR™	0.4			
1.0	Ni Based	2.2	25,000	n-butanol (oxo synthesis)	270,000
	CT-CO_2AR™	1.0			

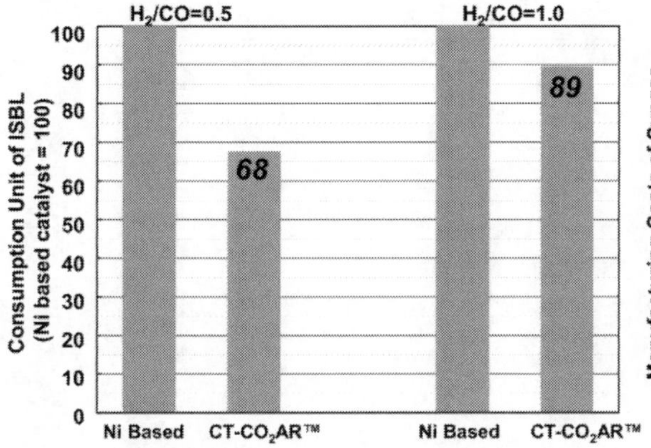

Fig. 7. Total Energy Input

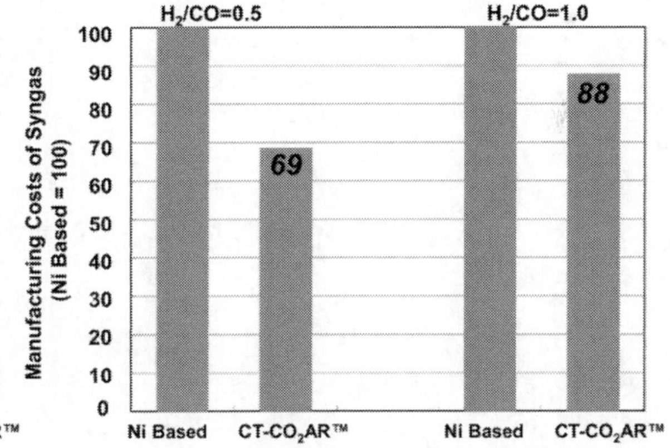

Fig. 8. Manufacturing Costs of Syngas

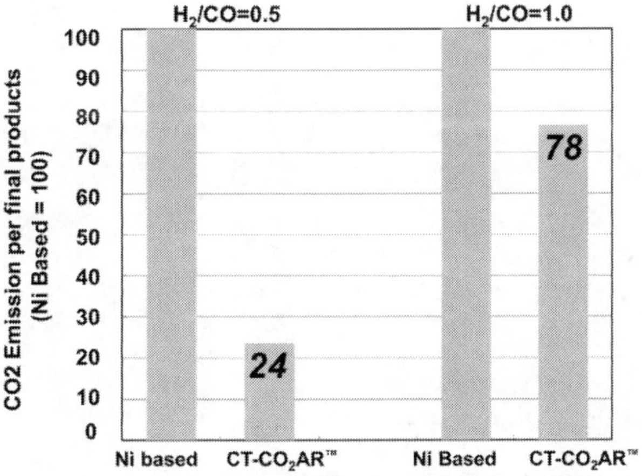

Fig. 9. CO_2 Emission per final products

4. Application to Commercial Plant

The CT-CO_2AR™ catalyst has been operated in a commercial plant in Japan since 2014. The information of the commercial plant is listed in Table 2. In this project, the reforming catalyst has been changed to the CT-CO_2AR™ catalyst and the S/C ratio at the inlet of the prereformer has reduced from 2.2 to 1.4 without changing the amount and specification of the produced synthesis gas.

Regarding the comparison between analysis result using the reaction model and the operation result, a pressure drop through the reformer tube is shown in Fig. 10 and the methane concentration of the synthesis gas is shown in Fig.11.

Since a significant increase in the pressure drop and methane concentration was not observed, a stable operation was achieved during the whole operation time.

In addition, our reaction model can accurately predict the performance of the CT-CO_2AR™ catalyst.

Table 2. Information about the commercial plant

Client's Name :	KH Neochem Co., Ltd
Plant Location :	Yokkaichi, Japan
Product :	Synthesis gas H_2/CO ratio=1.0
Product capacity :	5,000 Nm^3/hr (as H_2+CO)
Project Scope :	Catalyst supply to the existing plant
Feed Hydrocarbon :	Natural gas or Naphtha

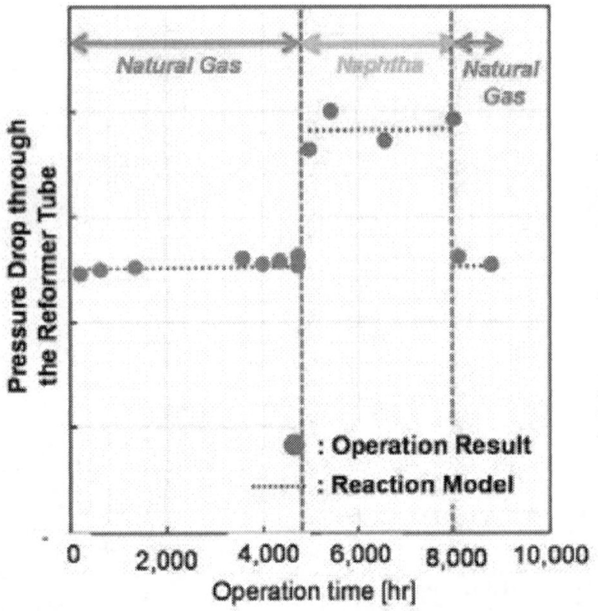

Fig. 10. Pressure difference of a reformer tube

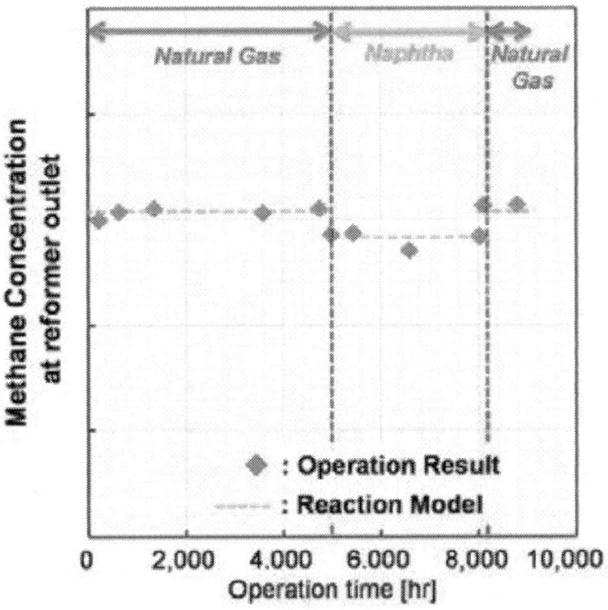

Fig. 11. Methane concentration at reformer outlet

Fig. 12 shows the comparison of the reformer duty and the amount of input natural gas between before and after replacing the CT-CO_2AR™ catalyst. By replacing the CT-CO_2AR™ catalyst, 10% of the natural gas input is reduced and this result matches the result of process study shown in section 3 (Advantages of the CT-CO_2AR™ catalyst).

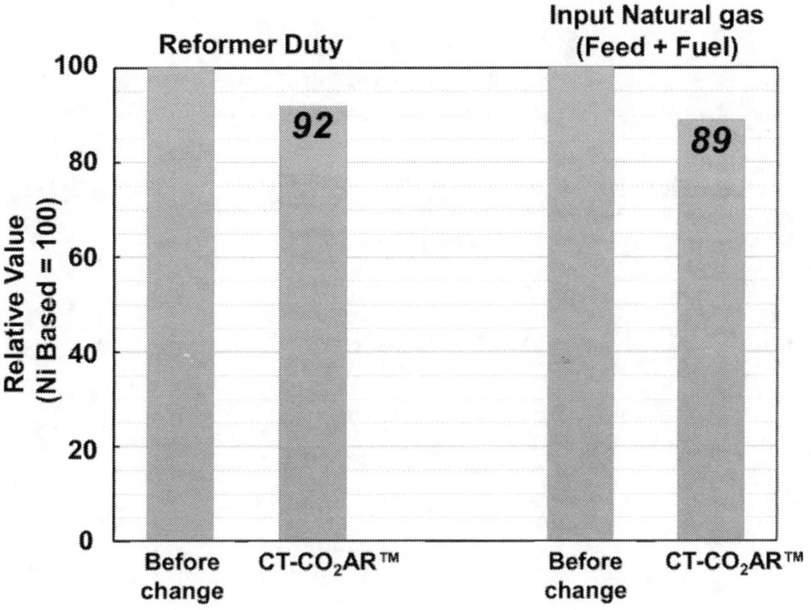

Fig. 12. Comparison result of commercial plant operation

5. Conclusion

The CT-CO_2AR™ catalyst enables a lower S/C condition operation because of its higher resistance to carbon formation.

Compared with a commercial catalyst, the CT-CO_2AR™ catalyst demonstrates high competitiveness in both manufacturing cost and CO_2 emission in the case of H_2/CO ratio = 0.5 and 1.0, especially, 0.5.

The CT-CO_2AR™ catalyst was installed in a commercial plant in 2014 and has been operating stably. 10% reduction of natural gas input was achieved in the commercial operation.

6. References

1) J.R. Rostrup-Nielsen, J. H. B. Hansen, *J. Catal.*, **144**, 38 (1993)

2) N.R.Udengaard, J.-H. Bak Hansen and D.C.Hanson, Oil & Gas J., Mar.9(1992)

Solar Thermochemical Splitting of H_2O and CO_2 using Nonvolatile Metal Oxides

(Principal Author)
Rahul Bhosale
Department of Chemical Engineering
College of Engineering
Qatar University, Doha, Qatar
rahul.bhosale@qu.edu.qa

(Other Authors)
Gorakshnath Takalkar
Department of Chemical Engineering
College of Engineering
Qatar University, Doha, Qatar

Anand Kumar
Department of Chemical Engineering
College of Engineering
Qatar University, Doha, Qatar

Fares A. AlMomani
Department of Chemical Engineering
College of Engineering
Qatar University, Doha, Qatar

Prepared for Presentation at
American Institute of Chemical Engineers
2017 Fall Meeting
Minneapolis, MN, USA
October 29 – November 3, 2017

Solar Thermochemical Splitting of H_2O and CO_2 using Nonvolatile Metal Oxides

Rahul R. Bhosale, Anand Kumar, Gorakshnath Takalkar, Fares A. AlMomani

Department of Chemical Engineering
College of Engineering
Qatar University, Doha, Qatar

Keywords: Thermochemical Reactions, Metal Oxides, Solar Fuels, H_2O/CO_2 Splitting

Abstract

In this study, a Co ferrite/ZrO_2 ceramic nanomixture was synthesized by the propylene-oxide-assisted sol–gel method. Cobalt, iron, and zirconium nitrate salts were used as the metal precursors, and ethanol and propylene oxide were used as the solvent and gelation agent, respectively. The prepared gels were aged, dried, and annealed at 1000°C for different dwell times. The annealed Co ferrite/ZrO_2 powder was analyzed by powder X-ray diffraction (PXRD), Brunauer–Emmett–Teller surface area analysis (BET), and energy-dispersive spectroscopy (EDS),. The characterization results are reported in detail.

1. Introduction

Metal-oxide-based solar thermochemical water and carbon dioxide splitting is considered a promising technology to satisfy future energy demands [1]. Zinc oxide [2-5], iron oxide [5-6], tin oxide [7-8], doped iron oxides (ferrites) [9-14], ceria-based redox materials [15-19], and perovskites [20-24] are used for splitting. Among these oxides, ferrites have been used extensively for solar thermochemical water splitting; however, their use is extremely limited for CO_2 splitting.

However, the major issue limiting the use of ferrites in solar thermochemical cycles is the increased size of the ferrite grains obtained as a result of high-temperature processing. In this study, a Co ferrite/ZrO_2 ceramic mixture was synthesized by the propylene-oxide-assisted sol–gel method [25-27], exhibiting mixing of Co ferrite and ZrO_2 with a uniform distribution. The effect of annealing time on the phase and chemical composition, crystallite size, specific surface area (SSA), pore volume, and morphology of the Co ferrite/ZrO_2 ceramic nanomixture was investigated, and the results were reported in detail.

2. Experimental

2.1 Sol–Gel Synthesis

Metal precursors of cobalt, iron, and zirconium were first mixed in a specific weight ratio and dissolved in ethanol using a sonication bath. Once a visibly clear solution was obtained, a predetermined amount of propylene oxide was slowly added, affording a gel. The gel was aged, dried, and annealed at 1000°C (at various annealing times) in air. Fig. 1 shows the flow chart for the synthesis of the Co ferrite/ZrO_2 ceramic nanomixture.

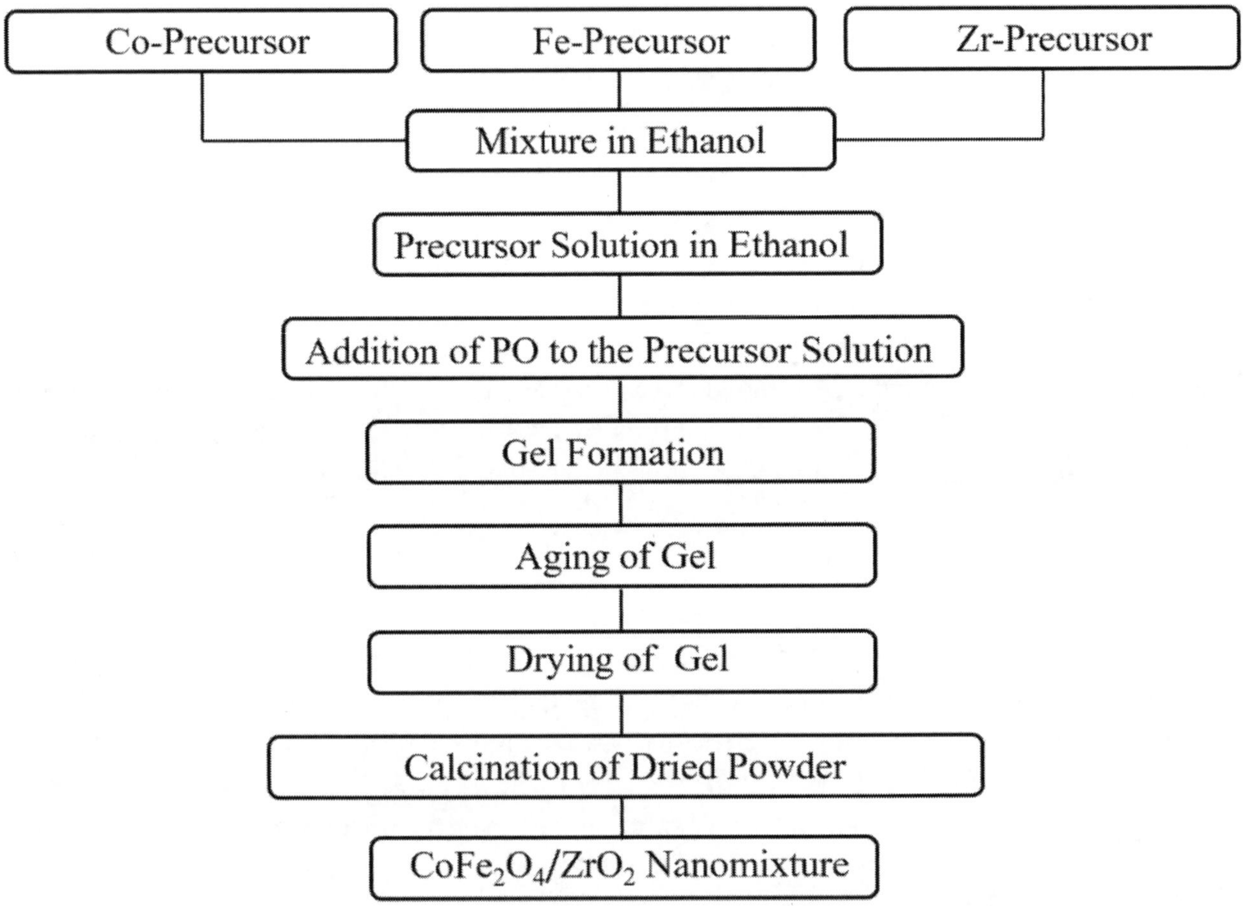

Fig. 1. Flow chart for the synthesis of the Co ferrite/ZrO_2 ceramic nanomixture by the sol–gel method.

2.2 Characterization

To investigate the effects of annealing on the physiochemical properties of the Co ferrite/ZrO_2 ceramic nanomixture, the Co ferrite/ZrO_2 powder obtained after annealing was characterized by powder X-ray diffraction (PXRD), Brunauer–Emmett–Teller (BET) surface area analysis, scanning electron microscopy (SEM) with a field-emission gun, X-ray energy-dispersive spectroscopy (EDS), and high-resolution transmission electron microscopy (HR-TEM).

3. Results & Discussion

The effect of annealing time on the physicochemical properties of the sol–gel-derived Co ferrite/ZrO_2 ceramic nanomixture was investigated. For this purpose, the Co ferrite/ZrO_2 gels (prepared using 20 mL of propylene oxide, aged for 24 h, and dried at 100°C) were annealed at 1000°C for 1–5 h. The calcined powders thus obtained were characterized by PXRD and EDS to determine the phase and chemical composition. The results shown Fig. 2 clearly indicated that the phase and chemical composition of the sol–gel-derived Co ferrite/ZrO_2 ceramic nanomixture is unaffected by the increase in the annealing time.

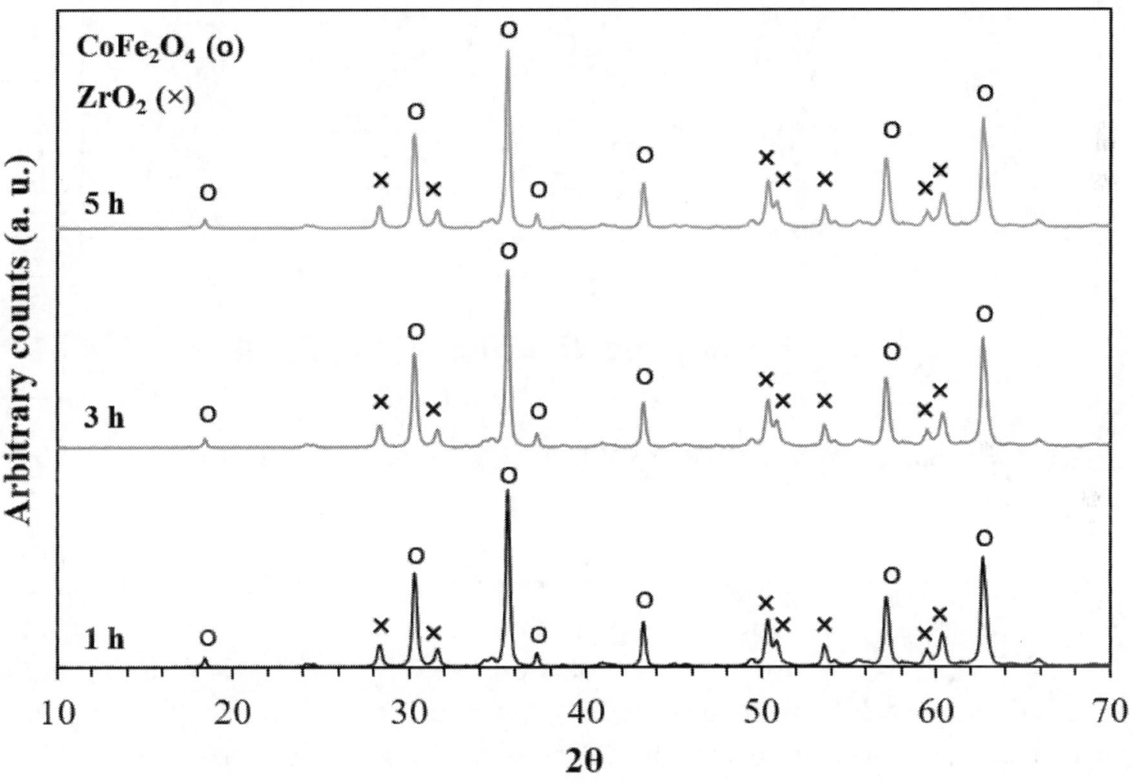

Fig. 2 – Effect of the annealing time on the phase composition of sol–gel-derived Co ferrite/ZrO_2 ceramic nanomixture.

The effect of the annealing time on the crystallite size of the sol–gel-derived Co ferrite/ZrO_2 ceramic nanomixture was also determined. Fig. 3 shows the results obtained. With the increase in the annealing time from 1 to 3 h, the crystallite size increases by 14.86 nm. With the further increase in the annealing time from 3 to 5 h, the crystallite size further increases by 22.83 nm. In summary, with the increase in the annealing time from 1 to 5 h, the size of the Co ferrite/ZrO_2 crystallites increases from 53.66 to 91.35 nm.

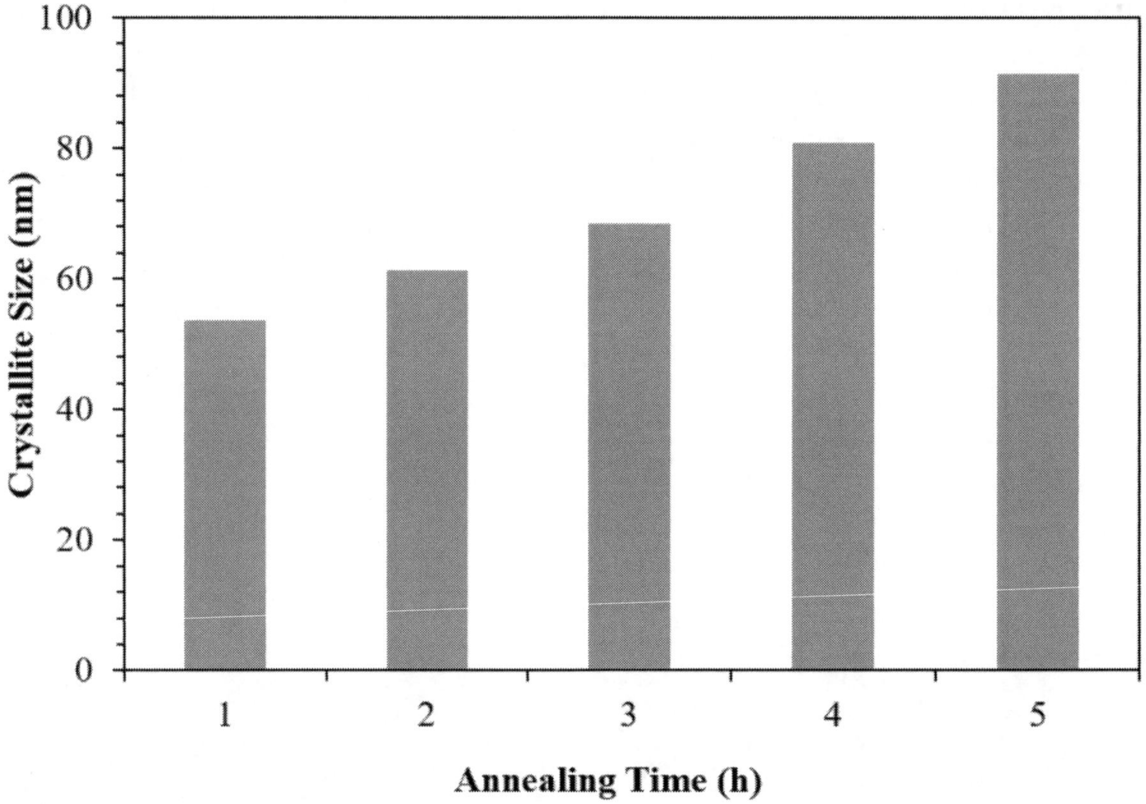

Fig. 3 – Effect of the annealing time on the crystallite size of the sol–gel-derived Co ferrite/ZrO_2 ceramic nanomixture.

In addition to the composition and crystallite size, the SSA and pore volume are also crucial parameters to be investigated. The results obtained by BET analysis indicated that with the increase in the annealing time from 1 to 3 h, the SSA decreases by 14.33%. Similarly, with the increase in the annealing time from 3 to 5 h, the SSA decreases by 13.57%. With the increase in the annealing time from 1 to 5 h, the pore volume decreases from 1.52×10^{-3} cm^3/g to 1.03×10^{-3} cm^3/g.

4. Conclusion

In this paper, a Co ferrite/ZrO_2 ceramic nanomixture, with a pure phase composition, high SSA, pore volume, nanoparticulate morphology, and a uniform distribution of Co ferrite and ZrO_2 in the mixture, was synthesized by the propylene-oxide-assisted sol–gel method. The phase and chemical composition of the Co ferrite/ZrO_2 nanomixture was not affected under all of the annealing conditions. With the increase in the annealing time from 1 to 5 h at an annealing temperature of 1000°C, the crystallite size increased by 37.69 nm. with the increase in the annealing time from 1 to 5 h, the SSA and pore volume decreased by 2.59 m^2/g and 0.484×10^{-3} cm^3/g, respectively.

5. Acknowledgments

This publication was made possible by the NPRP grant (NPRP8-370-2–154) from the Qatar National Research Fund (a member of Qatar Foundation). The statements made herein are solely the responsibility of author(s).

References

[1] J.R. Scheffe, A. Steinfeld, Oxygen exchange materials for solar thermochemical splitting of H_2O and CO_2: a review, Materials Today 17 (2014) 341-348.

[2] P.G. Loutzenhiser, A. Steinfeld, Solar syngas production from CO_2 and H_2O in a two-step thermochemical cycle via Zn/ZnO redox reactions: thermodynamic cycle analysis, International Journal of Hydrogen Energy 36 (2011) 12141-12147.

[3] A. Stamatiou, P. Loutzenhiser, A. Steinfeld, Solar syngas production from H_2O and CO_2 via two-step thermochemical cycles based on Zn/ZnO and FeO/Fe_3O_4 redox reactions: Kinetic analysis, Energy & Fuels 24 (2010) 2716-2722.

[4] R. R. Bhosale, A. Kumar, F. AlMomani, R. B. Gupta, Solar thermochemical $ZnO/ZnSO_4$ water splitting cycle for hydrogen production, International Journal of Hydrogen Energy (2017) in-press.

[5] P. Loutzenhiser, M. Gálvez, I. Hischier, A. Stamatiou, A. Frei, A. Steinfeld, CO_2 splitting via two-step solar thermochemical cycles with Zn/ZnO and FeO/Fe_3O_4 redox reactions II: Kinetic analysis, Energy & Fuels 23 (2009) 2832-2839.

[6] R. R. Bhosale, A. Kumar, van den Broeke, Leo JP, S. Gharbia, D. Dardor, M. Jilani, J. Folady, M.S. Al-Fakih, M.A. Tarsad, Solar hydrogen production via thermochemical iron oxide–iron sulfate water splitting cycle, International Journal of Hydrogen Energy 40 (2015) 1639-1650.

[7] P. Charvin, S. Abanades, F. Lemont, G. Flamant, Experimental study of $SnO_2/SnO/Sn$ thermochemical systems for solar production of hydrogen, AIChE Journal 54 (2008) 2759-2767.

[8] R. R. Bhosale, A. Kumar, P. Sutar, Thermodynamic analysis of solar driven SnO_2/SnO based thermochemical water splitting cycle, Energy Conversion & Management 135 (2017) 226-235.

[9] R. R. Bhosale, A. Kumar, F. AlMomani, U. Ghosh, P. Sutar, G. Takalkar, A. Ashok, I. Alxneit, Effectiveness of Ni incorporation in iron oxide crystal structure towards thermochemical CO_2 splitting reaction, Ceramics International 43 (2017) 5150-5155.

[10] C.C. Agrafiotis, C. Pagkoura, A. Zygogianni, G. Karagiannakis, M. Kostoglou, A.G. Konstandopoulos, Hydrogen production via solar-aided water splitting thermochemical cycles: Combustion synthesis and preliminary evaluation of spinel redox-pair materials, International Journal Hydrogen Energy 37 (2012) 8964-8980.

[11] R. R. Bhosale. A. Kumar, F. AlMomani, I. Alxneit, Propylene oxide assisted sol–gel synthesis of zinc ferrite nanoparticles for solar fuel production, Ceramics International 42 (2016) 2431-2438.

[12] M. Roeb, N. Gathmann, M. Neises, C. Sattler, R. Pitz-Paal, Thermodynamic analysis of two-step solar water splitting with mixed iron oxides, International Journal of Energy Research 33 (2009) 893-902.

[13] R.R. Bhosale, R.V. Shende, J.A. Puszynski, Thermochemical water-splitting for H_2 generation using sol-gel derived Mn-ferrite in a packed bed reactor, International Journal of Hydrogen Energy 37 (2012) 2924-2934.

[14] J.R. Scheffe, J. Li, A.W. Weimer, A spinel ferrite/hercynite water-splitting redox cycle, International Journal of Hydrogen Energy 35 (2010) 3333-3340.

[15] R.R. Bhosale, A. Kumar, F. AlMomani, I. Alxneit, Sol–gel derived CeO_2–Fe_2O_3 nanoparticles: Synthesis, characterization and solar thermochemical application, Ceramics International 42 (2016) 6728-6737.

[16] R.R. Bhosale, A. Kumar, F. AlMomani, U. Ghosh, S. Al-Muhtaseb, R. Gupta, I. Alxneit, Assessment of $Ce_xZr_yHf_zO_2$ based oxides as potential solar thermochemical CO_2 splitting materials, Ceramics International 42 (2016) 9354-9362.

[17] J.R. Scheffe, A. Steinfeld, Thermodynamic analysis of cerium-based oxides for solar thermochemical fuel production, Energy & Fuels 26 (2012) 1928-1936.

[18] W.C. Chueh, C. Falter, M. Abbott, D. Scipio, P. Furler, S.M. Haile, A. Steinfeld, High-flux solar-driven thermochemical dissociation of CO_2 and H_2O using nonstoichiometric ceria, Science 330 (2010) 1797-1801.

[19] P. Furler, J.R. Scheffe, A. Steinfeld, Syngas production by simultaneous splitting of H_2O and CO_2 via ceria redox reactions in a high-temperature solar reactor, Energy & Environmental Science 5 (2012) 6098-6103.

[20] J.R. Scheffe, R. Jacot, G.R. Patzke, A. Steinfeld, Synthesis, Characterization, and Thermochemical Redox Performance of Hf^{+4} , Zr^{+4} , and Sc^{+3} Doped Ceria for Splitting CO_2, The Journal of Physical Chemistry C 117 (2013) 24104-24114.

[21] J.R. Scheffe, D. Weibel, A. Steinfeld, Lanthanum–strontium–manganese perovskites as redox materials for solar thermochemical splitting of H_2O and CO_2, Energy & Fuels 27 (2013) 4250-4257.

[22] A.H. McDaniel, E.C. Miller, D. Arifin, A. Ambrosini, E.N. Coker, R. O'Hayre, W.C. Chueh, J. Tong, Sr-and Mn-doped $LaAlO_{3-\delta}$ for solar thermochemical H_2 and CO production, Energy & Environmental Science 6 (2013) 2424-2428.

[23] M. Gálvez, R. Jacot, J. Scheffe, T. Cooper, G. Patzke, A. Steinfeld, Physico-chemical changes in Ca, Sr and Al-doped La–Mn–O perovskites upon thermochemical splitting of CO_2 via redox cycling, Physical Chemistry Chemical Physics 17 (2015) 6629-6634.

[24] R. R. Bhosale, A. Kumar, A. Ashok, P. Sutar, G. Takalkar, M. Khraisheh, F. AlMomani, U. Ghosh, La-Based Perovskites as Oxygen-Exchange Redox Materials for Solar Syngas Production, MRS Advances (2017) in-press.

[25] R. R. Bhosale, R. Khadka, J. Puszynski, R. V. Shende, H_2 generation from two-step thermochemical water-splitting reaction using sol-gel derived $Sn_xFe_yO_z$, Journal of Renewable & Sustainable Energy 3 (2011) 063104-1 – 063104-12.

[26] R. Bhosale, R. Shende, J. Puszynski, H_2 generation from thermochemical water splitting using sol-gel derived Ni-ferrite, Journal of Energy and Power Engineering 4 (2010) 27-38.

[27] R.R. Bhosale, R.V. Shende, J.A. Puszynski, H_2 Generation From Thermochemical Water-Splitting Using Sol-Gel Synthesized Zn/Sn/Mn-doped Ni-Ferrite, International Review of Chemical Engineering 2 (2010) 852-862.

500493 Critical Issues in the Development of Commercial Natural Gas to Industrial Chemicals Bioprocesses

Abstract:

The low cost and abundant supply of natural gas make it America's greatest untapped energy resource, leading to the development of technologies for natural gas conversion to compounds of greater value. Intrexon has developed the first natural gas-to-liquids bioconversion platform powered by methantrophic fermentation, which applies synthetic biology to program bacteria and produce higher value materials. Our innovative genetic toolbox allows us to engineer the biology of the methanotroph to more efficiently convert natural gas into a wide range of valuable chemicals and fuels. These applications include isobutanol, farnesene, 1,4-butanediol, 2,3-butanediol, isoprene and isobutyraldehyde – the building blocks for industrial and consumer products like fuels, synthetic rubber, acrylics, resins and spandex.

Using a methanotrophic bioprocess to produce fuels and chemicals has several advantages. Natural gas bioconversion offers scaling opportunities beyond traditional gas-to-liquid (GTL) facilities, as well as biomass feedstock facilities. The theoretical yields of methanotroph/methane exceed that of yeast/sugar approaches by a wide margin. Biocatalytic processes that achieve high yields under milder processing conditions allow for reduced capital and operation expenditures, offering greater economic viability. To successfully bring this technology to commercial scale, consideration must be given not only to optimizing the bacterial metabolic pathway, but to key issues in the fermentor design, downstream processing, scale-up and overall plant economics.

At any scale, aerobic natural gas fermentation must overcome the challenge of gas-to-liquid transfer of low solubility gases to increase bioavailability of gas substrates to the bacteria. A low gas-to-liquid mass transfer rate reduces the bacterial productivity and process yield, which in turn increases fermentor size, capital cost, operating costs and may add additional equipment necessary to recycle gases. During commercial scale-up of gas fermentation processes, these factors can become cost prohibitive. Standard stirred tank reactors become unfeasible for use at larger scales, as the agitator power input and size become excessive. There are many fermentor designs currently available to increase the overall mass transfer coefficient of these gases. However, capital and operating expenditures of the reactor are critical metrics for evaluating commercialization feasibility. Unique fermentor configurations can greatly increase capital cost by raising the cost of fabrication and decreasing the ability to receive competitive bids. These unique fermentors also may not create a homogenous, well-mixed broth – which increases the cost of controls and instrumentation. The economic feasibility of gas transfer designs must be balanced with the biologic effects of reactor forces on the bacteria. Methods of increasing mass transfer often include increasing turbulence in the reactor, which can create detrimental effects on the bacteria through high shear forces, high pressure and vacuum areas, and dissolved gas substrate inhibition. Careful equipment selection and pilot trials are necessary to determine the various effects of gas transfer methods on mass transfer rates and cell viability.

Downstream process selection for commercialization is a balance of biologic limitations, technical feasibility, capital cost, operating cost and the risk of cutting-edge over well-established techniques. The developing bioprocess must be iterated upon the limitations of strain engineering and practicality of downstream processing techniques. Fermentation broth concentrations may be low to product inhibition on the cell, but this increases the size and energy requirement of downstream equipment. Cutting-edge technologies can increase the efficiency of low titer separation and reduce energy inputs, which can be capital intensive and unproven at commercial scales. Genetic engineering tools can increase the tolerance of the bacteria to products. However, the directed or natural evolution process requires time.

A minimum broth titer must be determined through techno-economic modeling of processes at each stage of process development and refinement. Bioprocesses also create metabolic byproducts which can add unit operations, and therefore cost, to the downstream purification process. Some byproducts may originate from the genetic pathway or from commercial feedstocks, as they are often lower in purity and have higher variability than those at the research scale. Additional downstream steps must be evaluated to determine if they are technically and economically feasible to meet final product specifications for sale, or if the byproduct genetic pathways can be manipulated to terminate production and not add cost to the overall process.

Furthermore, certain tools that aid in genetic engineering and strain development at the research and development scale must be eliminated or retooled to create commercially viable strains. During development, production pathways are often introduced to the bacteria through plasmids which carry antibiotic resistance. To avoid the material cost and downstream separation of the antibiotic at commercial scale, the genetic material must be integrated into the chromosome. This genomic integration may affect the strain's capability to produce product at previously observed yields, titers and rates. Another common development tool is that production pathways may be turned on or off using an externally applied chemical inducer. At scale, these chemical inducers can add material cost as well as affect downstream processes and product purity. Induction systems may be completely removed for commercial scale or based off an easily controllable environmental switch.

Favorable plant economics are necessary to move forward with commercialization of a technology. Scale is carefully selected based on a comparison of the production rate, fermentation productivity and yield targets to revenue, capital and operating costs. The plant location is an important selection factor, as costs are affected by utility rates in the area and/or co-location of the plant with shared infrastructure.

Traditional engineering workflows, specifically front-end loading strategy, were used to shape the commercial scale process design of Intrexon's natural gas to chemical plants. The biologic, technical and economic commercial feasibilities of the end-to-end natural gas to isobutanol process were critical to Intrexon's plant design. Laboratory scale learnings were applied to select technology that was commercially feasible at demonstration scale. This demonstration scale design was scaled down to pilot scale, where the operation will yield learnings into refinement of at-scale design, equipment selection, empirical data on heating and cooling, and controls strategies and feedback to further iterations on genetics.

495124 Direct, Single-Pass Thermocatalytic Upgrading of Biogas and Landfill Gas into Renewable Natural Gas over the Ultra-Low Loading Ru/ Î³-Al_2O_3 Catalyst

Abstract:

Direct, single-pass thermocatalytic upgrading of biogas and landfill gas into renewable natural gas over the ultra-low loading Ru/g-Al_2O_3 catalyst

Yichen Zhuang and David Simakov

Department of Chemical Engineering, University of Waterloo, Waterloo, ON N2L 3G1, Canada

Converting CO_2 into synthetic renewable fuels is an attractive pathway to decrease CO_2 emissions and to reduce our dependence on fossil fuels. Thermocatalytic hydrogenation provides advantages of fast reaction rates and high conversion efficiencies (as compared to electro-catalytic and photo-catalytic routes), thus allowing for compact, high-throughput operation. One option is to convert CO_2 into synthetic methane, so-called renewable natural gas (RNG) via the Sabatier reaction ($CO_2 + 4H_2 = CH_4 + 2H_2O$). Water electrolysis using renewable or surplus electricity can be used to provide a non-fossil source of H_2 required for the reaction. There are multiple sources of concentrated CO_2-containing streams including biogas and landfill gas (up to 50 mol% CO_2). It is highly desirable to develop a technology that can convert such CO_2-rich streams directly, without costly upstream CO_2 separation.

One of the most fundamental challenges is the development of a catalyst, which is highly active, selective to CH_4 formation, stable against coking, and low-cost at the same time. The selectivity becomes a major challenge when a mixed feed (biogas or landfill gas contain up to 50 mol% CH_4 in addition to CO_2) is used instead of purified CO_2 feed. In this study, a highly selective Ru-based catalyst with ultra-low Ru loadings (less than 0.5 wt%) is reported. A series of catalysts with Ru loadings ranging from 0.05-0.5 wt% were synthesized and evaluated for their catalytic activity in selective conversion of CO_2 in the mixed CH_4/CO_2 feed. CO_2 conversions up to 80% were achieved with nearly complete selectivity to CH_4 generation (left plot in the figure below). Importantly, the catalysts showed good stability, without any notable decline in performance over 60 hours on stream at very high space velocity of GHSV = 90,000 1/h (right plot in the figure below). A variety of analytical techniques were employed in order to investigate the mechanism of the superior selectivity and activity, including electron microscopy, thermal gravimetric analysis, temperature programmed reaction, chemisorption, and X-ray photon spectroscopy. Our results point out at the strong structure-activity relationship and the importance of the active phase interaction with the support.

492217 Energy Quality Factor and Exergy Destruction Processes Analysis for a Proposed Polygeneration System

Abstract:

A new polygeneration system was established through combining coal pyrolysis with semicoke combustion, semicoke gasification and a steam Rankine cycle, and then the calculation methodology was developed to calculate the system. The energy and exergy efficiencies of the system are 65.9 % and 75.1 %, respectively; and the standard energy quality factor *Î*±Î· values of the involved 17 components and the energy quality factor *Î*± of each system stream were obtained. By comparing the differences among the standard energy quality factor *Î*±Î· values of the related 17 components, the features of the energy grade of the components at the standard conditions were elucidated. Then the effect of streamÂ¢s composition and temperature on the actual energy quality factor was discussed. Combining the concept of energy quality factor, the processes of exergy destruction in system blocks were visually expressed via a kind of diagram; and the reasons of exergy destruction were tried to be explained from the perspective of energy quality factor.

500440 Large Eddy Simulation of Soot Formation in Oxy-Coal Combustion

Abstract:

Oxy-Coal combustion is a promising clean coal technology. Typical combustion burns coal in air, resulting in a product mixture of water, nitrogen, and carbon dioxide. Separating carbon dioxide from nitrogen is costly. In oxy-fired combustion, nitrogen is first separated from air and a mixture of coal, oxygen, and (optionally) recirculated flue gases is burned. The resultant products are water and carbon dioxide, which are easily separated for subsequent sequestration. Oxy-fired combustion often occurs at high temperatures and under flow and thermal conditions that differ somewhat from combustion in air. Radiative heat transfer is a key process in such combustion as it is a primary source of heat transfer in steam generation applications. Radiation is present from gaseous species, ash, coal char, and soot, with soot being a significant contributor. Unfortunately, soot formation is a complex process. While much research has been done on soot formation in general, relatively little research has been published on soot formation in coal combustion. In oxy-fire systems with high water and carbon dioxide concentrations, and at high temperature, soot consumption via gasification may be significant. (Normally so consumption is dominated by oxidation reactions.)

We present a series of large eddy simulations in an oxy-coal combustor in which sensitivity to the soot formation mechanism is presented. Differences is flow, temperature, and radiative heat transfer are shown. We compare results using a previously published coal soot model, as well as extensions that include optimized gasification and oxidation rates. An advanced physics-based soot formation from coal tar mechanism is presented.

THE UNIVERSITY OF UTAH

Ash aerosol and ash deposit formation during high temperature oxy-combustion of petroleum coke

Yueming Wang [a], Xiaolong Li [a], Jost Wendt [a]

[a] Department of Chemical Engineering, University of Utah, Utah, USA

Background

What is petroleum coke

- Peroleum coke (petcoke) is a byproduct of petroleum refining process

Advantages

- Petcoke is available in large amounts with low price
- Petcoke has low ash content and high heating value

Challenges

- The volatile content is very low in petcoke making it hard to be burned efficiently
- The sulfur content is very high and can cause corrosion and pollution

THE UNIVERSITY OF UTAH

2

Research Objectives

- ➢Assess the combustion efficiency of petroleum coke under pulverized coal furnace
- ➢Uncover the effect of flame temperature on combustion performance of petroleum coke
- ➢Study the formation mechanisms of ash aerosol and ash deposition (fouling)
- ➢Develop the relationship between ash deposition rates and concentration of the ash aerosol

Oxy-Fuel Combustor (OFC)

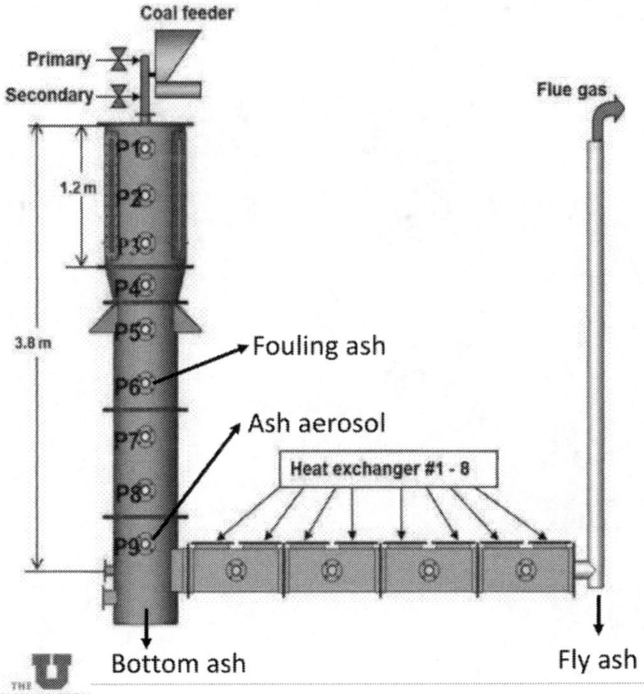

- ➢ 100 kW pilot scale
- ➢ Self-sustained down-fired
- ➢ Systematically controlled
- ➢ Temperature profile and particle concentration comparable with full scale units

Fouling ash sampling system

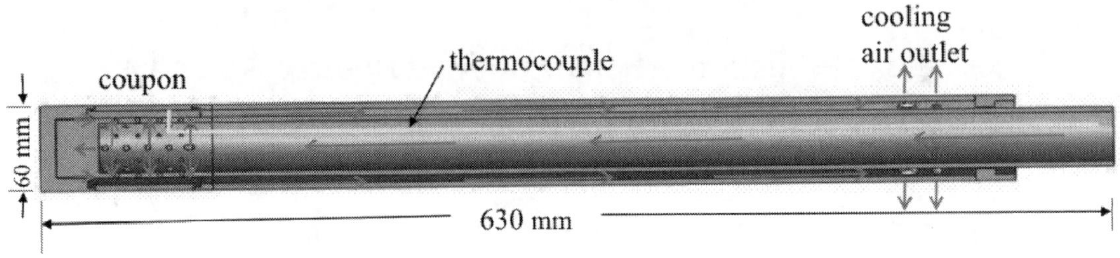

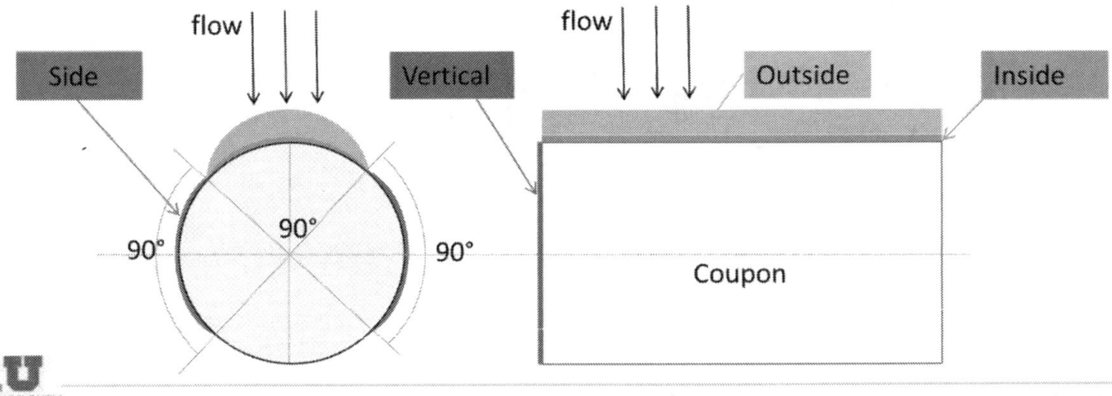

THE UNIVERSITY OF UTAH

5

Ash aerosol sampling system

BLPI:
0.0324 - 15.7 μm

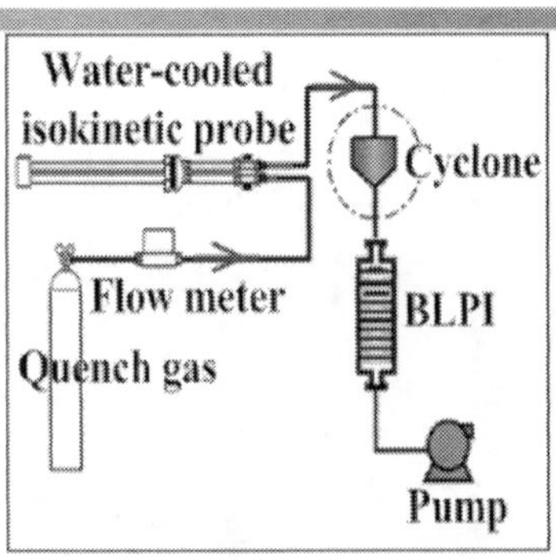

+

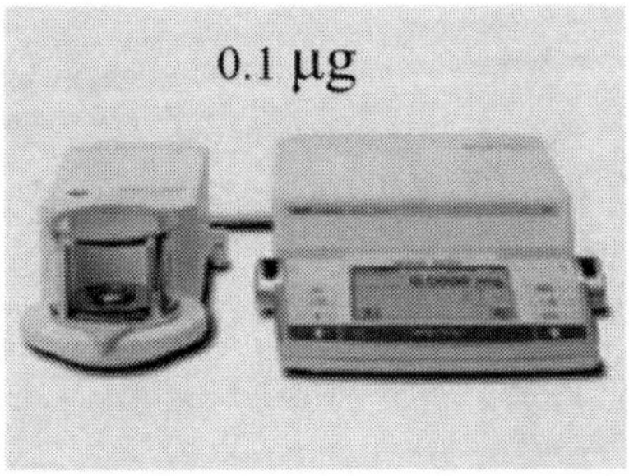

SMPS:
0.0143 - 0.6732 um
APS:
0.532 - 20 um

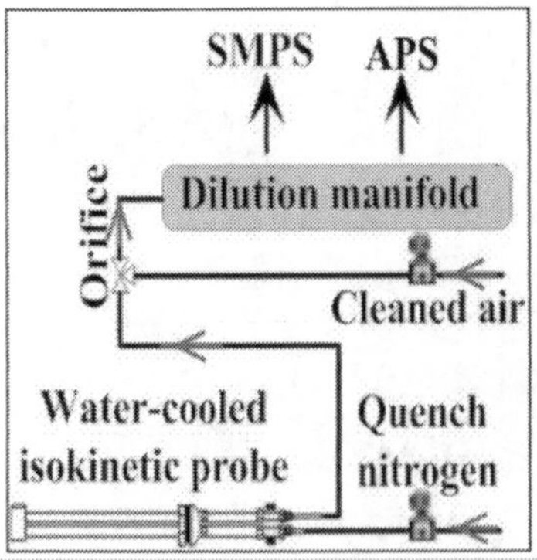

THE UNIVERSITY OF UTAH

6

Combustion parameters

Composition Analysis

ASH (%)	C (%)	H (%)	N (%)	S (%)	O (diff) (%)	H_2O (%)	Volatile (%)	Fixed Carbon (%)	HHV (kJ/kg)
2.99	82.51	6.02	1.71	5.65	0.49	0.57	10.18	86.26	34447.6

Al_2O_3 (%)	CaO (%)	Fe_2O_3 (%)	MgO (%)	K_2O (%)	SiO_2 (%)	Na_2O (%)	SO_3 (%)	TiO_2 (%)	NiO (%)	V_2O_5 (%)
19.4	4.22	7.02	0.66	1.17	46.7	0.72	3.77	0.63	1.26	8.24

Combustion conditions

Tests conditions	OXY70	Air
Feeding rate of petroleum coke (kg/h)	1.36	1.36
Feeding rate of natural gas (kg/h)	1.36	1.36
Excess O_2 (vol%_dry)	3	3
Adiabatic flame temperature (K)	3095	2312
Flow rate of flue gas (m^3/h_standard condition)	11.85	37.73

THE UNIVERSITY OF UTAH

7

Temperature profile of flue gas

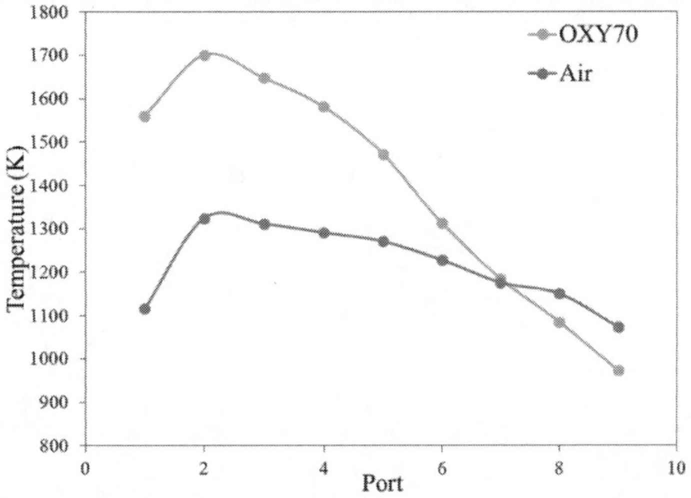

- Peak temperature of OXY70 (1701 K) is much higher than Air (1324 K)
- Peak temperatures for both cases are much lower than adiabatic flame temperatures.

THE UNIVERSITY OF UTAH

8

Aerosol particle size distribution (PSD)

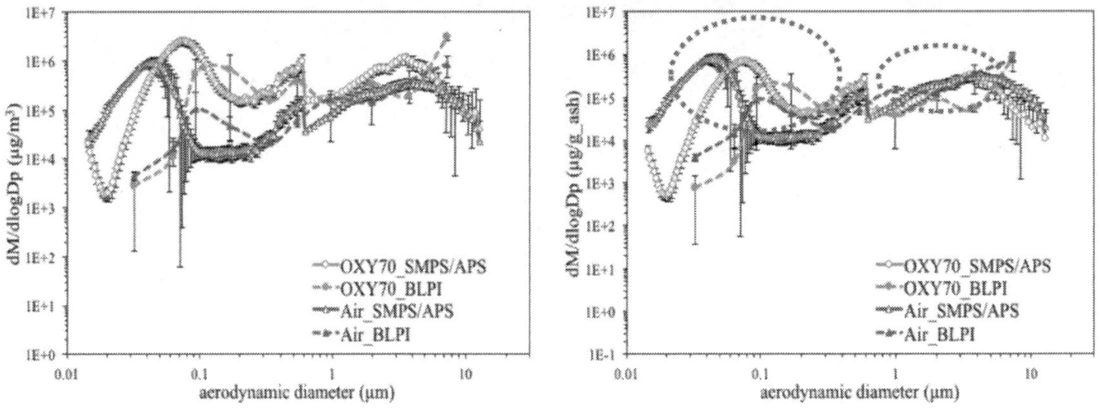

- BLPI data match with SMPS/APS data
- High temperature oxy-firing generates more submicron particles (99.1 mg/g_ash) than air-firing (72.4 mg/g_ash), but supermicron particles are not affected by temperature

THE UNIVERSITY OF UTAH

9

Aerosol size segregated composition

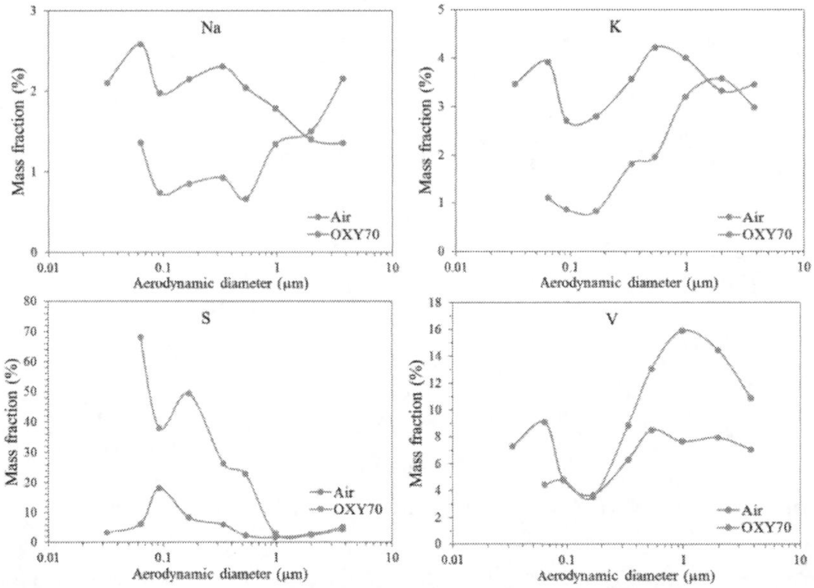

- More alkali metals are scavenged from submicron particles by alumina-silica under high temperature condition

THE UNIVERSITY OF UTAH

10

Composition of fouling ash

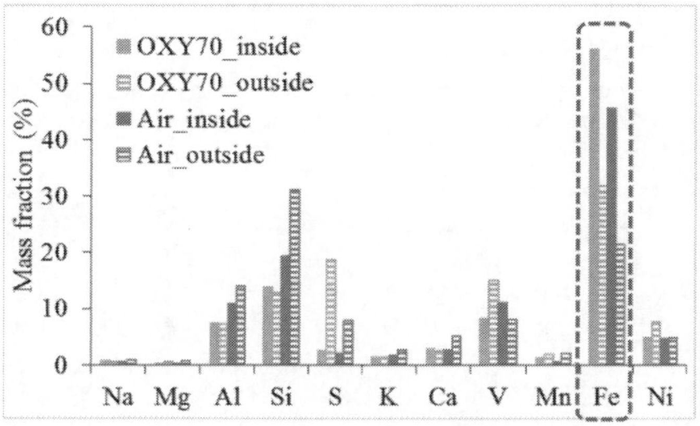

Al_2O_3 (%)	CaO (%)	Fe_2O_3 (%)	MgO (%)	K_2O (%)	SiO_2 (%)	Na_2O (%)	SO_3 (%)	TiO_2 (%)	NiO (%)	V_2O_5 (%)
19.4	4.22	7.02	0.66	1.17	46.7	0.72	3.77	0.63	1.26	8.24

- Composition of deposits are dominated by Fe

THE UNIVERSITY OF UTAH

11

SEM and EDS analysis for deposits

Na	0
Mg	0.17
Al	1.63
Si	2.30
S	1.05
K	0.33
Ca	0.82
V	1.33
Mn	0.61
Fe	**90.01**
Ni	1.75

THE UNIVERSITY OF UTAH

12

Composition of fouling ash

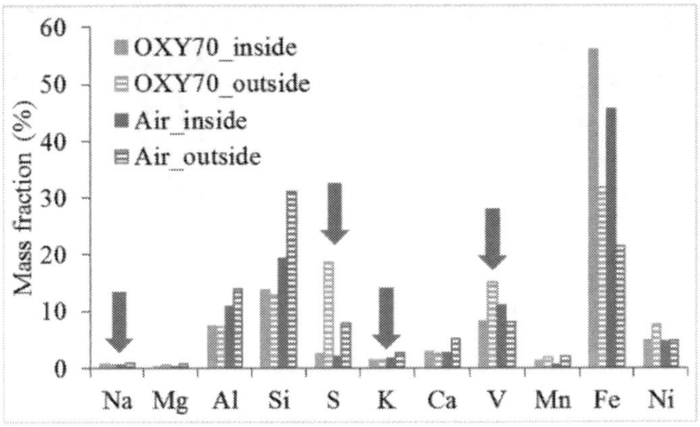

- Composition of deposits are dominated by Fe
- OXY70 deposits have more S and V, but less Na and K than Air, which are consistent with aerosol composition analysis

THE UNIVERSITY OF UTAH

13

Sulfur and carbon analysis

Condition	OXY70		Air	
Content	S (%)	C (%)	S (%)	C (%)
outside deposit	3.21	79.0	1.13	28.0
fly ash	3.43	11.9	1.55	13.3
bottom ash	3.12	68.8	1.89	36.8

- Air firing seems surprisingly more efficient to burn petcoke than OXY70
- Sulfur content in OXY70 is higher than Air case

THE UNIVERSITY OF UTAH

14

Correlation of deposition rates with PM1

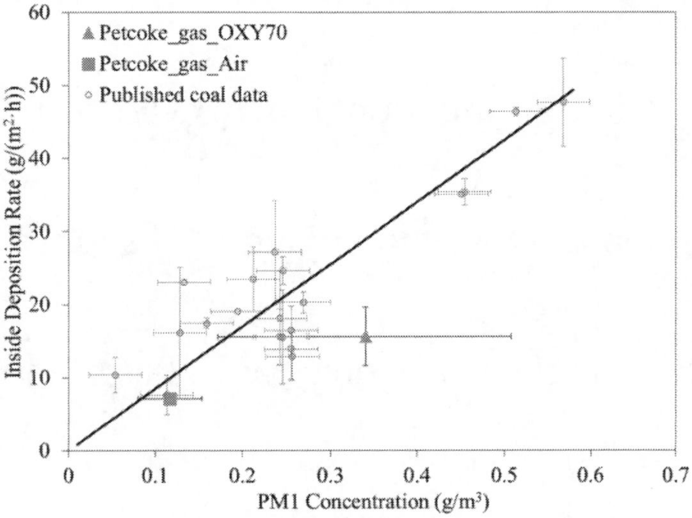

Published data from "Zhan, Z.; Wendt, J. O. L. Energy & Fuels **2017**, 31 (3), 2198–2202."

- Inside deposition rates of petcoke show positive relationship with concentration of PM1, which is consistent with previous findings in coal.

THE UNIVERSITY OF UTAH

15

Conclusions

- More alkali metals are scavenged from ultrafine fly ash by alumina-silica under high temperature condition
- The composition of fouling ash can be predicted by ash aerosol
- Fe is enriched in fouling ash, especially for high temperature condition
- Mixing between oxidant gas and petcoke is more important for carbon burnout than flame temperature
- Inside deposition rates can be predicted by the concentration of PM1

THE UNIVERSITY OF UTAH

16

Acknowledgements

- National Science Foundation under Award 1603249
- Praxiar for contributing the O_2 and CO_2 supply
- Brittany Walker, Dave Wagner, Paulo Perez and Brian van Devener

THE UNIVERSITY OF UTAH

17

Thank you!

THE UNIVERSITY OF UTAH

18

499074 Mercury Oxidation over Cu-SSZ-13 Catalysts under Flue Gas Conditions

Abstract:

Increasing concerns over mercury (Hg) emissions from coal-fired power plants have led the creation of new emission standards such as the recent Mercury and Air Toxic Standards, which requires a 90% reduction of Hg emissions. These new standards have led to the implementation of various novel technologies to control Hg emissions. Selective catalytic reduction (SCR) catalysts have shown promise as bifunctional catalysts for limiting both nitrogen oxide (NO_x) and Hg emissions through simultaneous NO_x reduction and Hg oxidation. The catalytic oxidation of elemental Hg, which is difficult to capture without dedicated Hg capture systems, to Hg^{+2} via HCl enables the capture of Hg^{2+} in the existing wet flue gas desulfurization unit as it is water soluble. Previous studies have investigated the possibility of utilizing current commercial SCR catalysts to fill this role, however these catalysts often have unsatisfactory performance when it comes to reducing Hg pollution to the levels required by current emission standards. To develop the next generation of SCR catalysts, a copper (Cu) exchanged small pore zeolite (Cu-SSZ-13) was investigated for its possible application in coal-fired power plants as a new bifunctional SCR and Hg oxidation catalyst. Cu-SSZ-13 has shown promise as a SCR catalyst for diesel exhaust, but has not been studied in coal-fired power plants.

In this study, the activity of Cu-SSZ-13 was investigated for its ability to oxidize Hg under a variety of flue gas conditions. The Cu-SSZ-13 catalyst was compared to SSZ-13 to understand the effect of Cu in the Hg oxidation mechanism. The Hg oxidation performances of both zeolites were compared to that of a commercial SCR catalyst (V-Ti-W), which served as a benchmark. All catalysts required hydrogen chloride (HCl) to exhibit any Hg oxidation with the zeolites outperforming the commercial catalyst. However, the activity of all catalysts significantly decreased in the presence of either sulfur oxides (SO_x) or SCR reactants (nitrogen oxide (NO)+ammonia (NH_3)) in the flue gas, with SO_x having a greater effect on Cu-SSZ-13 and SCR reactants having a greater one on the commercial catalyst. Additionally, it was determined that Cu was not the active site for Hg oxidation but was necessary for the SCR reaction. In fact, in the presence of SO_2, Cu actually contributed to the deactivation of the catalyst for Hg oxidation, likely due to the formation of sulfates as confirmed by XPS analysis.

502807 Mediated Oxycombustion with Integrated Uncoupled Oxygen Supply Â (MOBIUS)

Abstract:

Combustion systems that circumvent or reduce the energy penalty for carbon dioxide separation are being vigorously investigated. The approach we have pioneered aims at synergistically combining the basic objectives of three such technologies, namely oxy-combustion, chemical looping and liquid anode DCFC, and transforming them into a single lean, thoroughly optimized single unit process - Mediated Oxycombustion with Integrated Uncoupled Oxygen Supply (MOBIUS) - through innovations that overcome the bottlenecks of each of these technologies such as a) controlled combustion in a molten oxide to overcome the issues related to high temperatures (oxycombustion); b) obviating the need of solids transport between reactors (CLC); and c) avoiding operational issues such as electrical resistance, ash separation, impurity tolerance etc (DCFC). The system (simplistic concept in Figure 1) synergistically combines the effective aspects of the above mentioned technologies along with innovations to overcome the hurdles faced by these technologies. In MOBIUS, the solid fuel is combusted in the absence of any nitrogen as in oxy-combustion. However, the carbon is not combusted directly by oxygen but by a metal oxide as in a CLC. Unlike conventional CLC, the MeO_x and Me in the proposed Chemical Looping with Oxygen Uncoupling (CLOU) systems are in the liquid phase to improve the combustion rates and the mobility of the metal and the oxide. The metal is re-oxidized at the walls of the combustor with oxygen that is uncoupled from the reactor wall material. The individual subsystem studies will be presented and the integration of the various subsystems into a single reactor system will be discussed. Results from the subsystem will include the combustion rates, metal oxidation rates, oxygen uptake, transport and release rates under various conditions and composition of the membrane (during the membrane evaluation). The long terms operation of the integrated system will be demonstrated.

497932 Mobility Size and Effective Density of Carbonaceous Aerosols

Abstract:

Mobility Size and Effective Density of Carbonaceous Aerosols

Georgios A. Kelesidis, Eirini Goudeli, Sotiris E. Pratsinis

Particle Technology Laboratory, Institute of Process Engineering,

Department of Mechanical & Process Engineering, ETH Zurich, Switzerland

The mobility size estimation of evolving fractal-like soot aggregates is not trivial, as it depends on their structure as well as the number and size of their constituent primary particles. A scaling law describing the mobility size of agglomerates consisting of monodisperse primary particles in point contact (Sorensen, 2011) overestimates the mobility diameter of mature soot aggregates from diffusion flames for a given number of primary particles (Rissler et al., 2013). Rissler et al. (2013) attributed this difference to the primary particle polydispersity and chemical bonding induced by surface growth (aggregation), neglected in the above scaling law (Sorensen, 2011). Typically population balance models for soot formation estimate the mobility diameter assuming it is equal to the gyration diameter (Yapp et al., 2015).

Here, the Discrete Element Model of nascent soot agglomeration and surface growth by acetylene pyrolysis (Kelesidis et al., 2017) is extended to investigate the dynamics of mature soot formation at longer residence times and different flow regimes than nascent soot in the absence of soot oxidation. The evolution of soot mean gyration, mobility and primary particle diameters as function of the mean number of primary particles per aggregate are elucidated. The common assumption that the mobility is equal to the gyration diameter leads to 40 % underestimation of the nascent soot mobility size and 30 % overestimation of the mature soot aggregate mobility diameter.

Relationships for the soot mobility size and effective density are derived by following the evolution from nascent to mature soot morphology and compared to the correlation of Sorensen (2011) and mass-mobility measurements in quenched diffusion flames (Rissler et al., 2013; Yon et al., 2015). The scaling law for agglomeration in the absence of surface growth (Sorensen, 2011) overestimates the measured mature soot aggregate mobility size by 37 % and underestimates its effective density by 44 %. Accounting, however, for the primary particle polydispersity and chemical bonding or aggregation induced by surface growth reduces the deviation from the mass-mobility measurements down to 7 and 11 % for the mobility size and the effective density, respectively.

References:

Kelesidis, G.A., Goudeli, E., Pratsinis, S.E. (2017) *Proc. Combust. Inst.* **36**, 29-50.

Rissler, J., Messing, M.E., Malik, A.I., Nilsson, P.T., Nordin, E.Z., Bohgard, M., Sanati, M., Pagels, J.H. (2013)

Aerosol Sci. Technol. **47**, 792-805.

Sorensen, C.M. (2011) *Aerosol Sci. Technol.* **45**, 765-779.

Yapp, E.K.Y., Chen, D., Akroyd, J., Mosbach, S., Kraft, M., Camacho, J., Wang, H. (2015) *Combust. Flame* **162**, 2569-2581.

Yon, J., Bescond, A., Ouf, F.X. (2015) *J. Aerosol Sci.* **87** (2015) 28-37.

502531 Fossil Energy Policy in the Trump Administration

Abstract:

Fossil energy policy in the Trump administration

Those interested in a federal fossil energy policy are very alert to attempt to read the proper signals early in the Trump Administration. Several circumstances provide a sense of direction. Others will unfold over the June to January timeframe.

What we know now is:

- It appears that the Clean Power Plan is dead. This may be helpful because industry players perceive a somewhat arbitrary "number" representing emission reduction levels on a state-by-state basis. The energy industry is reducing emissions already and can likely meet climate goals more cost-effectively by not having to spend money on compliance costs and to use this capital to reduce emissions broadly instead on arbitrary emissions reduction targets.
- It appears that the Paris Accord is dead. Staying in or pulling out of the Paris Accord may be meaningless from an emission reduction standpoint. The U.S. energy industry is on track to realize the emission reductions represented in the accord. It can achieve emission reductions more cost-effectively without arbitrary reduction levels and regulatory compliance costs. Contributions that the United States would have made to the Green Climate Fund can be used for other purposes consistent with the President's "America First" campaign promise.
- And the U.S. Department of Energy Fossil Energy budget, this is trickier because while the "budget path" cuts fossil research it also cuts research in the energy efficiency and renewable energy budgets. The administration has strived to stress that the federal role should be basic research, not pre-commercialization research; which they suggest is more appropriately a private sector function.

Much more will be known over the next few months – impacting the fossil fuel research agenda perhaps for years.

502530 Emerging Technologies in Fossil Energy and Chemical Conversion

Abstract:

Advanced energy technologies are usually developed to meet emerging goals in stationary power, transportation, or aerospace applications. Research efforts typically focus on meeting energy goals such as efficiency, emissions, and cost. Less attention is given to possible “derivative” technologies where advances in energy conversion could also produce benefits for chemical conversion. This presentation will describe new developments in energy conversion that may also provide advantages for chemical production and process intensification. The talk will discuss emerging energy technologies—continuous detonation engines, cold plasmas, rotating fluid beds—and describe how they may benefit both energy and chemical conversion needs.

504579 University Coalition for Fossil Energy Research - a New Doe-NETL National Alliance for Clean Energy Solution

Abstract:

University Coalition for Fossil Energy Research - A New DOE-NETL National Alliance for Clean Energy Solution

Chunshan Song*

DOE-NETL University Coalition for Fossil Energy Research, the Pennsylvania State University, EMS Energy Institute, C211 Coal Utilization Lab, University Park, PA 16802

*E-mail: csong@psu.edu

The University Coalition for Fossil Energy Research (UCFER) was established in 2016 by the U.S. Department of Energy (DOE), National Energy Technology Laboratory (NETL) at Penn State after a nation-wide open competition in 2015. The Coalition brings together a multidisciplinary team of researchers from participating universities to address the fundamental challenges in developing clean carbon-based energy technologies. Currently there are 16 member universities in the DOE-NETL UCFER as of May 2017 including Carnegie Mellon Univ, Louisiana State Univ, MIT, Ohio State Univ, Penn State Univ, Princeton Univ, Texas A&M Univ, Univ of Kentucky, Univ of North Dakota, Univ of Pittsburgh, Univ of Southern California, Univ of Tulsa, Univ of Utah, Univ of Wyoming, Virginia Tech, and West Virginia Univ. Research performed by the coalition will directly support the mission of DOE NETL and NETLâ€™s Research and Innovation Center, by focusing on more efficient and more environmentally-friendly fossil energy utilization including diverse research programs such as advanced energy systems, CO_2 capture, storage and utilization, energy conversion, oil and gas technology, unconventional natural gas resources and infrastructure, crosscutting research and system analysis. The Coalition will facilitate basic and applied energy research and promote multidisciplinary collaboration among the member universities and NETL. The Coalition also seeks industry participation in research projects and promote technology transfer to the private sector. The research, completed throughout the six-year initiative, is expected to accelerate the development and deployment of fossil fuel-based technologies, enabling the continued use of our nationâ€™s abundant natural resources in a cost effective and environmentally responsible manner.

502533 Reliance Energy R&D

Abstract:

Reliance Energy R&D

Reliance industries is operating large scale refineries at its Jamnagar Manufacturing Complex. Reliance Industries has been continuously carrying out R&D work in Energy sector. We are at advance stage of implementation of Petcoke Gasification project at Jamnagar as value addition to Refinery and as a part of energy integration. Petcoke Gasification is carried out with entrained flow slagging Gasification technology.

Petcoke gasification is not a mature technology and a very few plants in the world are operating on 100% Petcoke. Therefore to understand the technology, Reliance has initiated an advance R&D projects in Petcoke Gasification mainly in the area of Slag Management, Petcoke Gasification Kinetics, Refractory Material Improvement in terms of life and alternate of Chrome Refractory.

Another area of research is recovery of V/Ni from Gasification Slag. Process has to be developed as there is no off-the- shelf Technology available for V/Ni recovery from Slag.

Reliance is also exploring next generation coal liquefaction technology and Bio conversion of syngas to higher hydrocarbons & petrochemicals.

Reliance scientists and technologist are engaged in indigenously developing Low Temperature & High Temperature PEM fuel cell.

Reliance is engaged with world class universities PSU, MIT, Frieberg, Kansas University to name a few working with us in field of Gasification, and environmentally beneficial technologies.

Reliance is carrying out research in the field of Bio-CBM i.e. bio-conversion of underground coal to methane, another potential breakthrough technology.

Another significant research is being carried out in the area of generating Oil from Algae (A2O).

501439 Short Contact Time Catalytic Partial Oxidation of Methane over Rhodium Catalysts Supported on 3-D Printed Ceramic Supports

Abstract:

Short contact time catalytic partial oxidation (SCT-CPO) is a process by which methane and other fuels can be converted to hydrogen and carbon monoxide at millisecond contact times without the addition of heat. To achieve such low contact times, structured supports with high void fractions. Since the reaction is limited by mass transfer, small pore diameters are desired at the cost of potential increases in pressure drop. 3-D printed ceramic supports (created via robocasting) can be produced with well defined and consistent pore structures and sizes with the desired high void fraction necessary for short contact time operation.

In this work, we have investigated the SCT-CPO of methane over 3-D printed ceramic supports experimentally demonstrating their ability to achieve high yields at low residence times and virtually no pressure drop. Supports of different compositions including alumina, silica, and zirconia have been investigated as well as differing pore size supports at constant void fraction. Experimental data at varying space velocities and feed ratios will be presented. Under the appropriate conditions, 3-D printed supports achieve methane conversions over 85% with hydrogen selectivities over 90% at space velocities over 400,000 hr-1.

499273 Characterization of Ni-Substituted Pyrochlore Catalysts for the Dry Reforming of CH_4

Abstract:

Pyrochlores are mixed metal oxides that have proven to be effective catalysts for the reforming of hydrocarbon fuels. These materials are thermally stable at the demanding conditions associated with fuel reforming, and have a modifiable structure which can incorporate various dopants to improve their catalytic activity. One example is a lanthanum zirconate pyrochlore (La-Zr). The Zr framework has been partially substituted by Rh to disperse the catalytic metal into small and stable clusters that are resistant to both carbon and sulfur poisoning [1]. Furthermore, alkaline metals (e.g. Ca) can be doped into the A-site for La to introduce oxygen vacancies which can increase lattice oxygen mobility, while also providing surface basicity that promotes the gasification of carbon.

Ni is an attractive catalytic reforming metal, because of its well-known specific activity, and low cost. However, Ni is problematic due to its low sintering temperature, and propensity to form carbon. This study will evaluate the substitution of different levels of Ni (1 vs 3 wt%) into a La-Zr pyrochlore to determine whether the pyrochlore structure is able to stabilize the Ni under conditions required for dry reforming of CH_4. The materials will be synthesized by a variation of the Pechini method [2], which is a well-known procedure to obtain mixed oxide solid solutions. The materials will be characterized by various methods to determine the location of the Ni, and how this will relate activity stability, and particularly carbon deposition. Traditional characterization methods like XRD, temperature programmed reduction, BET, and Raman are used. In addition, an analysis using by Cameca using a LEAP 5000 atom probe provided 3-D images with corresponding compositional analysis on each Ni containing material. The atom probe results (see Figure 1 for results on the 1 wt% substituted catalyst), and XRD results both show Ni is not soluble in the pyrochlore structure (for either weight loading). Instead, crystallization elicits the exsolution of Ni to the surface and grain-boundary regions of the pyrochlore, which forms NiO. At the elevated temperatures, and with high enough Ni loading (3 wt%), some of the Ni forms pocket of mixed-phases with the excess La (e.g. $LaNiO_3$). The Ni distribution at the surface and grain boundary regions suggests these enriched regions with metallic Ni would be the active component of the catalyst for dry reforming.

496600 Dry Reforming of Methane over Zr,Y-Modified Ni/Mg/Al Hydrotalcite Catalysts for Hydrogen and Carbon Monoxide Production

Abstract:

Dry reforming of methane over Zr,Y-modified Ni/Mg/Al hydrotalcite catalysts for hydrogen and carbon monoxide production

Katarzyna Swirk[a,b], Maria Elena Gálvez[b], Monika Motak[a], Teresa Grzybek[a], Patrick Da Costa[b]

[a] AGH University of Science and Technology, Faculty of Energy and Fuels, Al. A. Mickiewicza 30, 30-059, Krakow, Poland

[b] Sorbonne Universités, UPMC Paris 6, CNRS, UMR 7190, Institut Jean Le Rond d'Alembert, 2 Place de la Gare de Ceinture, 78210, Saint-Cyr-L'Ecole, France

1. Introduction

Dry reforming of methane (DRM) is a method of producing hydrogen and carbon monoxide while utilizing CO_2. The process has not been launched yet on an industrial scale due to lack of stable, active and selective catalyst [1]. Ni-based catalysts, used e.g. in steam reforming, seem to be a great alternative for expensive noble metal materials. However, the main drawbacks of nickel catalysts is their fast deactivation caused by carbon deposition and/or sintering [2].

The hydrotalcites (HTs, or layered double hydrotalcites LDHs, which exhibit brucite-like structure) appear to be promising support for catalysts due to their strong interrelated nanostructure which hinders Ni from sintering at high temperatures needed for DRM [3]. On the other hand, an addition of metallic promoters may positively influence catalytic activity and stability in DRM process [4].

Zr and Y species seems to be good components for the DRM catalyst, because only a slight addition of yttrium may stabilize the zirconia phase resulting in high thermal stability of catalysts [5]. Moreover, the yttrium as a promoter, enhances activity and stability of the used catalyst in DRM process, due to ability of oxygen vacancies creation [6].

Thus, the goal of this work was to examine the effect of the promotion of Ni/Mg/Al hydrotalcites with Zr and Y on their structure and performance in dry methane reforming.

2. Experimental

The HT sample was prepared by the co-precipitation method at pH of 10 ± 2, from the aqueous solution of nickel, magnesium and aluminum nitrates (Ni/Mg=0.33, M^{2+}/M^{3+}=3.0). Subsequently, the fresh HT material was modified with zirconium (5 wt.%) and yttrium (0.2 wt.%, 0.4 wt.%, or 0.6 wt.%), resulting in HT/Zr5 Y0.2, HT/Zr5 Y0.4, HT/Zr5 Y0.6 catalysts. The introduction of these two metals occurred through wetness impregnation method (co-impregnation). So synthesized materials were dried at 80°C overnight, and calcined in air at 550°C for 6 hours. The obtained catalysts were characterized by XRD, TPR-H_2, TPD-CO_2, and FTIR.

The dry reforming tests were carried out in a fixed-bed quartz reactor at temperature range of 850-600°C. Before each catalytic test, the samples were reduced in situ at 900°C for 1 hour with hydrogen (5% H_2 in Ar, flow 10°C/min). The gas composition was adjusted to 1/1/8=CH_4/CO_2/Ar, with total flow of feed gases of 100 cm^3/min (GHSV=20,000h^{-1}). The final products were analyzed by Varian GC-490 micro-GC.

3. Results and discussion

The X-ray analysis for the fresh samples confirmed successful synthesis of hydrotalcite-like materials (reflections at 2θ equal to ca. 11, 22 and 35°). Also, the periclase-like structure was registered after thermal treatment in air. On top of this, separate phase of Zr compound (ca. 31° and 51°) was observed. However, no separate phase was registered for yttrium in the XRD patterns of calcined and reduced samples.

The TPR-H_2 profiles show wide peaks with maximum value at ca. 830-860°C, which arise from reduction of Ni particles derived from mixed oxides in hydrotalcite structure. The reduction temperature is dependent on the presence of promoters. There was observed a shift of the wide peak towards higher temperatures in Zr, and Zr,Y-modified samples (847°C), with the highest temperature reached at 860°C for HT/Zr5 Y0.4 sample. This shift indicates decrease in reducibility, and stronger interaction of Ni with the support. The observed peaks at lower temperature, ca. 400°C, arise from NiO weakly bonded with the support.

The Zr,Y-modification influenced basicity of tested samples by expanding the number of medium sites while increasing Y content. Although, the total basicity decreased with promoters addition.

The FTIR measurements showed typical spectra for hydrotalcite-derived material.

The DRM tests (CH_4/CO_2/Ar=1/1/8) showed increase in CH_4 and CO_2 conversions while increasing temperature (Fig. 1 and Fig.2). The best results in the low temperature region (600-700°C) were observed for Zr,Y-modified samples. At 700°C the conversions obtained following sequence: HT/Zr5 Y0.6 > HT/Zr5 Y0.4 > HT/Zr5 > HT > HT/Zr5 Y0.2. Above this temperature the best performance was found for the catalyst impregnated only with Zr. The presence of side reactions, such as RWGS was deduced through CO_2 conversion greater than CH_4 conversion, and H_2/CO lower than 1.

Fig. 1 CH_4 conversion as a function of reaction temperature (DRM process CH_4/CO_2/Ar=1/1/8, GHSV=20,000h^{-1}) of hydrotalcite catalysts modified with Zr and Y

Fig. 2 CO_2 conversion as a function of reaction temperature (DRM process CH_4/CO_2/Ar=1/1/8, GHSV=20,000h^{-1}) of hydrotalcite catalysts modified with Zr and Y

4. Conclusions

All prepared catalysts were catalytically active in the DRM process. The presence of Zr and Y in Ni-containing hydrotalcite-derived catalysts resulted in a higher number of medium basic sites, which contributes to enhanced catalytic activities in DRM at low temperatures. The DRM tests show that the HT/Zr5 Y0.6 gives the highest CH_4 and CO_2 conversion, and the highest H_2/CO ratio in the temperature range of 600°C to 700°C. At temperatures higher than 700°C the presence of yttrium did not increase the catalytic activity of the Zr-modified sample. The effect of the reverse water gas shift reaction, on the hydrogen production and the molar ratio H_2/CO has been observed.

Acknowledgments

K. Swirk would like to acknowledge for financial support the French Embassy in Poland and InnoEnergy PhD School.

References

1. M. Rezaei et al., J Nat Gas Chem, 15 (2006) 327-334

2. A. S. A. A. Fatish et al., Appl Catal A, 364 (2009) 150-155

3. K. Mette et al., ChemCatChem, 6 (2014) 100-104

4. K. Asami et al., Catal Today 84, 27 (2003) 27-31

5. F. Legorreta-García et al., Bol Soc Esp Cerám V, 54 (2015) 119-123

6. Q. Wu et al., Fuel Process Technol, 89 (2008) 993-999

492067 A Multicore-Shell Catalyst Derived from Ni Phyllosilicate@SiO_2 for Low Temperature Dry Reforming of Methane: Confinement Effect Against Carbon Formation

Abstract:

Dry reforming of methane has been a popular research topic since it converts two â€œgreenâ€ gases into syngas (CO+H2). The main issue for Ni-based catalysts is the severe carbon formation which covers active sites, leading to deactivation. A lot of effort has been put to get stable Ni catalysts with less carbon formation, in which confinement effect has been proven to be a good strategy.

In this report, we manage to synthesize a novel multicore-shell catalyst. Firstly, Ni phyllosilicate is supported on silica nanospheres with ammonia evaporation method. Then Ni phylloslicate nanospheres are coated with a layer of mesoporous SiO2 to obtain a core-shell structure of Ni phyllosilicate@SiO2 via hydrolysis of TEOS. The thickness of shell can be tuned via varying the amount of TOES. After H2 reduction at high temperature, multiple small Ni nanoparticles (~3nm) are formed supported on the inner SiO2 core but also encapsulated within the outer mesoporous SiO2 shell.

This multicore-shell catalyst shows a high and stable conversion (~60%, GHSV=60,000ml/h g_{cat}) along 24h time of stream (TOS) at 600°C while the pristine Ni phyllosilicate suffers from heavy carbon formation and the reactor is blocked after 3h TOS. TGA and TEM of spent catalysts have shown that there is almost no carbon formation for this novel multicore-shell catalyst.

Compared with conventional Ni@SiO2 core-shell catalyst, our multicore-shell catalyst is much easier to synthesize and does not need any toxic organic solvent. And low reaction temperature (600°C) also helps to reduce the energy input. We believe this strategy of making multicore-shell catalyst can be extended and may be helpful to researchers in the same field.

489031 Mixed (dry & partial oxidation) Reforming of Methane with a New Catalyst Derived from a Negative Value Mining Residue Spinellized with Nickel

Abstract:

Mixed (dry & partial oxidation) reforming of methane with a new catalyst derived from a negative value mining residue spinellized with nickel

Frank DEGA[(+)], Mostafa CHAMOUMI, Nadi BRAIDY and Nicolas ABATZOGLOU[(*)]

Department of Chemical Engineering and Biotechnological Engineering, Université de Sherbrooke, Sherbrooke, QC, J1K 2R1, Canada

Key words: spinel, reforming, catalyst, nickel, carbon deposition, stability, selectivity.

[(+)] Presenting author: Frank.Blondel.Dega.Dongmo@USherbrooke.ca

[(*)] Corresponding author: Nicolas.Abatzoglou@USherbrooke.ca

This work belongs to a larger endeavor aimed at studying the performance and the stability of a hydrocarbons reforming spinellized nickel catalyst prepared from an ilmenite metallurgical residue. Said residue is a negative value upgraded slag oxide (UGSO) coming from a TiO_2 slag production unit, operated by Rio Tinto Iron & Titanium, Quebec, Canada. Recently published results from experiments carried out with this catalyst exhibited high hydrogen production selectivity during methane steam and dry (CO_2) reforming and stability over time-on-stream (TOS).

More specifically, this work report on the behavior of this catalytic formulation during methane partial oxidation (POX) regimes employing both CO_2 and O_2 as reactants. This has been done at 800-850°C, atmospheric pressure, molar ratios $CH_4/CO_2 = 3$, $CH_4/O_2 = 1, 2, 3$, and a space velocity between 4000 and 4600 $ml_{STP}/h/g_{cat}$. The conversion of CH_4 and CO_2 and the selectivity of $H_2$and CO have been assessed by means of mass balances whose closure is 100 +/-5%. The crystalline phases of the catalyst components as well as their changes over TOS in presence of different oxygen concentration have been studied. A first attempt to correlate the catalytic performances and the catalytic stability (mainly related to carbon deposition tendency) with the mechanisms and structural modifications of the catalyst over TOS is reported.

The promising results have shown that, within the experimental range tested, an optimum performance has been obtained at 850°C and at molar $CH_4/CO_2 = 3$, $CH_4/O_2 = 2$. At CH_4/O_2 lower and higher than 2 the catalytic

performance decreases. The catalyst exhibited a good stability for 2 days of reaction, with no detectable carbon deposition, high conversions (99% of CH_4, 62% of CO_2) and H_2 selectivity of 60%. The in situ X-ray diffraction (XRD) analysis showed that the crystalline structure of the catalyst is subject to modifications. The most significant observation is that the catalyst structure changes over, typically, the first 30min of TOS then stabilizes. This catalyst structure behavior has been correlated to its good stability, the absence of carbon deposition and the high methane conversion and hydrogen selectivity. This study sheds more light on the role of the reaction environment in the activity and the stability of this UGSO spinellized nickel catalyst.

500693 Design of Robust and Efficient Diesel Autothermal Reforming System Using Bi-Metallic Catalyst for the Mobile Applications

Abstract:

In contrast to the studies dealing with the hydrogen production from the light fossil fuels, the degradation problem of the catalyst for the diesel reforming has not been scrutinized thoroughly. Among the several reasons for the degradation problem, coke formation on the surface of the catalyst is known as the main cause of the catalytic failure and it encourages the production of the un-cracked hydrocarbons such as ethylene (C_2H_4), which may lead the severe poisoning problem on the anode of the fuel cell. In this study, bi-metallic catalyst consisted of platinum and ruthenium was introduced to prevent coke formation and ethylene production during the long-term autothermal reforming (ATR) of commercial diesel fuel. Every catalysts synthesized using a combustion method were demonstrated its long-term stability and catalytic activity over 200 hours. The rate of degradation for each time-on-stream experiment was evaluated by calculating its reforming efficiency containing the amount of hydrogen, carbon monoxide and methane produced from the diesel fuel. Through this study, we have found that the additive ruthenium performed as a great oxidation assistant on the catalytic surface and it could successfully oxidize the carbon species formed on the surface of the catalyst. Consequently, the prepared catalysts containing ruthenium showed great stability and the lowest rate of degradation among the samples. Several fresh and post-mortem characterizations were performed in terms of physical and redox properties using X-ray diffraction (XRD), N_2-isotherms, temperature programmed reduction (TPR) and temperature programmed oxidation (TPO) analyses. Based on the findings from the catalytic experiments and characterization, we could minimize the oxygen to carbon ratio (OCR) and steam to carbon ratio (SCR) for the diesel autothermal reforming reaction to enhance the overall efficiency without sacrificing the long-term stability. The bi-metallic catalyst containing ruthenium showed great performance and negligible degradation under the low OCR and SCR condition than conventional values with minimal production of ethylene species. The overall autothermal reforming efficiency was also dramatically enhanced without significant degradation for over 200 hours operation. Moreover, we suggested the total system design containing diesel reformer consisted of the bi-metallic catalyst and conventional solid oxide fuel cell using the commercial software and thermodynamic calculations. Finally, to demonstrate the possibility to use this new catalyst for the mobile application, we coated the bi-metallic catalyst on the cordierite monolith structure using a dip-coating method. The final structured catalyst containing only 0.4 g of active bi-metallic catalyst layer showed excellent reforming activity and efficiency near the thermodynamic limitation and presented no degradation signal over 1,000 hours under the commercial diesel autothermal reforming condition.

495591 Core-Shell Ni-Phyllosilicate@CeO2 Catalyst with Excellent Coke-Resistance in Steam Reforming of Biomass Tar

Abstract:

In this work, a highly active and coke-resistant core-shell structured Ni-SiO_2@CeO_2 catalyst has been developed for steam reforming of biomass tar. A major challenge in the production of hydrogen from biomass gasification is the removal of tar that is formed as a byproduct. Nickel-based catalysts have been shown to be highly active in catalytic steam reforming of tar, but suffer from a high tendency for forming coke that results in blockage of reactors and deactivation of catalyst. Here, we have explored novel core-shell structured catalysts with a Ni-phyllosilicate derived active core and a Cerium oxide shell, that shows high activity (> 60% conversion at 700°C) and stable performance in steam reforming of toluene as a model tar component. In a 20 hr catalytic test run, the Ni-SiO_2@CeO_2 catalyst showed negligible deposition of coke (ca 0.005 gC/ g-cat hr), whereas a reference Ni-SiO_2 catalyst without the CeO_2 shell showed extensive formation of filamentous carbon (ca 0.08 gC/g-cat hr) under the same reaction conditions, that resulted in blockage of the reactor. The excellent thermal stability and resistance to carbon formation of the Ni-SiO2@CeO2 catalyst is attributed to the high oxygen storage capacity and redox nature of CeO_2, and the confinement effect provided by the CeO_2 shell on the metallic nickel nanoparticles. The ceria shell hinders the spatial movement and sintering of Ni nanoparticle at the high reaction temperature, and thus reduces the tendency of carbon formation, which has been shown to be highly dependent on the Ni particle size. At the same time, the CeO_2 shell can also catalyze the oxidation of the coke precursors deposited on the nickel surface by supplying necessary oxygen species. Different synthesis routes have been explored and synthesis conditions have been optimized to tune the morphology of the core-shell catalyst and maximize the activity by increasing Ni dispersion and minimizing diffusion resistance through the shell. Characterization of catalyst has been done by TEM, XRD, XPS, TPR, and TGA analysis. The influence of reaction temperature on the catalyst performance has also been investigated. Due to its high activity and carbon resistance, the application of this catalyst can be extended to other significant H_2 producing reactions such as dry reforming of methane, steam reforming of ethanol etc.

500502 Improved Activity and Stability of Ni-based Anode for Direct Methanolâ€"Fueled Solid Oxide Fuel Cells

Abstract:

Ethanol steam reforming (ESR) is a promising route for sustainable hydrogen production. Recently, $LaNiO_3$ has received a lot of attention as the precursor of the catalyst in ESR reaction due to its interesting properties, such as good redox ability, accessibility and coke resistance. The major drawback of traditional $LaNiO_3$ is its low surface area, which limits its catalytic application. Therefore, it is highly desirable to develop an effective strategy for generating porous $LaNiO_3$. In recent years, three-dimensionally ordered macroporous (3DOM) perovskites have been reported to show better activity for soot combustion, methane combustion and CO oxidation. In this work, high-surface-area 3DOM $La_{1-x}Ce_xNiO_{3-\delta}$ (x=0, 0.1, 0.2, 0.3) (LCNO) catalysts are successfully synthesized and firstly used in ESR reaction. The 3DOM perovskite structure was examined in terms of the catalyst structure and reactivity. After the calcination at 850 °C, all the LCNO samples exhibit a 3DOM structure, whereas the $LaNiO_3$ sample shows an irregular morphology with a particle size of 200–600 nm. The incorporation of Ce has almost no effect on the morphology of the 3DOM structure. The 3DOM structure catalysts show significantly higher H_2 yields and lower C_2H_4 yield than that of LaNiO3 at 600 °C. In addition, the addition of Ce enhances the stability of the catalyst and effectively suppresses coking in the reaction.

495447 The Modeling and Validation of CO Preferential Oxidation (CO PROX) Multi-Reactor System Using CuO/Rare Earth-Doped Ceria Catalysts for Commercial Diesel Reforming Applications

Abstract:

The multi-reactor system using copper-ceria mixed metal oxide catalysts was proposed for the CO preferential oxidation (CO PROX) in H_2-rich streams. Although the CO PROX reactor system using precious metal-based catalysts has been widely adopted in commercial applications, it has crucial drawbacks such as severe exothermic hydrogen oxidation, unwanted methanation and narrow operating temperature window due to its intrinsic reaction mechanism. To solve the problems, researchers have introduced the redundant inter-coolers and liquid coolant pumps between the CO PROX reactors to keep the operating temperature for each reactor, however the additional heat exchangers lead the over-sized system. In this study, we demonstrated the possibility of copper-ceria catalysts for CO PROX system as an alternative of current precious metal-based catalysts and we suggested optimized multi-reactor design to reduce physical size of the CO PROX system without redundant heat exchangers and coolers. Rare-earth elements doped cerium oxides synthesized using a glycine nitrate combustion method were prepared as a support material for the catalytic system. Various rare-earth elements such as La, Pr, Nd, Sm, Gd, and Tb were introduced in the ceria lattice and the final catalysts containing copper oxides were demonstrated to find its operating temperature for CO PROX reactions. Several characterizations of the prepared catalysts in terms of structural and redox properties using inductively coupled plasma atomic emission spectrometer (ICP-AES), N_2-isotherms, X-ray diffraction (XRD), transmission electron microscope (TEM) and temperature programmed reduction (H_2-TPR) analyses. The industrial CO PROX reactants simultaneously containing H_2, CO, CO_2 and H_2O were blended to demonstrate the CO PROX activity and selectivity of the prepared catalysts. The introduction of rare-earth dopants in the ceria support showed significant effect on the CO PROX activity, selectivity and operating temperature. Consequently, we could utilize these various copper-rare earth doped ceria catalysts to construct the multi-reactor CO PROX system in proper temperature regions for each reactor. The computational fluid dynamics (CFD) modeling using commercial software was performed to build the complete multi-reactor CO PROX system with different copper-ceria catalysts for each reactor canister and laboratory scale validation was also conducted. Moreover, techno-economic evaluation was followed to compare the current precious metal-based CO PROX system with the newly proposed copper-ceria based system in the case of commercial diesel reforming application.

499097 Conceptual Design and Techno-Economic Analysis of a Solar Hybrid Sulfur Process

Abstract:

The hybrid sulfur (HyS) cycle is the only practical two-step thermochemical water-splitting cycle with all fluid reactants. It is also one of the most advanced of the thermochemical water-splitting cycles, with both reaction steps having been experimentally validated and with detailed system designs having been proposed for hydrogen production from a continuous heat source like an advanced, high-temperature reactor. This presentation describes the conceptual design of a HyS process based on a variable heat source, specifically concentrated solar energy, which is only available for part of the day. The presentation will address how the diurnal variation was accommodated by the design of the plant and will lay out a techno-economic analysis for a hypothetical 50-MT/d solar hydrogen plant in the southwestern US. It will conclude with a development path toward hydrogen production at $2/kg H_2, the renewable hydrogen production cost target set by the DOE Office of Energy Efficiency and Renewable Energy.

NREL | 40
NATIONAL RENEWABLE ENERGY LABORATORY | 1977 - 2017

Potential Size of and Value Proposition for H2@Scale Concept

Mark F. Ruth, Paige Jadun,

Amgad Elgowainy,* Bryan Pivovar

2017 AIChE Annual Meeting

November 1-3, 2017

Minneapolis, Minnesota

NREL/PR-6A20-70457

* Amgad Elgowainy is an employee of Argonne National Laboratory

H2@Scale Energy System

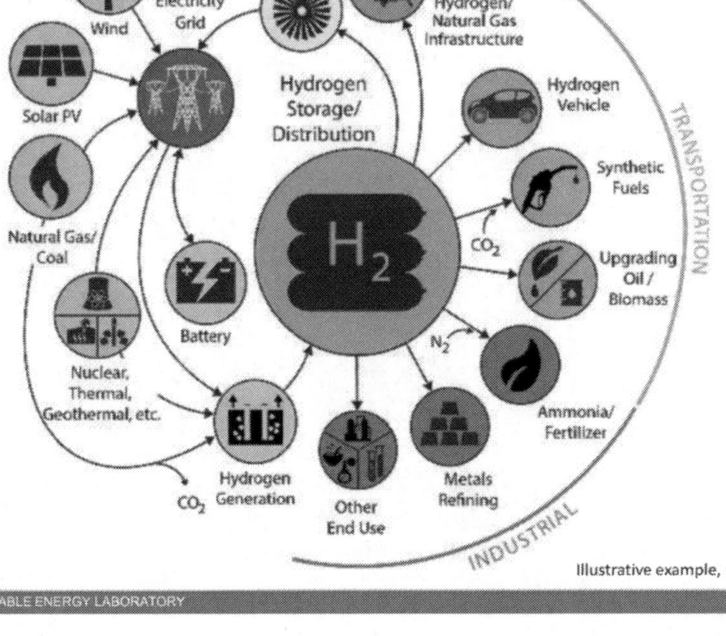

Illustrative example, not comprehensive

Technical Potential: Definition

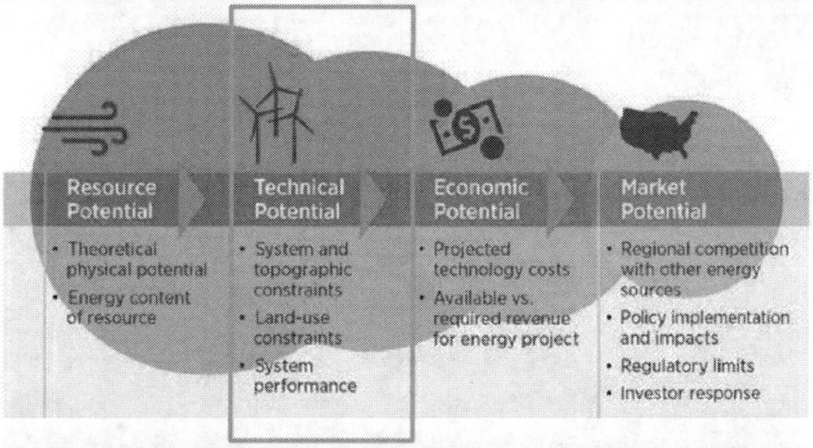

Technical potential is the subset of the available resource potential that is constrained by real-world geography and system performance but not economics

Brown, A., P. Beiter, D. Heimiller, C. Davidson, P. Denholm, J. Melius, A. Lopez, D. Hettinger, D. Mulcahy, and G. Porro. 2015. *Estimating Renewable Energy Economic Potential in the United States: Methodology and Initial Results.* Golden, CO: National Renewable Energy Laboratory. NREL/TP-6A20-64503

Technical Potential Hydrogen Demand

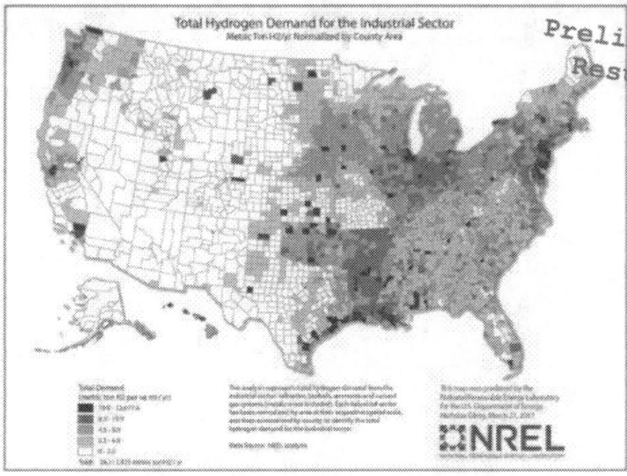

Use	Market potential (million metric tonne H_2 / year)
Industrial Use	
Refineries & CPI[$]	8*
Metals	5
Ammonia	5
Natural Gas	7
Biofuels	4
Light Duty Vehicles	28
Other Transport	3
Total	**60**

Total market potential: 60 MMT/yr

Current U.S. market: ≈ 10 MMT/yr

Near-term Outlook for Hydrogen Production Volume: 5% CAGR (2014-2019)[1]

[$] CPI: Chemical Processing Industry not including metals, biofuels, or ammonia
* Current potential used due to lack of consistent future projections
Light duty vehicle calculation basis: 190,000,000 light-duty FCEVs from http://www.nap.edu/catalog/18264/transitions-to-alternative-vehicles-and-fuels
1. Global hydrogen Generation Market by Merchant & Captive Type, Distributed & Centralized Generation, Application & Technology- Trends & Forecasts (2011-2016)

Technical Potential: Impact on Renewable Resources

	EIA 2015 current consumption (quads/yr)	Required to meet demand of 60 MMT / yr (quads/yr)	Technical Potential (quads/yr)
Solid Biomass	4.7	15	20
Wind Electrolysis	0.7	9	170
Solar Electrolysis	0.1	9	1,364

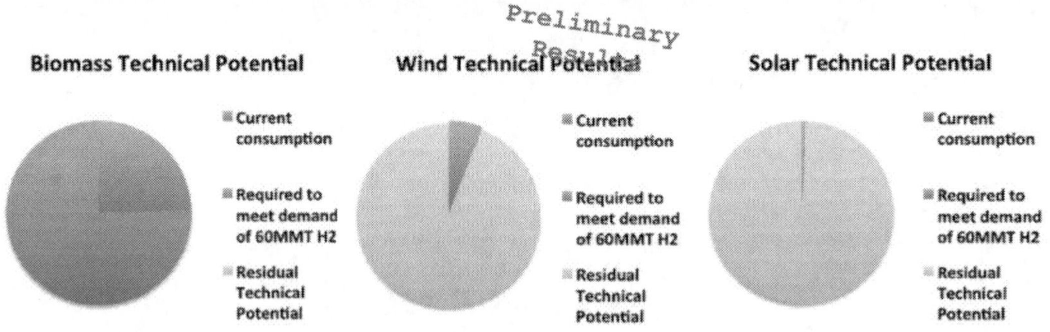

Total demand including hydrogen is satisfied by ≈6% of wind, <1% of solar, and ≈100% of biomass technical potential

Technical Potential: Impact on Fossil & Nuclear Resources

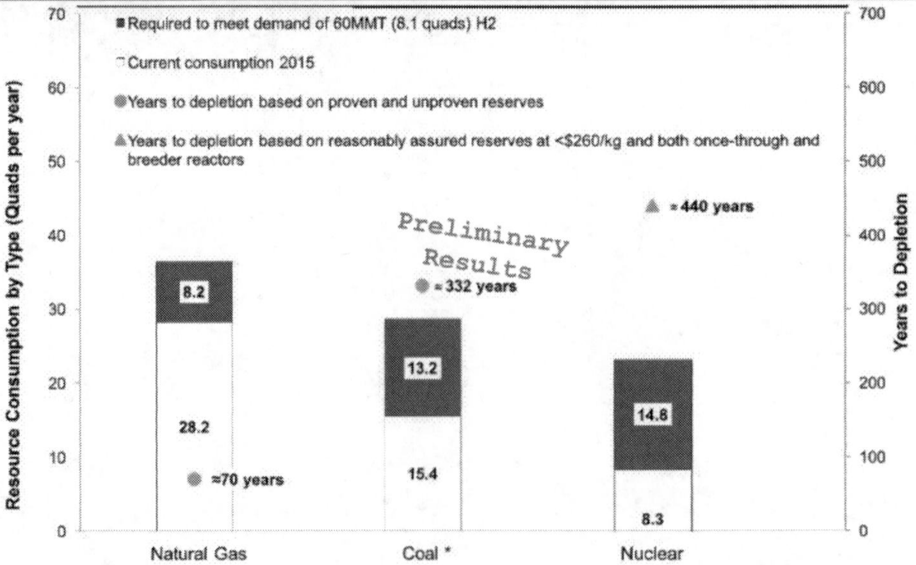

Hydrogen can be produced from diverse domestic resources to meet aggressive growth in demand

* Based on estimated recoverable reserves

Technical Potential: Impacts on Resources

Use	H_2 Consumed MMT / yr	Resource Savings: Petroleum (bbl/yr)	Resource Savings: Natural Gas (quad btu/yr)	Emissions Reduction: CO2 (million metric ton/yr)
Refineries	8	900,000	1.332	87
Metals	5	0	0.365	78
Ammonia	5	500,000	0.833	54
Natural Gas System	7	700,000	0.923	63
Biofuels§	4	77,500,000	-0.026*	28
Light Duty Vehicles	28	1,017,600,000	0.629	469
Other Transport	3	113,400,000	0.051	50
Total	**60**	**1.2 Billion bbl**	**4.1 Quads**	**830 Million MT**

Preliminary Results

~17% of U.S. petroleum consumption in 2016

~14% of U.S. natural gas consumption in 2016

~16% of U.S. energy-related emissions in 2016

Growth in electrolytic hydrogen using renewable electricity can reduce petroleum and natural gas utilization by ≥15%

*Negative values represent increase in use due to fertilizer production

§ 12% of the benefits of hydrogenated biofuels are credited to hydrogen

Economic Potential: Definition

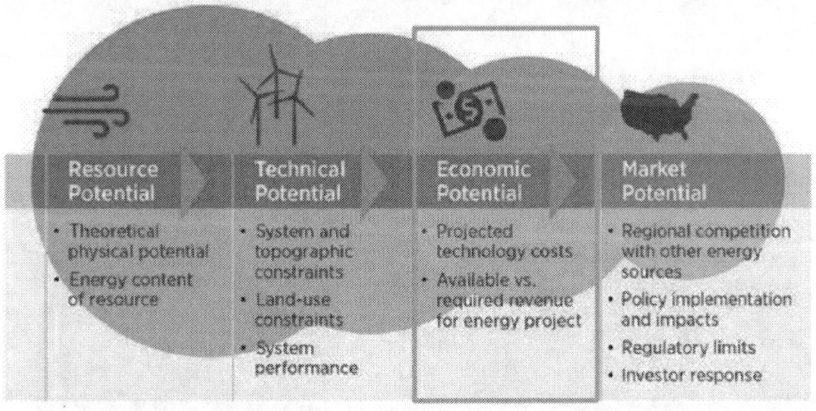

Economic potential is the **subset of the technical potential** where the **cost required** to produce hydrogen **is below** the **revenue available**

Brown, A., P. Beiter, D. Heimiller, C. Davidson, P. Denholm, J. Melius, A. Lopez, D. Hettinger, D. Mulcahy, and G. Porro. 2015. *Estimating Renewable Energy Economic Potential in the United States: Methodology and Initial Results.* Golden, CO: National Renewable Energy Laboratory. NREL/TP-6A20-64503

Improvements Enabling Use of Low-Cost Electricity

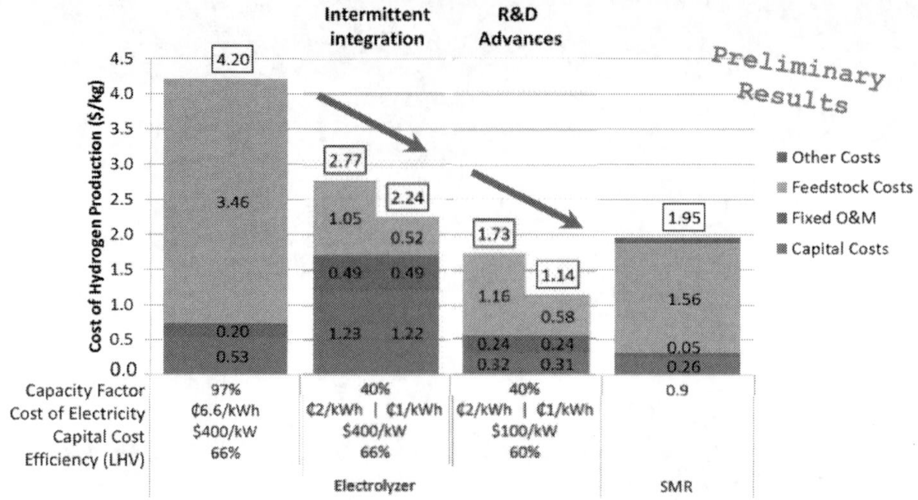

Leveraging of intermittent low-cost electricity can enable low-cost hydrogen production and also support grid stability.

Methodology for Estimating Otherwise Curtailed Electricity

Value of Otherwise Curtailed Electricity

Scenario Assumptions

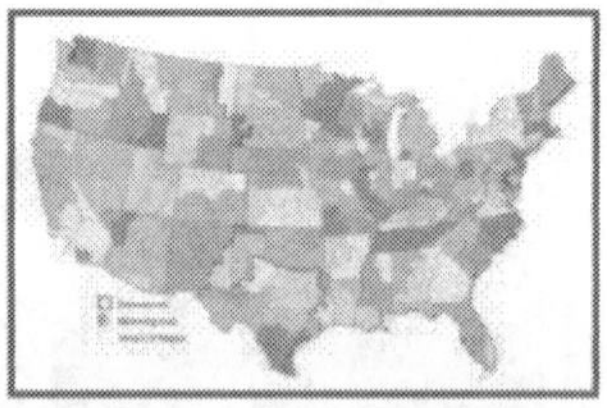

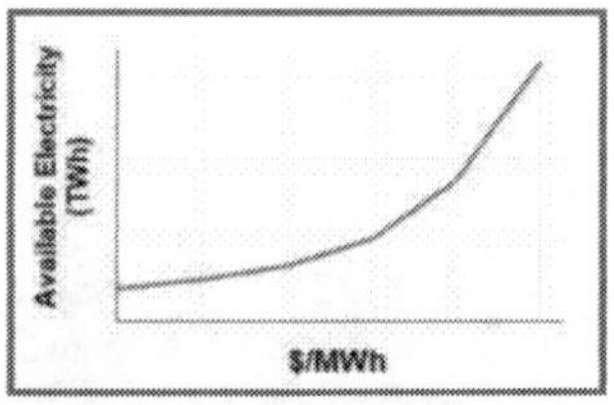

Key Assumptions

- ReEDS input parameters to achieve high renewable energy penetration:
  - Lower renewable energy technology costs
  - Higher natural gas prices

ReEDS

- Capacity expansion model
- Competes conventional, renewable, and storage technologies
- Projects capacity, generation, and transmission portfolio for 136 U.S. balancing areas out to 2050

Supply Curve

- Value of otherwise curtailed electricity vs. amount of curtailed electricity calculated by ReEDS
- Converted to a low-temperature electrolysis hydrogen supply curve using H2A

Economic Potential: Supply Curves

Developed supply curves using three resources

- Steam methane reforming (SMR)
- Nuclear generation
- Otherwise curtailed electricity with high penetrations of variable renewable generators on the grid and no transmission costs

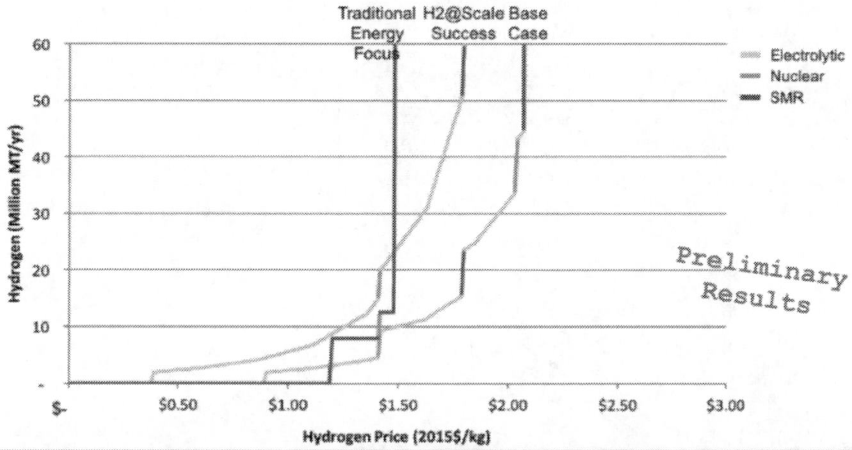

Economic Potential: Demand Curves

Developed demand curves under three paradigms

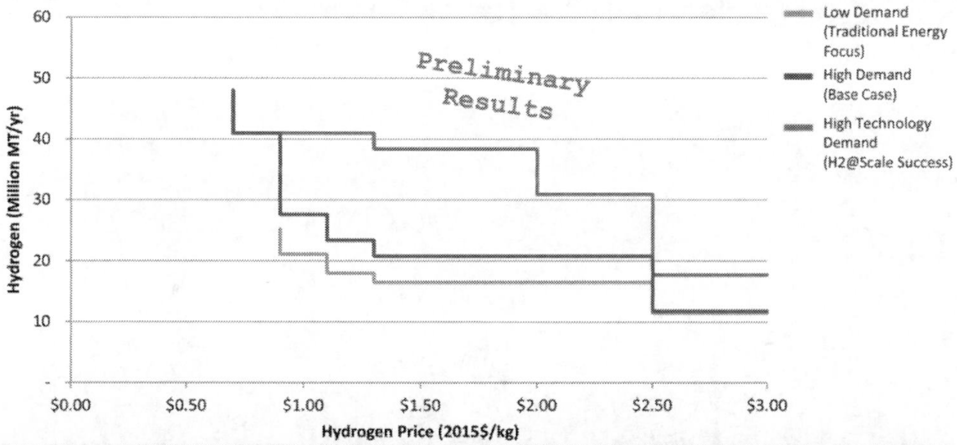

Demand	Base Case (High)	H2@Scale Success (High Tech.)	Traditional Energy Focus (Low)
Metals Reshoring	Economically competitive	Willingness to pay for H2 for metals	Economically competitive
LDV	Economically competitive	Full potential at $2.50/kg	Economically competitive
Synthetic Fuels	Economically competitive	Full potential at $2.00/kg	Economically competitive

Economic Potential: Scenario Summary

Preliminary

	Base Case	H2@Scale Success	Traditional Energy Focus
H_2 Use	21 MMT/yr	38 MMT/yr	17 MMT/yr
H_2 Price	$1.80/kg	$1.70/kg	$1.49/kg
Demand (MMT/yr)	• Refining (8), • Ammonia (3), • Synthetic fuel (1), • LDVs (9)	• Refining (8), • Ammonia (3), • Synthetic fuel (9), • Metals (6), • LDVs (13)	• Refining (8), • Ammonia (3), • Synthetic fuel (1), • LDVs (5)
Supply (MMT/yr)	• Low-temperature electrolysis (11), • Existing nuclear plants (5), • Existing NG reforming (5)	• Low-temperature electrolysis (33), • Existing nuclear plants (5)	• Existing nuclear plants (5) • NG reforming (8 MMT/yr from existing and 4 MMT/yr from new)
Electrolysis	10% curtailment, $19/MWh wholesale price	25% curtailment, $26/MWh wholesale price	No grid electrolysis

Economic Potential: Energy Use and Emissions Summary

H2@Scale can reduce emissions by up to 20% on top of baseline electricity sector emission reductions

Preliminary

Reduction Metric	Base Case	H2@Scale Success	Traditional Energy Focus
NO_X (Thousand MT)	130 (1%)	230 (2%)	61 (1%)
SO_X (Thousand MT)	33 (1%)	170 (5%)	13 (0%)
PM_{10} (Thousand MT)	10 (0%)	59 (2%)	4.0 (0%)
Crude Oil (Million Barrels)	470 (7%)	800 (12%)	280 (4%)
CO_2 (Million MT)	280 (9%)	590 (19%)	110 (4%)

Conclusions and Future Work

H2@Scale can transform our energy system by providing value for otherwise-curtailed electricity and a clean feedstock for numerous industries

- Technical potential:
  60MMT H_2/ yr can reduce emissions by 15%

- Economic potential:
  17-38 MMT H_2/ yr can be produced, given R&D advancements and access to low-cost intermittent power

Further analysis into regional issues and temporal issues is warranted to better quantify H2@Scale potential

www.nrel.gov

NREL
NATIONAL RENEWABLE ENERGY LABORATORY

NREL is a national laboratory of the U.S. Department of Energy, Office of Energy Efficiency and Renewable Energy, operated by the Alliance for Sustainable Energy, LLC.

Economic Potential Analysis Caveats

- Ideal markets
- Hydrogen storage and transportation requirements and costs are negligible
- Technology improvement and market assumptions in ReEDS result in a high VRE penetration and quantity of curtailed electricity
- Electrolyzer capacity factors are estimated using coarse temporal resolution of ReEDS (16 time slices representative of entire year)
- Low-temperature electrolyzer system costs achieve a target of $100/kW
- Nuclear-generated hydrogen costs do not include potential value to the grid in capacity and/or flexibility
- Economic rebound effects are negligible
- Additional hydrogen markets outside of our analysis may develop

504346 An Overview of H2@Scale: Energy System-Wide Benefits of Increased H2 Implementation

Abstract:

Supplying all of societyâ€™s growing energy demands is hard and has been limited to date to the fossil fuel based energy system in place since the start of the industrial age. Hydrogen, as a flexible, clean energy carrying intermediate, has the potential to be a centerpiece of a future energy system where aggressive market penetration of renewables (wind and solar) are coupled with renewable hydrogen production to meet societyâ€™s energy demands using clean, renewable resources and processes.

A number of key recent changes are bringing hydrogen back into the spotlight: 1) Renewable energy is getting cheaper, and penetration levels are increasing at an exponential pace; 2) the costs of GHG emissions (climate change) and criteria pollutants (health concerns) are being acknowledged and are reaching a point where society is demanding change (renewable portfolio and zero emission vehicle standards); and 3) commercial viability of fuel cell vehicle technology has been demonstrated (commercial vehicles being sold). Hydrogen, as an area of R&D interest (from either the federal or private sectors) has had its ups and downs over the past few decades. Hydrogen has suffered at times from being over-hyped or misunderstood. Claims of fuel cell commercialization, in particular, were in the past (early 2000â€™s) overly optimistic to be polite. The safety concerns of hydrogen have also often been poorly understood or overblown. These situations have arisen, in part, due to a lack of appreciation for the science and data of hydrogen and hydrogen systems.

This presentation will focus on the role of hydrogen at (grid-)scale and the efforts of a large, national lab team assembled to evaluate the potential of hydrogen to play a critical role in our energy future. Facts about hydrogen will be shared, as will the vision of how it will fit into our future energy system.

502524 Highly Efficient Visible Light Photocatalytic Production of H_2

Abstract:

Intensive effort has been made to develop photocatalytic processes to convert solar energy to clean fuels. So far, however, energy conversion efficiency for the visible-light photocatalytic splitting of water to hydrogen is less than 1%. In this talk, I will highlight recent findings for visible light photocatalysis in my group, including: (1) demonstrating that surface-diffuse-reflected-light can be 2 orders of magnitude more efficient than incident light for photocatalysis, (2) revealing that the inefficiency of absorbed visible light for the photocatalytic H2 production from water with a sacrificial agent is due to its kinetic limitation, and (3) inventing a novel temperature-induced visible light photocatalytic H2 production from water steam with a sacrificial agent to reach a high photohydrogen yield of 497 mmol/h/gcat with a large apparent quantum efficiency (QE) of 65.7% for entire visible light range.

498247 Dynamic Simulation of a Solar Powered Hybrid Sulfur Process for Hydrogen Production

Abstract:

Hydrogen can be produced using a thermochemical Hybrid Sulfur (HyS) Process. Using solar energy as the high temperature energy source allows for efficient environmentally-friendly production of hydrogen. The proposed route is an alternative to traditional photovoltaic plus electrolysis for hydrogen production. Making the process economically competitive is a major challenge. Operating the process with changes in the availability of solar energy also increases process complexity. The HyS flowsheet was modeled in Aspen Plus. The process includes electrolyzers, H2S04 decomposition, and O2 separation of the SO2/O2 product from the H2SO4 decomposition. Flowsheets for this process have demonstrated a thermal efficiency in excess of 50% higher heating value basis. The current work will discuss the process flowsheet, present results from a dynamic process simulation, and consider hydrogen production economics.

491322 Hydrogen Generation from Hydrogen Sulfide Decomposition

Abstract:

We propose a new method for the processing of hydrogen sulfide to create elemental sulfur and hydrogen gas. The process is based on a thermochemical cycle that leverages differences in temperature and reaction conditions to accomplish the thermodynamically unfavorable hydrogen sulfide decomposition over two reaction steps. First, hydrogen sulfide is reacted with iodine in aqueous conditions to produce dissolved hydroiodic acid and elemental sulfur. The sulfur is filtered out and the concentrated hydrogen iodide and water mixture is vaporized. The hydrogen iodide is then catalytically decomposed to hydrogen gas and iodine, which are flash separated. Overall, the two reaction and separation steps combine to accomplish the decomposition of hydrogen sulfide to hydrogen and elemental sulfur. This proposed process is similar to ideas put forth in the 1980s and 1990s by Kalina, Chakma, and Oosawa, but makes use of more recently discovered thermochemical hydrogen iodide decomposition methods and catalysts rather than electrochemical or photoelectrochemical methods.

This process is especially useful when the hydrogen sulfide was created by the reaction of hydrogen with sulfur-rich hydrocarbons during hydrodesulfurization. By recovering and recycling the hydrogen gas originally used to desulfurize the hydrocarbon, the amount of hydrogen gas needed to treat a sulfur-rich hydrocarbon is dramatically reduced. As larger quantities of sulfur-rich resources are extracted, methods to efficiently desulfurize hydrocarbons are increasingly important. Further, this process is inherently carbon neutral which is an advantage over steam reforming, the current main source of hydrogen gas.

A process model was created to explore the viability of the process. The model suggests that hydrogen could be profitably generated in processes with a maximum temperature of just 300 degrees Celsius. This implies that the heat required for the decomposition of hydrogen sulfide could be supplied by solar concentrators or heat integration with other process units. Process units in the model were defined by literature and experimental results.

Adsorptive on-Board Desulfurization of Liquid Fuels: High Efficiency in Desulfurization and Full Thermal Regeneration Via Hot Exhaust Gas

Raphael Neubauer, Institute of Thermal Engineering, Graz University of Technology, Austria
Norbert Kienzl, Bioenergy 2020+, Graz, Austria
Christoph Hochenauer, Institute of Thermal Engineering, Graz University of Technology, Austria

Introduction

Desulfurization of fuels is mandatory to fulfill environmental regulations and to protect the environment. Sulfur levels of liquid fuels are in the range of 10 to 5000 ppm and depend on the type of fuel and their correlated legal requirements (1). Hydrodesulfurization (HDS) is used for crude oil desulfurization to achieve the country-specific legal requirements. This HDS is the most prevalent and industrially relevant desulfurization process which operates at temperatures above 300 °C and more than 35 bar of H_2 operating pressure. The HDS efficiency decreases with increasing molecular size of the refractory sulfur compounds and thus is inefficient for removing benzothiophene (BT) and especially dibenzothiophene (DBT) and its derivatives.

Fuel cells are one of the most effective tools to convert chemical to electrical energy. This technology is not only suitable for stationary approaches but is also a promising source of on-board electricity supply for all kinds of vehicles, ships, and aircraft. The advantages of fuel cells are in particular higher efficiency, reduced emissions, and lower noise generation in comparison to conventional combustion engines. This technology becomes very attractive when on-board fuel is used to produce hydrogen-rich syngas via reforming, which is subsequently fed to the fuel cell. However, the sulfur threshold limit for the reformer and the fuel cell are < 50 and < 1 ppm of total sulfur, respectively (2, 3). Consequently, fuel cells as one of the most efficient tools for energy conversion are excluded from using the most common types of fuels.

Adsorptive desulfurization is a promising approach to provide the possibility of on-board desulfurization and thus operating fuel cells with commercial fuels. Several adsorbents have been investigated and published in recent years. However, adsorbents regeneration is still a major issue; especially after adsorption of dibenzothiophene (DBT). The aim of this work is to identify and prepare a highly thermal stable adsorbent for thermal on-board regeneration via hot exhaust gases.

Experimental Section

A silver based adsorbent was prepared by incipient wetness impregnation of γ-Al2O3. The impregnated support was further dried and calcined in order to get Ag-Al2O3. This adsorbent was characterized via different methods including N2-physisorption, XRD, SEM, EDX, and ICP-AES. The adsorbent was further used in different experiments including equilibrium saturation, kinetics, and breakthrough experiments. The adsorptive desulfurization performance of Ag-Al2O3 was tested under different conditions and fuels such as jet fuel and diesel fuel. These fuels contained different concentrations of poly aromatic sulfur heterocycles (PASHs) in the range of 300 to 900 ppmw of total sulfur. In situ regeneration experiments were carried out with preheated air and simulated exhaust gas from a solid oxide fuel cell driven auxiliary power unit

Experimental Section

The overall adsorption mechanisms of PASHs were studied by relating experimental data with theoretical principles. High breakthrough adsorption capacities were observed for different PASHs like BT, DBT, and 4,6-dimethyldibenzothiophen dissolved in commercial fuels. The breakthrough adsorption capacity was in the range of 0.9 to 2.3 mg/g at 20 °C and atmospheric pressure depending on the type of PAHS. Within these investigations, the role of the acid base interaction (S-H) on the overall adsorption mechanism was identified. The results showed, that the weak S-H interaction is a third mechanism beside the known π-complexation (π-Ag) and the direct metal sulfur interaction (S-Ag). Both π-Ag and S-Ag interactions have a selectivity order of DBT > BT which is caused by a higher adsorption energy of DBT in comparison to BT. The S-H interaction is thus responsible for the shift of the final selectivity order to BT > DBT at equilibrium saturation conditions. This result is confirmed by density functional theory (DFT) calculations reported in the literature as well as by the hard and soft acid base (HSAB) theory proposed by Pearson. Consequently, π-Ag and S-Ag interactions are the main adsorption mechanisms at breakthrough. In these two interactions, the silver cation (soft acid) interacts with the thiophenic ring or directly with the π lone-pair of electrons of the sulfur atom (soft base) (4).

Additional equilibrium saturation experiments were carried out at -10 and 60 °C. The adsorption temperature showed no influence on the equilibrium saturation capacity in the case of DBT. For BT, the equilibrium saturation capacity at -10 °C stayed at the same level as at 20 °C. At 60 °C, the equilibrium saturation of BT decreased to the level of DBT and thus confirmed that the weak S-H interaction is only relevant for BT at equilibrium stage. At increased temperatures (60 °C) the S-H interactions is to week to contribute in the overall adsorption mechanism of BT. The stronger adsorption of DBT in particular makes it difficult to achieve full thermal regeneration without thermal destruction of the adsorbent. These are important findings in order to develop highly efficient regeneration strategies.

The lack of solvents requires thermal regeneration of the adsorbent in the case of on-board desulfurization units. This is the reason, why we studied the desorption mechanisms of PASHs from $Ag-Al_2O_3$ in the presence of air; and in particular of N_2, O_2, CO_2, and H_2O. In these investigations, thermal regeneration was carried out in a preheated gas mixture with increasing temperature. A final temperature of 450 °C was used in the first regeneration experiments. This temperature was sufficient for full thermal regeneration in flowing air after adsorption of BT on $Ag-Al_2O_3$. The same experiment was carried out with DBT and showed only 60% recovery of the breakthrough capacity after thermal regeneration. This result confirms the stronger adsorption of DBT in comparison to BT. Additional regeneration experiments were carried out at a final temperature of 525 °C. The results showed full thermal regeneration in flowing air after adsorption of DBT. This is an excellent result as full thermal regeneration after DBT adsorption has not been reported so far. The regenerated $Ag-Al_2O_3$ was further analyzed and showed no thermal induced destruction.

One problem of thermal regeneration is the source of hot regeneration gas as electrical heating would be inefficient and cost-intensive. To overcome this drawback, hot exhaust gases were identified as a source of hot regeneration gases. This is the reason why regeneration experiments were performed by using hot off-gas from a solid oxide fuel cell based auxiliary power unit; this gas is referred to as APU. The flow sheet of this novel regeneration strategy is illustrated in Fig. 1 (right).

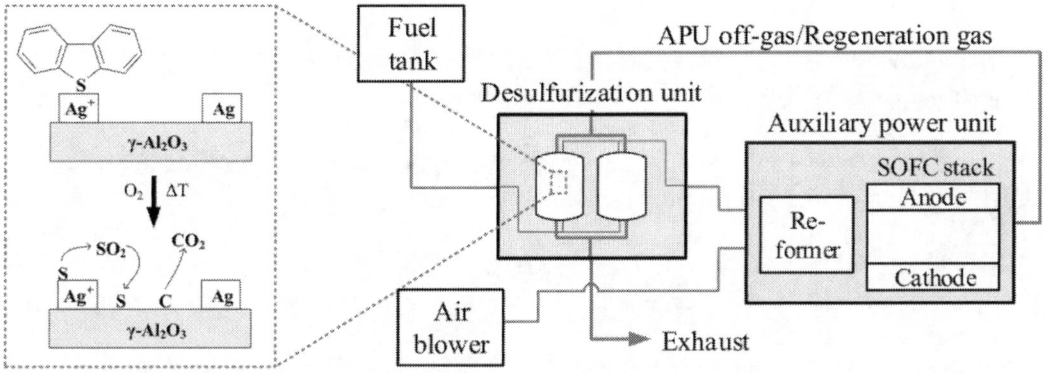

Fig. 1 Flow sheet diagram of novel regeneration strategy using hot APU off-gas for thermal regeneration of the adsorbent (right) and simplified scheme of the overall desorption mechanism of PASHs (left).

Excellent results were obtained with APU as regeneration gas. In these experiments, 100% regeneration was achieved over 5 cycles at a final temperature of 500 °C. The identified desorption mechanisms (Fig. 1 left) showed that CO_2 in the regeneration gas has no negative influence on the overall regeneration experiment as it desorbs already at temperature above 100 °C. In addition, H_2O was identified to have a positive influence on the thermal decomposition of adsorbed PASHs. This positive effect was proven by achieving full thermal regeneration at a final temperature of 450 °C with increased H_2O content in the regeneration gas (12 vol.%). These findings even highlight the possibility to use hot exhaust gas from a combustion engine as regeneration gas in a hybrid vehicle based on low temperature fuel cells.

Conclusion

Adsorptive desulfurization of liquid fuels is mandatory to provide the possibility to operate fuel cell with commercial fuels. $Ag\text{-}Al_2O_3$ was identified as excellent adsorbent for on-board desulfurization units. Investigations on the overall adsorption mechanism showed that the silver cations are the main active adsorption sites. Adsorption of DBT via π-Ag and S-Ag interactions is stronger in comparisons to BT. Consequently, regeneration of these silver sites after adsorption of DBT is essential to achieve full regeneration.

Hot exhaust gas was identified as ideal regeneration gas for thermal in situ and on-board regeneration. The identified desorption mechanism showed that CO_2 has no negative influence on the regeneration performance. Full thermal regeneration was reported for the first time after adsorption of DBT at a final temperature of 500 °C over 5 cycles. The final regeneration temperature could even be reduced to 450 °C by increasing the H_2O content in the regeneration gas. All these findings showed excellent adsorption and regeneration performances of $Ag\text{-}Al_2O_3$ under a wide range of conditions including different fuels. Full on-board regeneration via hot exhaust gas, as reported, highlights the possibility to operate fuel cell with commercial sulfur containing fuels and thus achieve high efficiency in energy conversion combined with lowest emissions.

References

1. A. Samokhvalov, B.J. Tatarchuk, Catal. Rev. 52 (2010) 381-410. http://dx.doi.org/10.1080/01614940.2010.498749.

2. X. Xu, S. Zhang, P. Li, Y. Shen, Fuel Process. Technol. 124 (2014) 140–146. http://dx.doi.org/10.1016/j.fuproc.2014.03.001.

3. C. Hulteberg, Int. J. Hydrogen Energy 37 (2012) 3978-3992. http://dx.doi.org/10.1016/j.ijhydene.2011.12.001.

4. R. Neubauer, M. Husmann, C. Weinlaender, N. Kienzl, E. Leitner, C. Hochenauer, Chem. Eng. J. 309 (2017) 840-849. http://dx.doi.org/10.1016/j.cej.2016.10.094.

495589 Thermodynamic Simulations and Techno-Economic Analysis on the Utilization of CO_2Â and a Novel Modularization Strategy for Chemical Looping Based Gtl Processes

Abstract:

The utilization of carbon dioxide (CO_2) as a feedstock in a chemical looping process that converts methane to syngas and a variety of downstream products has the ability to be impactful in the field of carbon capture, utilization, and sequestration (CCUS). The Ohio State University Chemical Looping methane to syngas (MTS) processâ€™ novelty comes from the coupling of carbon capture and utilization strategy, which captures and utilizes CO_2 to create a synergy atypical to many processes that are designed to deal with the challenges of carbon constraints. CO_2 utilization allows for increased flexibility of H_2:CO molar ratios that may be required for variations in the downstream processing while producing a comparable quantity of syngas with reduced natural gas input. In this study, thermodynamic models for chemical looping reactor systems are examined using ASPEN Plus. The chemical looping reducer reactor considered employs steam, natural gas and CO_2 as feedstock along with metal oxide oxygen carriers. This presentation will initially present system optimization based on thermodynamic Gibbs free energy minimization for the desired syngas composition required for downstream gas to liquid (GTL) processes. With CO_2 co-injection in the MTS process, the thermodynamic optimization indicates a natural gas savings of more than 20% over that of the baseline partial oxidation processes for GTL applications. Experimental verification using Thermo-gravimetric analyzer and bench-scale moving bed reactor that compared well with the thermodynamic model simulations will also be presented. A comprehensive techno-economic analysis on an energy, balanced (water balance, steam balance, heat and electricity balance) thermodynamic performance model of the MTS process with and without CO_2 utilization & modularization strategy was developed. The use of techno-economic analysis model for studying sensitivities associated with parameters including capital charge factor, first year versus leveled costs of fuel, crude oil equivalent prices and cost of chemical looping reactors will be presented. The comprehensive techno-economic analysis findings exemplify the importance of a CO_2 utilization and modularization strategy in chemical looping GTL processes and encourage further optimization in natural gas to commodity chemical conversion processes including gas to methanol, gas to ammonia, and gas to ethanol, amongst others.

534C:

Gasification of Residue from the Solvent Extraction of Powder River Basin Coal

William C. Schaffers
Ying Wang
David A. Bell (speaker)

American Institute of Chemical Engineers National Meeting

November 1, 2017

UW UNIVERSITY OF WYOMING

Shenhua Direct Coal Liquefaction (DCL)

- Commercial plant in China that makes liquid fuels by hydrogenating coal.
- Residue from plant contains undigested coal + 1st stage catalyst
- This project:
  - Investigate feasibility of gasifying residue
    - Provide hydrogen for plant
    - Eliminate residue waste stream
  - Sponsored by U.S./China Clean Energy Research Center for Coal

UW

Project Assumptions

- Unable to contact Chinese team
- Assumed:
  - Shenhua DCL plant or similar built in U.S.
  - Powder River Basin coal feed.
  - Need to recommend gasification technology for residue.

Fluidized bed gasifiers

- Lower cost than entrained flow gasifiers
- Requires reactive feedstock
  - Biomass
  - Lignite
  - Sub-bituminous coal, such as PRB
- Bituminous coals require more severe conditions in entrained flow gasifier.

Experimental strategy

- Approximate coal liquefaction process by extraction with tetralin, a hydrogen donor solvent

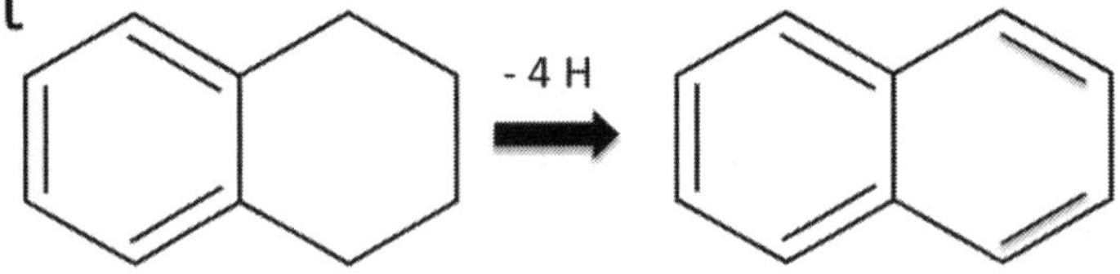

- Compare gasification reactivity of extraction residue to parent coal.

UW

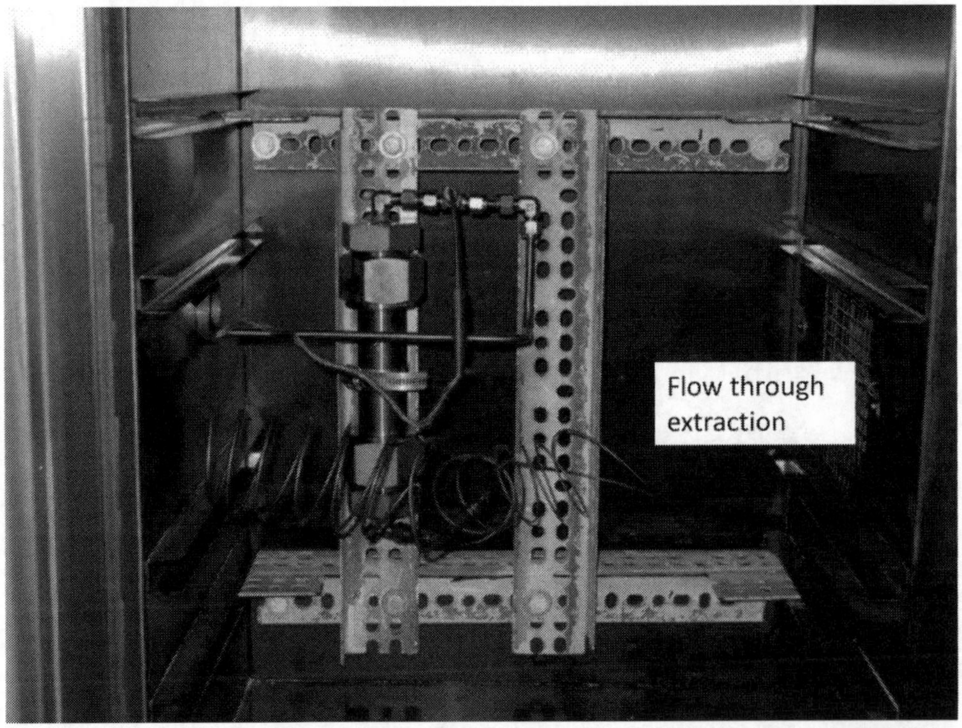

UW

UW

Extraction at 360 C

Run #	As Received Weight (g)	Dry Weight (g)	Moisture Loss (wt %)	Extraction Residue Weight (g)	Material Extracted (wt %)
3	54.38	46.54	14.42	25.31	45.62
5	57.88	50.19	13.29	30.01	40.21
6	60.27	51.38	14.75	28.39	44.75

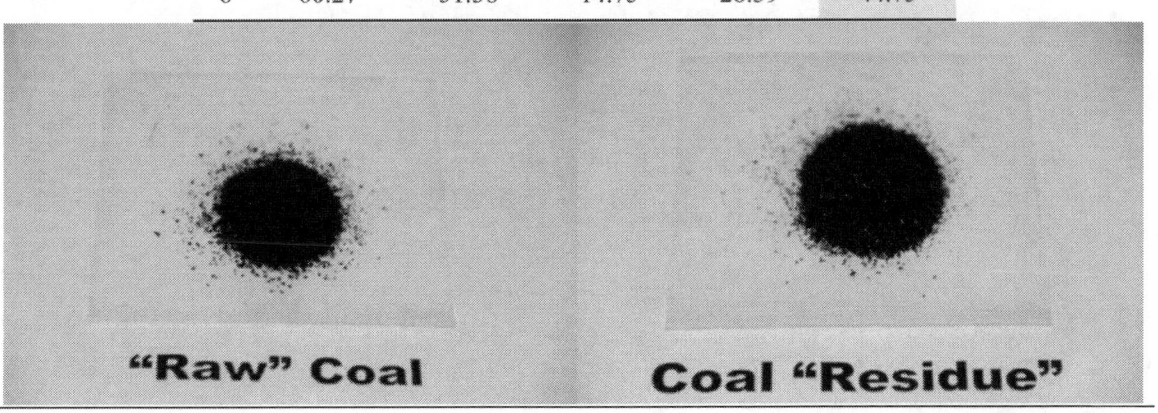

UW

Proximate & Ultimate Analyses

	As Received Basis (wt %)		Moisture Free Basis (wt %)		Moisture & Ash Free Basis (wt %)	
	Decker Coal	*Extraction Residue*	***Decker Coal***	*Extraction Residue*	***Decker Coal***	*Extraction Residue*
Proximate Analysis (Method: ASTM D-5142; D-7582)						
Moisture	24.29	*3.27*	****	****	****	****
Ash	3.62	*9.78*	4.78	*10.11*	****	****
Volatile Matter	29.42	*31.48*	38.86	*32.54*	40.81	*36.20*
Fixed Carbon	42.67	*55.47*	56.36	*57.35*	59.19	*63.80*
Total	**100.00**	*100.00*	**100.00**	*100.00*	**100.00**	*100.00*
Ultimate Analysis (Method: ASTM D-5142/5373; D-7582/4239)						
Moisture	24.29	*3.27*	****	****	****	****
Hydrogen	2.88	*3.80*	3.80	*3.93*	3.99	*4.38*
Carbon	56.48	*68.28*	74.59	*70.59*	78.33	*78.53*
Nitrogen	0.87	*1.05*	1.15	*1.09*	1.21	*1.21*
Sulfur	0.34	*0.28*	0.45	*0.29*	0.47	*0.32*
Oxygen	11.52	*13.54*	15.23	*14.00*	16.00	*15.57*
Ash	3.62	*9.78*	4.78	*10.11*	****	****
Total	**100.00**	*100.00*	**100.00**	*100.00*	**100.00**	*100.00*

UW

Gasification tests

- Modified drop tube reactor
- Preheated, gas flow established
- ~1 g of coal or residue dropped into reactor
- Lands on glass frit in center
- Effluent gases measured using mass spectrometer

UW

Typical mass spectrometer results

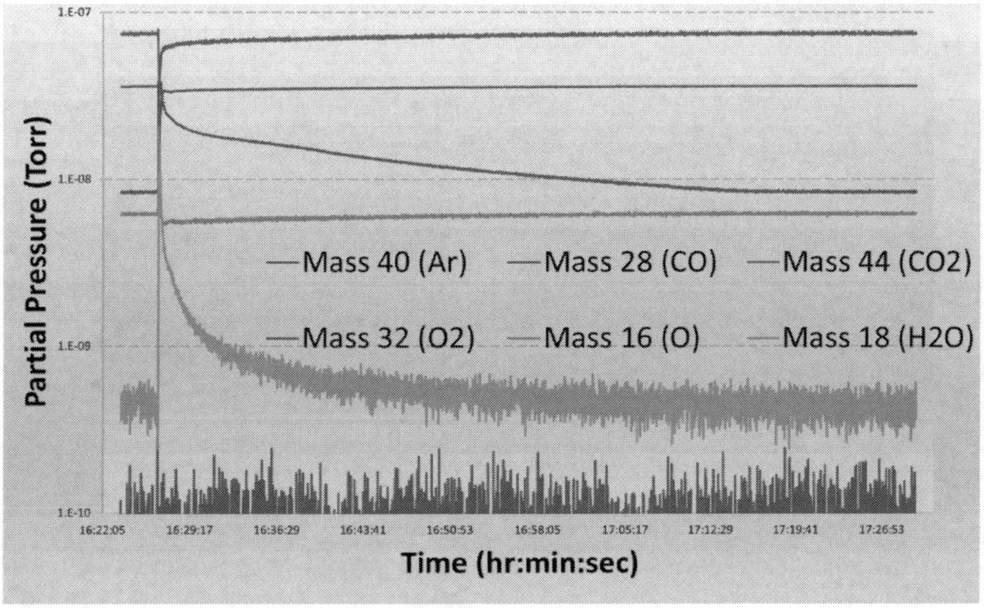

UW

Data fit

- Reaction complete when mass spectrometer signal returns to baseline
- Integration of signal versus time used to measure extent of reaction, *X*.
- Data fit to random pore model (assumes reaction on pore walls):

$$dX/_{dt} = \boldsymbol{k_p}(1 - X)\sqrt{1 - \psi ln(1 - X)}$$

Where $\boldsymbol{k_p}$ is the rate parameter, ψ is dimensionless structural parameter.

UW

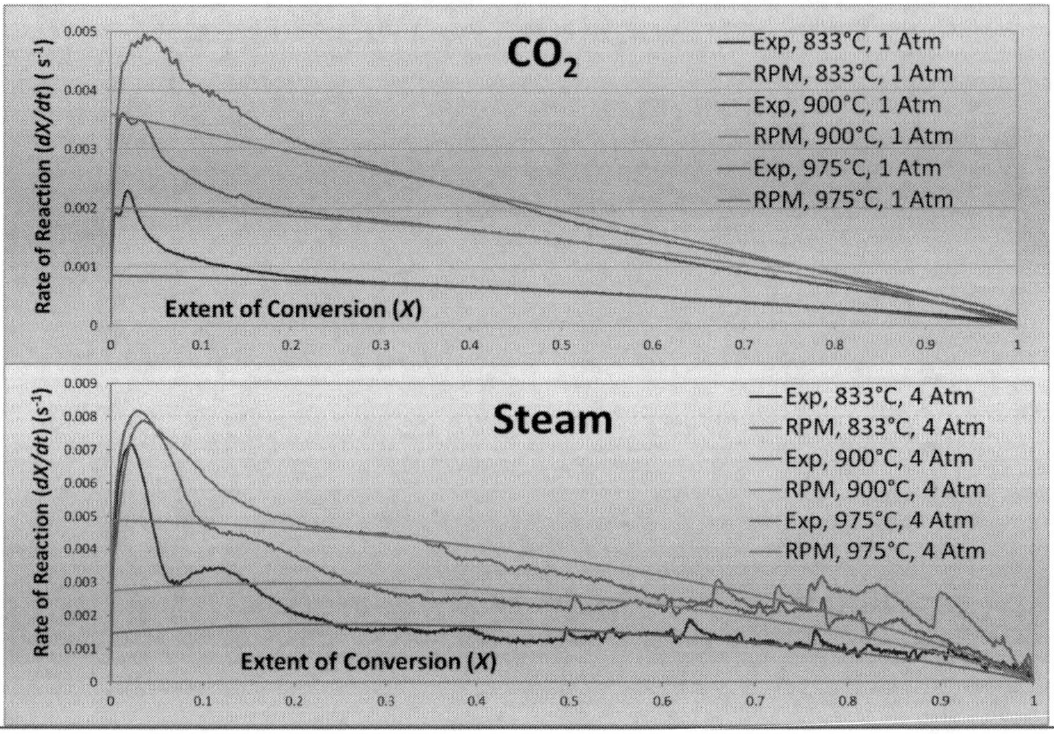

Coal & Residue Gasification in CO_2

Pressure	Temp.	Coal Time (hr:min:s)	Residue Time (hr:min:s)	*Coal* k_p *(s⁻¹)*	*Residue* k_p *(s⁻¹)*	*Coal* ψ	*Residue* ψ
1 atm	833 °C	1:06:12	0:47:45	4.10E-04	8.75E-04	3.37	1.50
	900 °C	0:25:00	0:22:52	1.20E-03	1.81E-03	3.12	1.51
	975 °C	0:12:06	0:15:28	2.75E-03	3.44E-03	2.45	0.63
4 atm	833 °C	1:01:36	1:00:02	5.00E-04	8.17E-04	2.65	0.92
	900 °C	0:23:00	0:28:36	1.28E-03	1.95E-03	2.63	0.28
	975 °C	0:11:00	0:17:11	2.50E-03	3.88E-03	3.25	0.10

Time is time to complete gasification.

Residue gasification is usually faster than gasifying parent coal.

Coal & Residue Gasification in Steam

Pressure	Temp.	Coal Time (min:s)	Residue Time (min:s)	*Coal* $k_p (s^{-1})$	*Residue* $k_p (s^{-1})$	*Coal* ψ	*Residue* ψ
1 atm	833 °C	17:41	13:19	1.28E-04	2.01E-03	5.25	4.69
	900 °C	9:26	9:34	2.75E-03	3.95E-03	5.25	2.18
	975 °C	4:31	7:07	6.30E-03	4.67E-03	3.50	3.39
4 atm	833 °C	19:53	19:19	1.28E-04	1.43E-03	3.63	4.27
	900 °C	13:15	10:02	2.60E-03	3.00E-03	2.95	3.65
	975 °C	6:47	07:31	5.85E-03	4.89E-03	2.23	2.88

Steam gasification faster than CO_2 gasification.

Residue gasification usually faster than coal gasification.

UW

Surface Areas of Chars from Coal & Residue

Method	Surface Area (m^2/g)					
	Coal Char (833°C)	Residue Char (833°C)	Coal Char (900°C)	Residue Char (900°C)	Coal Char (975°C)	Residue Char (975°C)
BET (N_2) (Mesopores)	0.36	10.81	1.49	24.39	34.54	40.25
D-R (CO_2) (Micropores)	408.51	382.13	341.44	382.61	344.68	404.13

UW

Calcium and Sodium on Char Particles

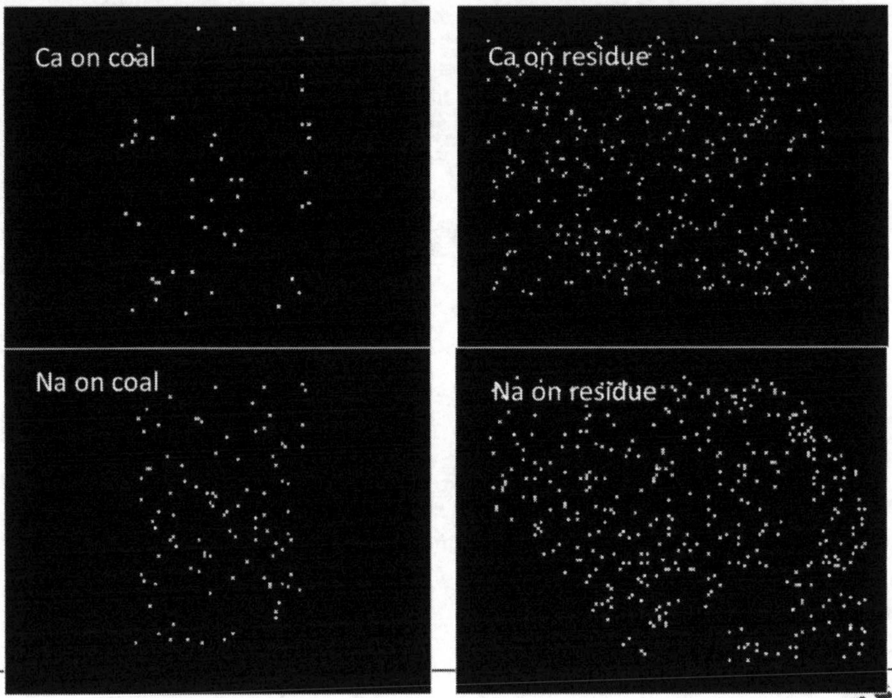

Conclusions

- Residue is more reactive than parent coal
- Effect may be due to
  - Increase in mesoporous surface area
  - Concentration of naturally-occurring catalysts, sodium and calcium
- Residue can be gasified in fluidized bed gasifier.

Development of a Modular Coal/Biomass to Liquids Research Facility

DOE Project Number: FE0010482

Andy Placido

University of Kentucky, Center for Applied Energy Research

http://www.caer.uky.edu/powergen/home.shtml

Facility Overview

- Integrated coal/biomass-to-liquids (CBTL) facility at a capacity of 1 bbl./day
- Purposely designed for maximum flexibility: modular, skid-mounted, anticipating frequent change-outs; "plug and play;" and future re-purposing

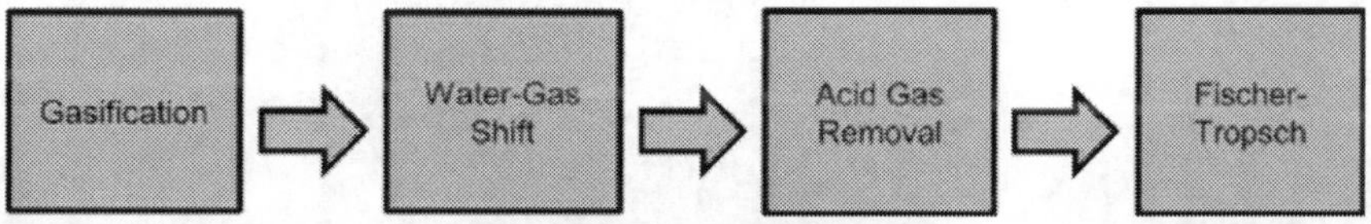

OMB Gasification System

- Entrained flow gasifier
  - Coal/water slurry
  - Oxygen blown
  - Molten slag
  - 1 ton/day coal consumption
- Includes biomass/coal feed preparation tower
  - Hydro-vibrating Ball mill

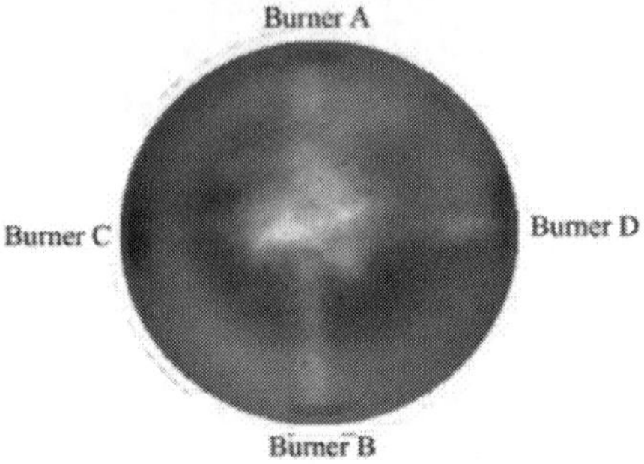

2017 AIChE Annual Meeting

November 1st, 2017

caer.uky.edu

Center for Applied Energy Research

Feed Preparation

- Ash fusion temperature generally limiting factor of feed
- Capable of utilizing wide range of coals and blends
- Designed to handle torrefied biomass
- Important Final Slurry Properties
  - Particle size
  - Viscosity

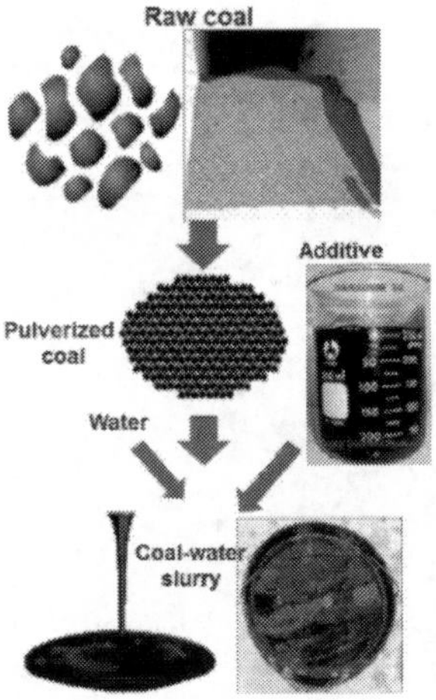

Typical Properties of CWS

Average particle size (μm)	Mass concentration	Viscosity (mPa·s)
<50	<60%	<250

2017 AIChE Annual Meeting

November 1st, 2017

caer.uky.edu

Center for Applied Energy Research

UK Center for Applied Energy Research

caer.uky.edu

WGS

$CO + H_2O \Leftrightarrow CO_2 + H_2$ Tunable H_2/CO ratio

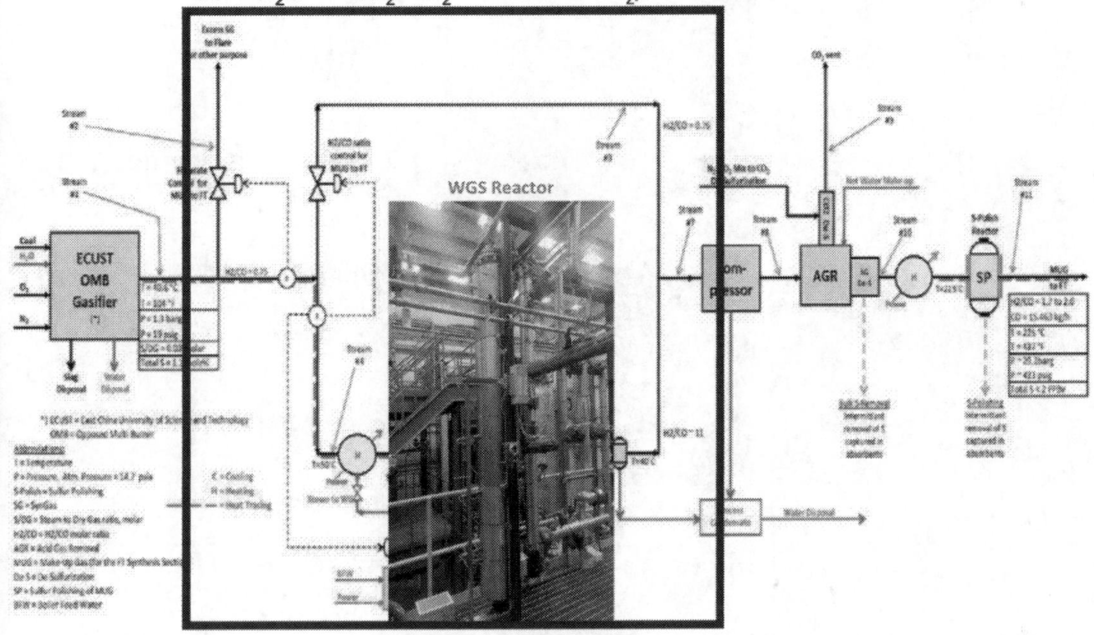

2017 AIChE Annual Meeting

November 1st, 2017

UK Center for Applied Energy Research

caer.uky.edu

Acid Gas Removal

Aqueous Solvent System

- CAER has 10+ years Experience with Similar Technology
- Thermal and Pressure Swing Absorption
- Howden Burton Corblin metal diaphragm compressor
  - Pressures up to 450 psi
- Wide range of Amine Solvents
- Sulfur Treatment on Both Rejected Acid Gas and Treated Gas
  - Activated carbon
- 95+% removal of CO_2
- 99.99% removal of sulfur (less than 1ppm)

2017 AIChE Annual Meeting

November 1st, 2017

Center for Applied Energy Research
caer.uky.edu

Downstream - FT

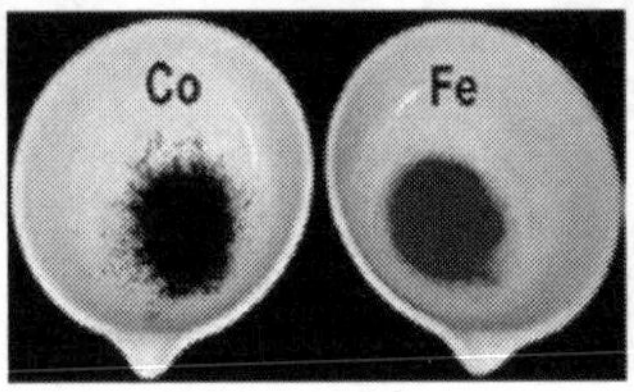

- Chart Energy microchannel reactor with production capacity of 1bbl./day
- FT Reactor is aluminum heat exchanger inside a steel kettle
- Capable of utilizing either Fe or Co based catalysts
  - Difference in H_2/CO molar ratio
- Wide ranging fuel production
- Water Gas Shift and FT modules were designed and built by Zeton Inc. with assistance from Stovlbaek Consulting

2017 AIChE Annual Meeting

November 1st, 2017

Center for Applied Energy Research
caer.uky.edu

Balance of Plant

2017 AIChE Annual Meeting

November 1st, 2017

Center for Applied Energy Research

caer.uky.edu

Experimental Data from Coal Only

- So far Gasifier test operation only
- Typical Operation data for 24 hour period
- Operationally stable for entire testing run
- Feed details
    - 30-40% coal/60-70% water
    - Co-fire small amount of natural gas (~15% heat input) to maintain heat

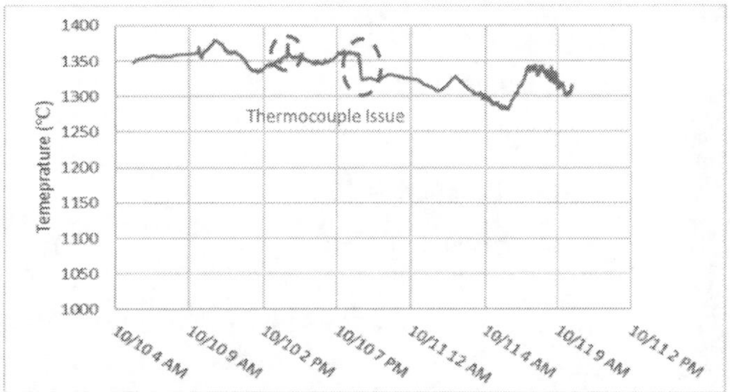

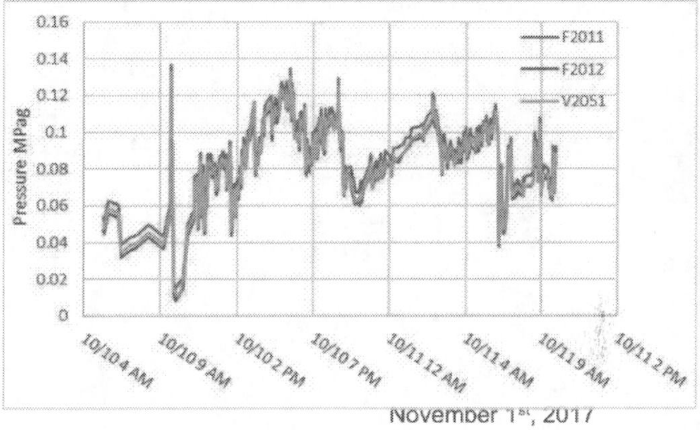

Center for Applied Energy Research

caer.uky.edu

Experimental Data from Gasifier Operation

- Overall availability ~50%
    - First operation for syngas quality control testing
    - GC learning curve (filters clogging, learning software and hardware)
- H_2/CO important syngas quality measure for downstream
    - Ratio stable around 1.1
    - Higher than expected due to co-firing with natural gas

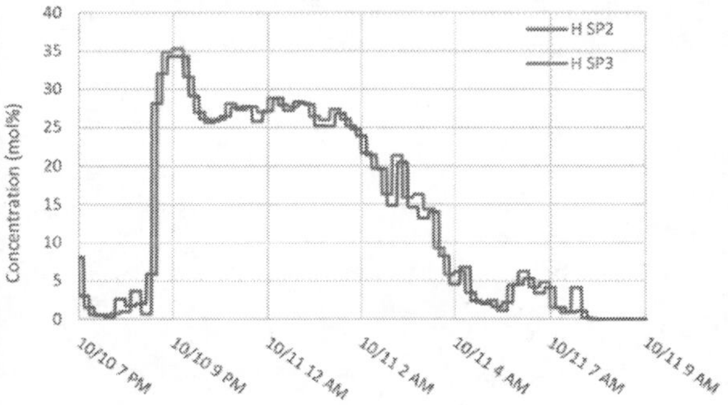

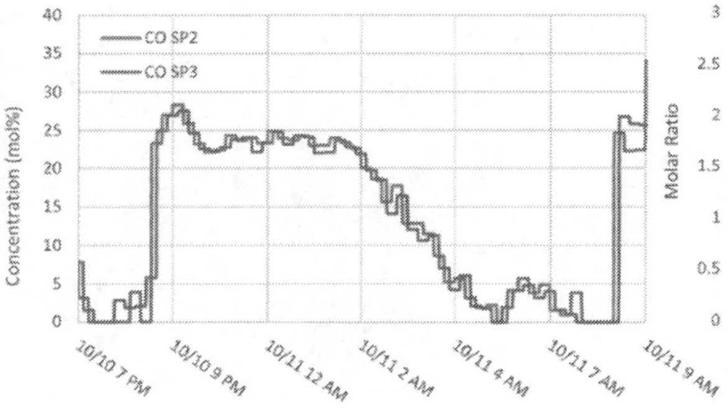

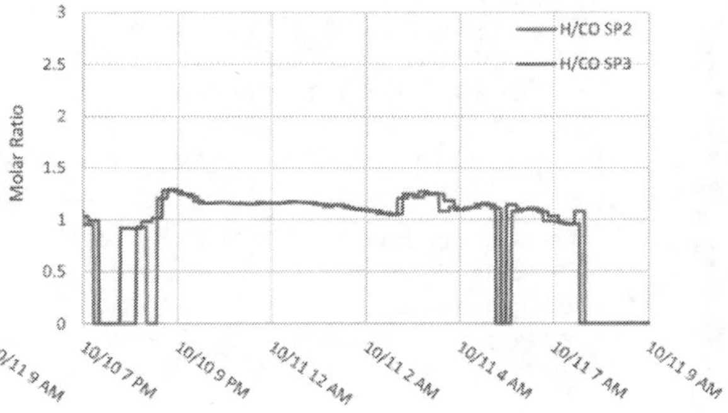

Experimental Data from Gasifier Operation

- Coal Conversion
  - CO and CO_2 produced
  - Total carbon fed (coal and natural gas)
- Low Conversion
  - First trial run
- Burner Atomization
- Slurry Concentration

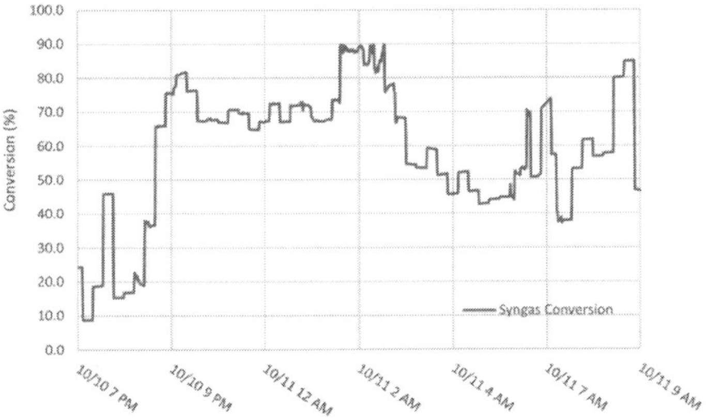

caer.uky.edu

2017 AIChE Annual Meeting

November 1st, 2017

Lessons Learned

- Finding pilot scale process equipment within budget and scope often require exhaustive and creative search
- The Project Benefited Greatly from Strong Collaborations in China in the area of Coal Gasification while also posing some unique challenges
- Slurry concentration has a dramatic affect on gasifier operation and gas composition
  - Challenge to maintain constant concentration (density)
  - Blend based on density and determining acceptable density range limits
  - Additive for settling and pumping
- Refractory on this small scale has caused some integrity issues with thermocouples
  - Thermal expansion/contraction
  - Wrong initial design
  - New thermocouples to meet our gasifier needs
  - Development/utilization of temperature probes without physical insertion (IR, Pyrometer, etc)

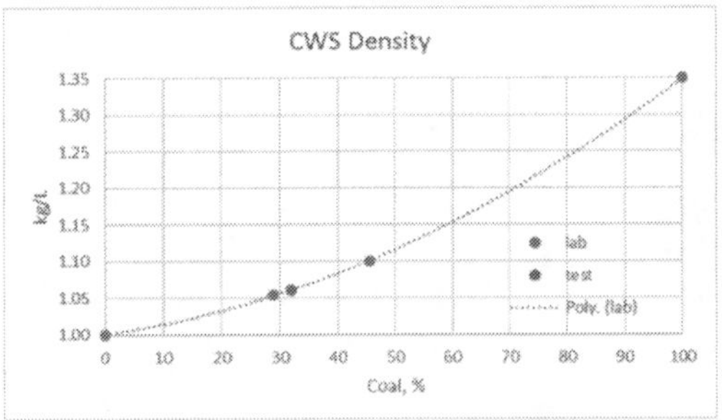

Gasifier Thermocouple Well

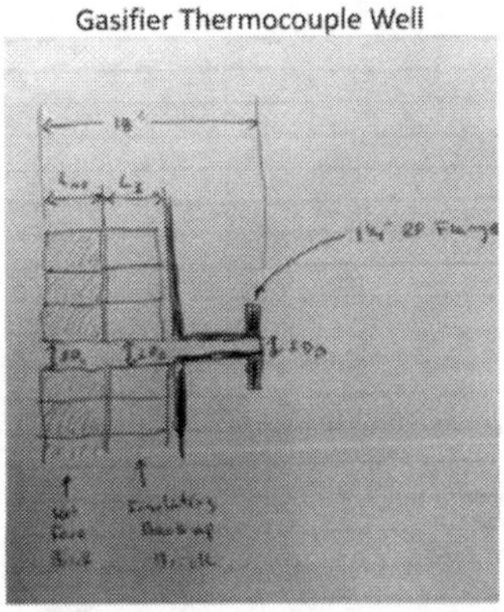

caer.uky.edu

2017 AIChE Annual Meeting

November 1st, 2017

Center for Applied Energy Research

caer.uky.edu

Summary

- Modular nature of the facility provides wide range of capabilities and flexibility for future use
- Whole facility ready for operation
- Facility is available for testing

2017 AIChE Annual Meeting   November 1st, 2017

Center for Applied Energy Research

caer.uky.edu

Acknowledgements

- DOE-NETL: Arun Bose, Jenny Tennant and Jason Hissam
- UK: Rodney Andrews, Don Challman, Kunlei Liu, Burt Davis, Dennis Sparks, Steve Summers, Tate VanHoose, Len Goodpaster, Otto Hoffman, Marshall Marcum, Brad Irvin, Jimin Zeng and Jonathan Pelgen
- Zeton: Ravin Dave, Danielle De Sousa, Erik Gaspar
- Stovlbaek Consulting: Peter Pedersen
- ECUST: Qinghua Guo, Yan Gong, Jianliang Xu
- CEEDI: Wang Qiuming and Tao Li

2017 AIChE Annual Meeting   November 1st, 2017

507270 Production of Fuels from Coal-Biomass Feedstock using Fischer-Tropsch

Abstract:

The Fischer-Tropsch synthesis is actually composed of four sub-processes: gasification, gas clean-up, Fischer-Tropsch synthesis and product upgrading. The impact of feedstock (coal, natural gas, biomass) and mixtures of these feedstocks upon each sub-process will be discussed. The major impact of the feedstock is on the first two sub-processes: gasification and gas clean-up.

499845 Coal Power Plants with Enhanced Profitability and No Carbon Dioxide Emissions

Abstract:

Coal Power Plants with Enhanced Profitability and No Carbon Dioxide Emissions

A zero carbon dioxide emission process is developed, in which electrical power and formic acid are coproduced from coal. The overall reaction for the proposed process uses coal, oxygen, and water as reactants to generate formic acid as the only product. A thermodynamic analysis is carried out, so that feasible operating limits for the proposed process are identified. Since the process cannot be realized within one single step, a reaction cluster is proposed which consists of feasible chemical reactions. These are: gasification of coal to syngas; CO_2 hydrogenation to formic acid; hydrogen combustion; and CO combustion. The principal technologies employed in this process network include combustion, gasification of coal to syngas, gas compression, supercritical fluid pumping, hydrogen separation using pressure swing adsorption, and formic acid production from CO_2 and H_2. The software UniSim [TM, Honeywell Inc] is employed to simulate the flowsheet. Two Gibbs reactors act as the gasification reactor and a combustor, while a conversion reactor carries out the formic acid reaction. Coal, water, and part of oxygen are fed into the gasification reactor; off-gas from the hydrogen PSA and oxygen are fed into the combustor; pure hydrogen from hydrogen PSA and CO_2 which is compressed to supercritical phase are fed into the formic acid reactor. By adjusting the oxygen delivery ratio, the coal gasification is designed to be energetically self-sufficient, so that the whole process produces the largest amount of power. Heat exchangers, phase separators, compressors, pumps, etc. are also employed to realize the process. Temperature and heat load information for each process stream is obtained from the software simulation and used in carrying out a heat and power integration study. Through the use of heat engine and heat pump subnetworks, a feasible operating envelope for the considered process is identified. A preliminary technicoeconomic analysis is carried out, which establishes the superiority of the proposed process to current coal power plant technology.

491886 Reforming of Hydrocarbons for Hydrogen Generation - Hyres

Abstract:

Steam methane reforming (SMR) is a commercial technology that is used at large scales to generate syngas and hydrogen, which can be used as fuel directly, or used as a chemical feedstock in petroleum refining and chemicals production. TDA Research is developing a reforming process capable of converting nearly any hydrocarbon feedstock into H_2. Our process has the potential to use as little as 1/4 of the energy required per pound of H_2 produced (energy required to operate the process) compared to SMR. We call our process *HyRes (Hydrogen from Residuum),* and we can steam-reform anything from shale derived natural gas to bottom-of-the-barrel atmospheric and vacuum residuum at low steam to carbon ratios. *HyRes* uses circulating fluidized bed reactors, one for steam reforming and one for catalyst regeneration. This reactor configuration permits us to maintain catalyst activity by continuously removing carbon deposited on the catalyst during the reforming process.

In addition to using less energy, the use of circulating fluidized beds for the reformer and regenerator means that the process no longer requires the use of the extremely expensive high alloy tubes such as those used in conventional fixed bed steam reforming. Also, because the catalyst is regenerated before enough carbon can build up to damage the catalyst, there is no irreversible catalyst deactivation due to coking or sulfur deposition. We discuss results where we have demonstrated our fluidized bed reforming process at the bench scale with several catalysts and hydrocarbon feedstocks.

497979 Syngas Production on Thermally Conductive SiC Catalyst

Abstract:

A new catalyst support technology has been developed to increase the thermal conductivity of a steam reforming catalyst system. The technology (*HeatPath*^{TM}) is designed to improve the heat transfer of the catalysts during the production of hydrogen from hydrocarbons. By allowing heat to move easily into the reaction zone, the new catalyst system reduces the amount of heat that must be generated, significantly lowering the cost of hydrogen produced.

The core technical achievement of this study is the feasibility validation of the high thermal conductivity supports with protective coating, and a washcoated catalyst layer on an application-suitable geometry. Packed-bed silicon carbide (SiC), along with structured metal foams and metal monoliths were all evaluated in the early stage to confirm the suitability of the coating processes and performance as steam reforming catalysts. Both metal and SiC supports showed promise for steam reforming reactions, though SiC has significant advantages in cost, weight and mechanical strength. Nexceris has optimized the integration of washcoated catalysts and protective layers on high thermal conductivity SiC supports. Testing has been performed to assess long term performance stability and to confirm agreement between the experimental and modeling results.

Through computational simulation of steam reformer operation, Nexceris determined that the best path for and reducing cost of syngas from natural gas is to reduce heat loss in the reformer. The identified approach was to establish coating technology that will enable the use of high thermal conductivity catalyst supports. Combining a protective coating technology and catalyst formulations designed for the reforming of natural gas, the *HeatPath*^{TM} platform offers three times better heat transfer than conventional catalyst support materials, which will reduce energy consumption during syngas production by more than 15 percent.

Results of our modeling and experiment work suggest that these newly developed supported catalysts will provide a leap forward in the design of hydrogen production systems.

497796 Thermodynamic Simulations, Experiments and Techno-Economic Analysis of Iron-Based Chemical Looping Systems for H_2 Production

Abstract:

The iron-based moving bed based chemical looping technology developed at the Ohio State University (OSU) is a promising candidate to yield cost benefits in natural gas utilization for a carbon-constrained scenario. OSU has developed three commercially relevant, distinct approaches for high efficiency H2 production from natural gas[1,2,3]. The first approach focuses on a three-reactor scheme that produces CO2 and H2 in separate reactors, providing process intensification benefits[1]. The second approach utilizes a membrane based separation scheme to surpass the thermodynamic limits associated with the H2 production system in the first approach[2]. The third approach retrofits the methane to syngas based co-current reaction system in a steam methane reformer and considers co-feed of steam and CO2 for high efficiency H2 production[3]. This presentation will initially examine the thermodynamic limits using Gibbs Free Energy minimization the three approaches. The thermodynamic performance will be further quantified by a cold-gas efficiency and an energy efficiency comparison between the three approaches and a baseline Steam Methane Reforming case. Relevant experiments that include high-pressure kinetic data and sub-pilot scale experimental demonstration will be used as an input for a base techno-economic analysis model for the three approaches. Sensitivity studies for each approach that identify the critical parameters to be de-risked for being competitive or better than steam methane reforming plants will be presented.

References

1. Kathe, M. V., Empfield, A., Na, J., Blair, E., & Fan, L. S. (2016). Hydrogen production from natural gas using an iron-based chemical looping technology: thermodynamic simulations and process system analysis. *Applied Energy*, *165*, 183-201.
2. Nadgouda, S. G., Kathe, M. V., & Fan, L. S. (2017). Cold gas efficiency enhancement in a chemical looping combustion system using staged H 2 separation approach. *International Journal of Hydrogen Energy*,Â *42*(8), 4751-4763.
3. Kathe, M., Empfield, A., Sandvik, P., Fryer, C., Zhang, Y., Blair, E., & Fan, L. S. (2017). Utilization of CO 2 as a partial substitute for methane feedstock in chemical looping methaneâ€"steam redox processes for syngas production.Â *Energy & Environmental Science*.

496174 Development of Thermo Neutral Reforming Catalyst for on-Board Hydrogen Production

Abstract:

This study investigated the development of a reforming catalyst for on-board generation of hydrogen for transportation applications. The development of a new reforming catalyst was aimed for high throughput system for syngas production for on-board transportation applications such as in auxiliary power units. A new reforming catalyst using a monolith support was prepared for liquid hydrocarbons based fuel processor. For mobile applications, liquid fossil fuels like diesel are suitable due to their high energy density, their easy storage and their already existing distribution infrastructure. However, diesel reforming is known as a major challenge due to its high carbon content, aromatics, and sulfur, which play a major role in catalyst deactivation. In this study, a new fuel delivery technique (atomization) was investigated for diesel reforming. An ultrasonic injector (instead of using conventional pre-heating) was used to mix all feedstock (fuel, air and steam) before it reaches the catalyst bed. The new method has improved the thermo neutral reforming (TNR) catalyst performance and stability with diesel. Monolith-based TNR catalysts of various sizes were prepared and their performance was evaluate in a fixed bed reactor with heavy naphtha and diesel as feedstocks. These catalysts were prepared by successive impregnation of active metals on a monolith with a wash coat of alumina on monolith. Compared to any other catalyst structure such as pellets or extrudates, a significantly lower pressure drop is observed when monolithic catalysts are used. The newly developed TNR catalyst showed stable performance with heavy naphtha as well as diesel for hydrogen production.

500939 1 kW_e Diesel Autothermal Reformer Integrated with Catalytic Burner for Exhaust Gas Oxidation

Abstract:

Heavy-duty trucks and recreational vehicles have to idle their engines when extended duration of using energy is required. However, idling causes the environmental issues and performance degradation of vehicles. Auxiliary Power Unit (APU) is a device on a vehicle that provides deficient energy for functions other than propulsion using the fuel from the vehicle. Solid Oxide Fuel Cell (SOFC) based auxiliary power units are expected to be about 15-20% more efficient compared to commercially available diesel based APUs resulting in significantly lower CO_2 footprint and lower emissions of criteria pollutants with low noise. The fuel reforming process, which is integrated with a SOFC for APU application, also plays an important role in an APU system. The synthesis gas consisting mainly of hydrogen, carbon monoxide, water and carbon dioxide generated as a result of hydrocarbon (i.e., diesel) reforming is sent to an SOFC stack for electrochemical oxidation operating at temperatures around 800°C. Moreover, the recent studies have reported a possibility to use SOFC as a CO_2 conversion system. Diesel autothermal reforming (ATR) is used to produce hydrogen-rich gas for SOFC application. 1 kW_e diesel autothermal reforming system is developed and tested for SOFC APU application with consideration of heavy-duty truck integration.

A catalytic burner is integrated into 1 kW_e diesel autothermal reformer to oxidize exhaust gases from the reformer. During the reformer operation including startup process, produced hydrogen and other combustible gases such as carbon monoxide must be fully oxidized before exhaust. Also, heat produced from the burner during the oxidation is utilized into the system to heat up the cathode air supply. In this study, the air supply rate of the catalytic burner is controlled to ensure full oxidation of the exhaust gases. With the burner temperature limitation of 800°C, hydrogen and carbon monoxide are fully oxidized with the catalytic burner. Generated heat is also utilized to increase temperature of cathode input air. As a result, stable startup and operation profile with integration of the catalytic burner can be obtained.

497030 CFD Study of a Pilot-Scale Multi-Tube Palladium Membrane Module for Hydrogen Separation

Abstract:

CFD Analysis of Pilot-Scale Multitube Pd Membrane Module for Hydrogen Purification from Coal-Derived Syngas

Rui Ma[1], Bernardo Castro-Dominguez[2], Ivan P. Mardilovich[1], Nikolaos K. Kazantzis[1], Anthony G. Dixon[1], Yi Hua Ma[1]

[1.]Center for Inorganic Membrane Studies, Department of Chemical Engineering, Worcester Polytechnic Institute, 100 Institute Road, Worcester, MA 01609, USA.

[2.] University of Limerick, Sreelane, Castletroy, Co. Limerick, Ireland

Syngas production and hydrogen separation technologies are very mature, and also extremely important for the energy and chemical industries. Enhancing or intensifying these mature technologies is extremely challenging, but on the other hand would have tremendous impact on performance and economics of many processes. At this moment, most research on Pd membrane technology has been conducted at lab scale, and membrane module scale up and commercial unit design are still challenges that need to be addressed in order to commercialize this promising technology. Compared with the one shell, one membrane tube module, many questions need to be answered for a pilot scale, multitube membrane module, such as the study of tube to tube variation in hydrogen permeation due to the position in the module, the exploration of the optimal shell side operating conditions and the evaluation of mass transfer limitations such as concentration polarization. For this purpose, a three dimensional CFD simulation model was developed and used to study the aforementioned aspects of a pilot scale membrane module.

The geometry of a seven-tube module with a total permeable area of 1050 cm^2 is shown in Figure 1a; coal-derived syngas was introduced to the shell side (retentate side) and pure hydrogen was collected from the tube side (permeate side). The asymmetrical fluid flow (Fig 2a) resulting from the outlet location leads to an uneven hydrogen permeation rate in each membrane tube (Fig 3). By adding a manifold for the membrane tubes in between the membrane tubes and the outlet (Fig1b), fluid flow tends to be more symmetrical (Fig 2b.), hydrogen permeance flux of each outer tube appears to be more even (Fig 3), thus the membranes are used more efficiently. In addition, due to the competition effect of the center tube with the six tubes around it, the center tube has the lowest hydrogen permeance flux. This phenomenon was addressed more in detail by the study of a further scaled-up module with 19 membrane tubes and a total permeable area of 2850 cm^2 (Fig 4). Concentration polarization was also observed from the model and indicated by the second Damkohler Number. Furthermore, hydrogen recovery appears to be favored by high pressure and low temperature while membrane usage benefits from moderate pressure and high temperature, the trade-off between these

two most important parameters of membrane modules is investigated and a thorough process optimization study was carried out using Box-Behnken method.

Figure1. Geometry of the 7 membrane tube module: (a) without manifold (b) with manifold

Figure2. Velocity field of the 7 membrane tubes module: (a) without manifold (b) with manifold

Figure 3. Hydrogen permeance flux of the outer tubes with different geometry

Figure 4. Cross sectional view of the 19 membrane tubes module

495931 Novel SAPO-34 Hollow Fiber Membrane for H_2 Separation Under Various Gas Atmospheres at High Temperature

Abstract:

Abstract

In the current research, a novel SAPO-34 hollow fibre membrane for hydrogen separation was fabricated using seeded secondary growth method. As compared to the conventional tubular membranes, hollow fibre membranes have a great advantage of higher surface to volume ratio, higher gas permeation due to reduced wall thickness etc. To obtain high quality SAPO-34 hollow fibre membranes, zeolite seeding and growing conditions were optimized to achieve a thin and defect-free zeolite layer for high gas permeability and selectivity. In addition, it was found out that membranes calcined under flowing oxygen had significantly higher gas permeability than those calcined under stagnant air. In this research, the membranes were subjected to permeation test of binary gas mixtures such as H_2/CH_4, H_2/CO_2, and H_2/N_2 at both room temperature and elevated temperatures ranging from 25 to 500 â„ƒ. Remarkable H_2 separation performance was achieved in separating a H_2/CH_4 mixture through the SAPO-34 hollow fiber membrane with a maximum H_2 selectivity of 45 at 100 â„ƒ and a H_2 permeation flux of $10.1*10^{-8}$ mol. $m^{-2}.s^{-1}.Pa^{-1}$. In addition, H_2 permeation flux increased to $28.9*10^{-8}$ mol. $m^{-2}.s^{-1}.Pa^{-1}$ with at a temperature of 500â„ƒ while the selectivity dropped to around 25. Due to the good H_2 permeation performance at high temperature, SAPO-34 hollow fiber membrane can be integrated with catalysts to form catalytic membrane reactors for various H_2 production reaction such as water gas shift (WGS) and dehydrogenation of hydrocarbons like propane. This novel catalytic membrane reactor can simultaneously remove hydrogen product, and increase the productivity of the reaction beyond the equilibrium yield.

502033 Kinetics of the thermal decomposition of sulfur compounds

Abstract:

Abstract

In this contribution the kinetics of the thermal decomposition of sulfur compounds will be discussed both experimentally and theoretically under typical steam cracking and pyrolysis conditions. The experimental investigation has been conducted in a new 85 cm long annular quartz flow reactor, having an outer and inner tube diameters of 2.1 and 1.5 cm respectively. Both pure sulfur compounds pyrolysis in nitrogen (0.04 mol% S in N2) as well as in the presence of hydrocarbons (0.04 mol% S and 13.7 mol% of HC in N2) were studied covering a wide temperature range of 600 to 850 °C, at a fixed pressure of 1.5 bara and residence time of ~0.5 s. The reaction effluent is sampled on-line at 300 °C. The unique on-line sampling system consists of valve-based sampling manifold and uniformly heated transfer lines connected to a comprehensive two-dimensional gas chromatograph (GC × GC) equipped with a flame ionization detector (FID) and a sulfur chemiluminescence detector (SCD). This dedicated set-up allows to detect and quantify both pure (with the FID) and sulfur-containing hydrocarbons (with the SCD) using heptane and 3-chlorothiophene as internal standards, respectively. The GC × GC-FID/-SCD settings are optimized for trace analysis of sulfur compounds in olefin-rich (ethylene- and propylene-) hydrocarbon matrices produced during steam cracking.

To understand the underlying decomposition mechanism of sulfur compounds in the presence of a complex hydrocarbon matrix formed by its decomposition, a group additive kinetic model was constructed via an in-house automated reaction mechanism generation software called Genesys [1]. It relies on ab-initio CBS-QB3 based kinetic coefficients for the reference/ parent reactions for each elementary reaction family, and for the others, contribution of groups to the single event Arrhenius parameters. No adjustment of the Arrhenius parameters to the experimental data is performed. The kinetic model is validated by comparison with the set of experimental data gathered in the annular quartz flow reactor by providing a satisfactory representation of the latter.

[1] Nick M. Vandewiele, Kevin M. Van Geem, Marie-Françoise Reyniers, Guy B. Marin, Genesys: Kinetic model construction using chemo-informatics, Chemical Engineering Journal, Volumes 207-208, 2012, Pages 526-538.

500300 A Cylindrical Formulation of the One Dimensional Turbulence (ODT) Model for Turbulent Jet Flames

Abstract:

The One Dimensional Turbulence (ODT) model has been successfully applied to a wide range of turbulent flows including homogeneous turbulence, jets, wakes, mixing layers, channel flows, Rayleigh-Taylor, buoyant wall flows, and others. ODT solves a flow by temporally-evolving one-dimensional transport equations for mass, momentum, and (optionally) reacting scalars. Advection is modeled through stochastic eddy events that are implemented as triplet maps. The size, location, and rate of these maps are drawn from an eddy rate distribution in a mamner that this consistent with turbulent inertial range advection. The model is an extension to the Linear Eddy Model (LEM), and differs from LEM in that, in ODT, momentum components are solved in addition to scalar quantities, and these momentum components are used to define eddy rate distribution function. Hence, ODT is a dynamic turbulence model and can accurately capture velocity and mixing profiles in nonuniform turbulence. Because the model is one dimensional, it is limited to boundary-layer like flows, but such flows are important and widely studied, with a wealth of experimental data available. ODT is unique in that it is able to resolve a full spectrum of turbulent time and length scales (in one dimension). Only direct numerical simulation is able to resolve all scales of a turbulent flow, but this is done at a computational cost that is prohibitive for engineering flows at realistic Reynolds numbers. Large Eddy Simulation directly captures large eddies, but models small subgrid scale motions. Conversely, ODT resolves fine scale structures, allowing diffusion processes in the physical space coordinate, but models the large-scale advection. Often the fine scales are more difficult to model since nonlinear multi-component diffusion and reaction processes occur at dissipation scales. Large-scale advection is often not too difficult to model in boundary layer flows.

ODT has been recently validated quantitatively against several DNS in compatible configurations, namely, temporal, planar jets. This has been done for soot formation, and flame extinction and reignition using ethylene and syngas fuels. Premixed flames have also been studied. As the model has been extended to more and more complex flows, there has been a desire to use ODT as a surrogate DNS, due to its low cost compared to DNS. Such validation studies comparing ODT and DNS are promising in this area. However, DNS is limited to relatively few simulations and relatively low Reynolds number. Application of ODT to experimental flows is limited by the configuration. Experimental jets are usually cylindrical and evolve spatially, whereas existing ODT simulations are planar and (normally) evolve temporally. (There is a spatial formulation of the model in which a one-dimensional flow evolves spatially in a down-stream direction like a parabolic boundary layer formulation.)

We extend the ODT model to a cylindrical formulation. The formulation is presented, along with challenges and interpretations. We present comparisons of the cylindrical ODT formulation to round pipe flow, round nonreacting jets at high Reynolds number, and to reacting nonpremixed jet flames. The new formulation is expected to improve previous predictions of jet flows, including, for instance, particle flows. We discuss such expected improvements.

AUTOMATED TRANSITION STATE THEORY CALCULATION OF HYDROGEN ABSTRACTION FROM NOVEL BIOFUELS

Nathan D. Harms, Richard H. West
Northeastern University, Boston MA 02115

Abstract

Understanding and predicting the combustion properties of novel fuels allows for development of greener and more efficient fuels and engines. To predict the detailed combustion kinetics of a single novel fuel requires the construction of a reaction mechanism consisting of thousands of elementary reactions and their kinetics. The reaction rates can be determined by performing transition state theory (TST) calculations on each elementary reaction. However, with dozens of novel biofuels being considered, and each fuel having its own reaction mechanism, the TST calculations must be automated. We present a fully-automated method to determine reaction rates via automated Transition State Theory (AutoTST) calculations, coupled to the Reaction Mechanism Generator (RMG) [1, 2]. We focus here on the abstraction of hydrogen by six common radicals present in combustion ($HO\cdot$, $H\cdot$, $HOO\cdot$, $O\cdot$, $H_3C\cdot$, and O_2) from several biofuels being considered by the U.S. Department of Energy "Co-Optima" program. Hydrogen abstraction reactions are one of the primary reaction families that facilitate combustion and is arguably the most important class of reactions in the combustion process; therefore, accurately determining kinetic parameters for these reactions is crucial. Our approach uses RMG to determine the possible hydrogen abstraction reactions for each fuel, and the reaction kinetics are determined using our AutoTST method. The AutoTST method estimates the transition state geometry using a group contribution and distance geometry method, performs DFT geometry optimization, performs intrinsic reaction coordinate calculations to verify the transition state, and uses CanTherm [3], a canonical transition state theory calculator, to determine kinetic parameters. This method has previously been applied to hydrogen abstractions and results have been compared to published values and benchmark calculations [4] The application of AutoTST is broad and can be applied to all families of combustion reactions given proper resources.

Introduction

Long term energy, environmental, and economic needs have motivated the study of renewable energy sources to replace fossil fuels. Currently, the United States is responsible for 25 % of the world's CO_2 production despite the size of its population. [5] In order to limit the United State's production of CO_2 from motor vehicle transport, the Department of Energy (DOE) developed the "Co-Optima" program to simultaneously optimize the selection of novel biofuels and the design of efficient engines, hoping co-optimization of both will maximize performance, efficiency, and adoption. Designing an engine for a fuel that has not yet been made in significant quantities, however, requires a *predictive* kinetic model, as does evaluating a fuel's performance in an engine that does not yet exist.

From this investigation, 23 biofuels were identified as potential replacements for fossil fuels or blend-stocks (Figure 1). Few of these fuels have accurate kinetic models.

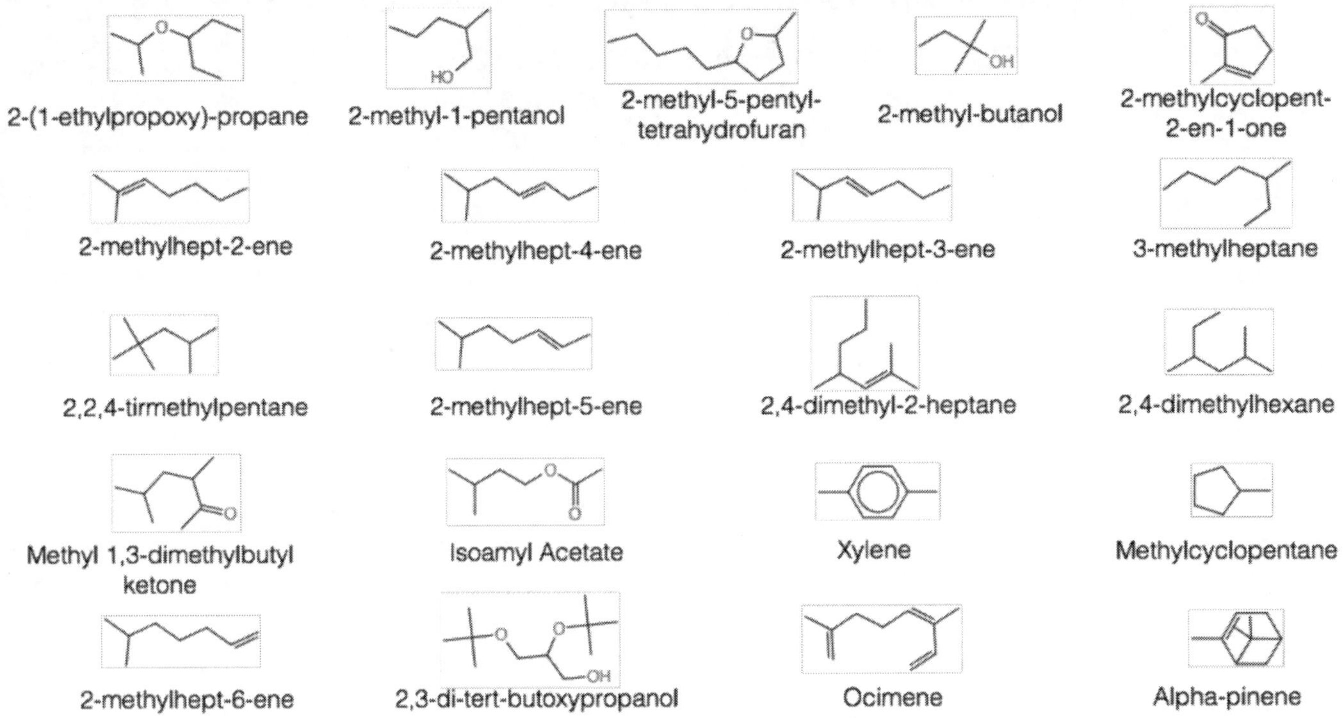

Figure 1: *The 23 novel biofuels being investigated in this study.*

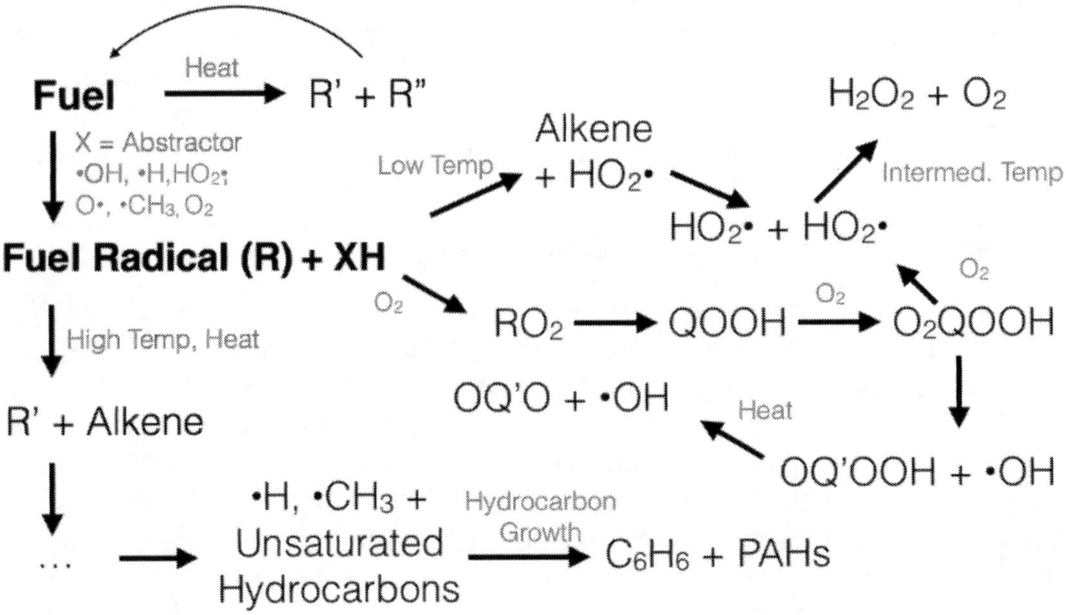

Figure 2: *Overview of a generic combustion process*

Reaction mechanisms often follow a typical road map outlined in Figure 2. For this work, we focus on the abstraction of hydrogen by six common radicals present in combustion: $HO\cdot$, $H\cdot$, $HOO\cdot$, $O\cdot$, $H_3C\cdot$, and O_2. Hydrogen abstraction reactions are some of the most important reactions in the combustion of fuel, and abstractions from the fuel molecule are the first in a set of thousands of reactions that allow for the rest of combustion to occur.

When constructing a reaction mechanism, rates are usually generated in three ways: estimates and intuition, rate rules, or transition state theory on the bases of quantum chemistry calculations [6]. Decent estimation require decades of prior knowledge in the combustion field and, even then, these estimates will be just that – estimates. Rate rules are an efficient method to determine kinetics for reactions, essentially codifying the estimates an experienced scientist might make. These methods are orders of magnitude faster than quantum calculations but their accuracy suffers when applied outside of the range of "chemical space" where the rules were developed and trained: These rate rules can be misapplied (i.e. applied to a reaction that is too dissimilar to the reactions from which it was derived), or the chemistry of a reaction might not follow the rules. Because of this, quantum calculations are often used to develop more accurate reaction rates.

Although experiments are essential to test and validate an overall kinetic model, it is often challenging to isolate an individual elementary reaction in an experimental study, and with the latest computational techniques some reactions can be calculated *ab initio* with an accuracy exceeding experiments, and at a fraction of the cost. Because of this, theoretical calculations are often utilized to determine kinetics in combustion mechanisms.

Utilization of quantum packages like Gaussian [7] allows one to optimize geometries of reactants and products by walking down the potential energy surface (PES) until a minimum energy has been found. This has been automated for on-the-fly calculation of thermochemical properties of intermediate species in RMG [8]. For transition state geometries, an initial guess of the transition state will undergo a similar search, but the program will search for a saddle point on the PES instead: a stationary point with negative curvature in one direction (the normal mode corresponding to the reaction coordinate) and positive curvature in all other directions. The search algorithm is usually only successful if the initial guess is close enough to the correct saddle point geometry that the direction of negative curvature is correct and pointing towards the saddle point. In other words, it needs a very good initial guess, which often requires significant input from the user. In order to automate the search for saddle points, Bhoorasingh et al. developed a procedure to provide a guess for the transition state geometry based on the reactants, products, and the type of reaction [9]. The algorithm used group-additivity to determine inter-atomic distances for the atoms participating in the reaction, and distance geometry to convert these into a 3D structure. Successful optimization of the TS geometry is validated using Intrinsic Reaction Coordinate (IRC) calculations. These calculations use the TS as a starting point and walk down the PES until a local minimum energy is attained. If the reactants and products of these local minima match the actual reactants and products of the reaction, the TS is validated.

Successful optimization of these geometries allows one to determine the the canonical partition functions from the results of the DFT calculations and subsequently kinetics using transition state theory (equation 1).

$$k(T) = \frac{k_B T}{h} \frac{Q^{TS}}{\Pi_i^n Q_i} \exp\left(\frac{-\Delta E_0}{k_B T}\right) \qquad (1)$$

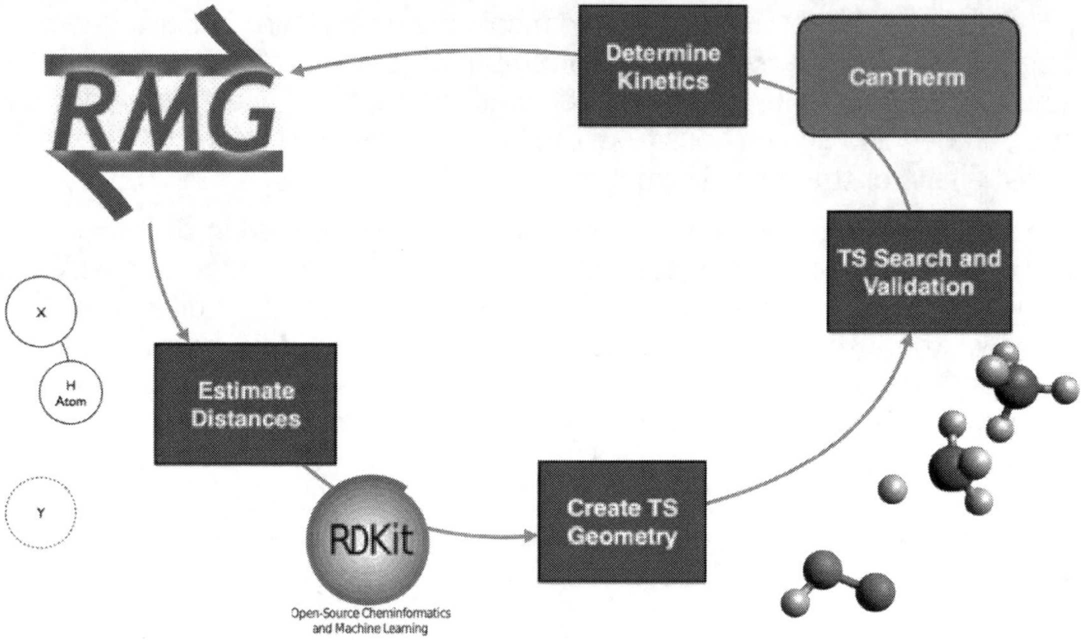

Figure 3: *Overview of the AutoTST process developed by [10]*

$k(T)$ represents the reaction rate, k_B is the Boltzmann constant, T is temperature, h is Plank's constant, Q^{TS} is the canonical partition function for the TS, $\Pi_i^n Q_i$ represents the product of the canonical partition functions for the reactants and ΔE_0 is the activation energy. This highlights that the temperature-dependent reaction rate can be determined by the partition functions and activation energy, which can all be determined from the electronic structure or DFT calculations of the reactants and transition state.

This geometry-producing algorithm was coupled to a canonical TST calculator to provide a fully automated AutoTST workflow (procedure outline in Figure 3) [4, 10].

Methods

Identification of possible hydrogen abstraction reactions

Identification of the hydrogen abstraction reactions of each of the proposed biofuels was accomplished by using Reaction Mechanism Generator (RMG) [1]. RMG identified 692 hydrogen abstraction reactions that were subsequently run using AutoTST.

Perform automated transition state theory calculations

The overall workflow is described in the recent paper by Bhoorasingh et al. [4] but a brief overview is given here. Reactant and product geometries were created using algorithms in RMG developed by Magoon and Green [8]. Transition state geometries were estimated using a group contribution method developed by Bhoorasingh and West [9]. This group contribution method identifies functional groups in each of the reactants and products, which are used to estimate the distances between atoms at the reacting center using previously identified transition states as training data. The transition state geometry estimate was optimized to an energy minimum with the reaction

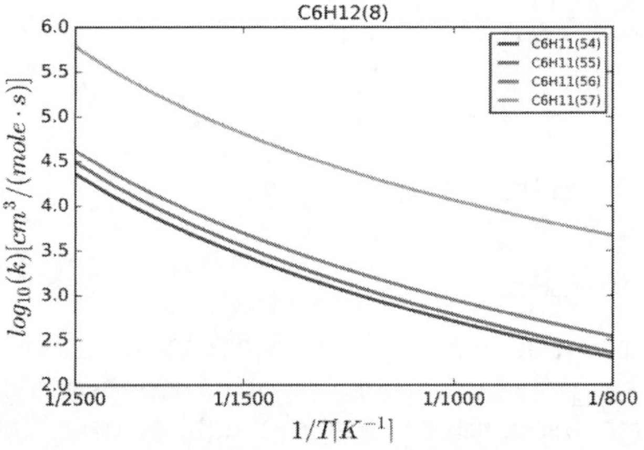

(a) *Arrhenius plot of HOO· abstracting a hydrogen from methylcyclopentane. Colors correspond to subfigure (b)*

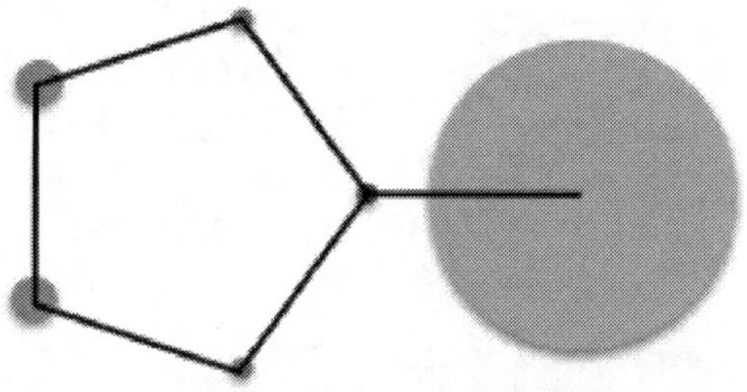

(b) *Depiction of methylcyclopentane. Colors correspond to subfigure (a). Area of each circle proportional to H abstraction rate at that site by HOO·*

Figure 4: *An Arrhenius plot and heat map of the hydrogen abstraction reactions from methylcyclopentane by HOO·.*

center distances frozen followed by a saddle point search with all distances frozen except for the reaction center. The resulting geometry is then fully optimized to a saddle point. If a saddle point is found, an intrinsic reaction coordinate (IRC) calculation is performed to verify the reactants and products. If the reactants and the products from the IRC calculations, determined from the Cartesian coordinates using a connect-the-dots algorithm, match the reactants and products from the initial reaction, the transition state validation is complete. All the geometries were optimized and validated with the M06-2X DFT functional with the MG3S basis set in the Gaussian 09 quantum chemistry package.

Following this validation, the CanTherm software package [3] was used to determine kinetic parameters using TST. Symmetry numbers for the rate calculations were determined using the SYMMETRY software [11]. SYMMETRY takes the optimized transition state geometry, allows for small deviations, and calculates the point group. Product properties were also calculated so that tunneling corrections could be applied via the Eckart model.

Results

Through our presented work flow, we have identified 692 hydrogen abstraction reactions on 23 novel biofuels, started calculations on 540 of these reactions, and successfully calculated kinetics for 106 of them. At time of preparation of this manuscript, the completion rate is 19.6%, which is substantially lower than the success rate previously reported for this method of ~70% [4].

Although the current success rate of this work flow is low, we have successful identified the reaction rates of HOO· abstracting a hydrogen from methylcyclopentane at all unique sites, for example. In figure 4, we see an Arrhenius plot of the hydrogen abstraction rate by HOO· and a corresponding "heat map" of methylcyclopentane. This has told us that the methyl group of this proposed biofuel is most likely to lose a hydrogen.

Total Reactions Identified	692
Reaction Calculations Started	540
Overall Successes	106
Success Rate	19.6%

Table 1: *Success rates at time of manuscript preparation.*

Conclusions

We are still investigating the currently low success rate. Some failure modes include: not finding the lowest energy conformer for a reactant so that the barrier height is negative; failed TS optimizations in Gaussian; converging upon a saddle point that is not the required transition state (eg. a conformational change); and long computational time associated with IRC calculations leading to interrupted calculations. Many abstraction reactions will succeed in one direction but not the other, which is not reflected in these results.

In order to obtain the lowest energy conformers, we intend to implement Fafoom, a conformer searcher that utilizes a genetic algorithm technique [12]. We have tested this approach on reactant and product molecules, and plan to extend the approach to transition states.

A number of our calculations resulted in a non-existent barrier height. This could have occurred because tolerances on calculated energies of reactants, products, and TSs can overlap if the optimization is not exact enough. We propose running a stricter calculation on these geometries to minimize their tolerances and obtain barrier heights.

Some of our TS optimizations failed to converge as well. This could be because the guess provided by AutoTST was too far from the actual TS. One method to try and correct this is to retrain the data which the distance geometry estimates in AutoTST are based on. This will result in more accurate reaction coordinates for the TS guess and should give a higher success rate. Another avenue to explore is performing a looser optimization on the geometry in order to get convergence, then performing a tighter optimization on the loosely optimization geometry.

Finally, we would like to find a faster method of validation for AutoTST calculations. Currently, the IRC calculations are the most computationally expensive calculation. We have identified Quick Reaction Coordinate calculations [13] and nodal analysis as possible avenues to explore.

References

[1] Gao CW, Allen JW, Green WH, West RH. Reaction Mechanism Generator: Automatic construction of chemical kinetic mechanisms. *Comput Phys Commun*. 2016;203:212–225. https://doi.org/10.1016/j.cpc.2016.02.013.

[2] Green WH, Allen JW, Barlow J, Bhoorasingh PL, Buesser BA, Class CA, Gao CW, Han K, Khanshan FS, Lambert V, Merchant SS, Slakman BL, Troiano S, Vandeputte A, Vandewiele NM, West RH, NathanYee. Reaction Mechanism Generator;.

[3] Allen JW, Green WH. CanTherm: Open-source software for thermodynamics and kinetics. Included in: Reaction Mechanism Generator, v2.0.0. 2016. URL http://reaction.mechanismgenerator.github.io

[4] Bhoorasingh PL, Slakman BL, Seyedzadeh Khanshan F, Cain JY, West RH. Automated transition state theory calculations for high-throughput kinetics. *J Phys Chem A*. 2017; https://doi.org/10.1021/acs.jpca.7b07361.

[5] U.S. Energy Information Administration - EIA - Independent Statistics and Analysis. URL https://www.eia.gov/tools/faqs/faq.php?id=32&t=6

[6] Van de Vijver R, Vandewiele NM, Bhoorasingh PL, Slakman BL, Seyedzadeh Khanshan F, Carstensen HH, Reyniers MF, Marin GB, West RH, Van Geem KM. Automatic Mechanism and Kinetic Model Generation for Gas- and Solution-Phase Processes: A Perspective on Best Practices, Recent Advances, and Future Challenges. *Int J Chem Kinet*. 2015;47(4):199–231. https://doi.org/10.1002/kin.20902.

[7] Frisch MJ, et al. Gaussian 09. Gaussian, Inc., Wallingford, CT. 2009.

[8] Magoon GR, Green WH. Design and implementation of a next-generation software interface for on-the-fly quantum and force field calculations in automated reaction mechanism generation. *Comput Chem Eng*. 2012;52:35–45. https://doi.org/10.1016/j.compchemeng.2012.11.009.

[9] Bhoorasingh PL, West RH. Transition state geometry prediction using molecular group contributions. *Phys Chem Chem Phys*. 2015;17(48):32173–32182. https://doi.org/10.1039/C5CP04706D.

[10] Bhoorasingh P, Slakman B, Khanshan FS, Cain J, West R. Kinetic data for manuscript describing the AutoTST algorithm for automated Transition State Theory calculations of chemical reaction rates. *figshare*. 2016;https://doi.org/10.6084/m9.figshare.4234160.v1.

[11] Patchkovskii S. SYMMETRY software, University of Calgary (2003). 2003. URL http://www.cobalt.chem.ucalgary.ca/ps/symmetry/

[12] Supady A, Blum V, Baldauf C. First-Principles Molecular Structure Search with a Genetic Algorithm. *J Chem Inf Model*. 2015;55(11):2338–2348. https://doi.org/10.1021/acs.jcim.5b00243.

[13] Goodman JM, Silva MA. QRC: a rapid method for connecting transition structures to reactants in the computational analysis of organic reactivity. *Tetrahedron Letters*. 2003; 44(45):8233 – 8236. https://doi.org/10.1016/j.tetlet.2003.09.074.

490269 First Principles Based Automated Kinetic Model Generation Using on-the-Fly *Ab Initio* Calculations

Abstract:

Introduction

Many kinetic models for combustion and pyrolysis processes have been published in the past decade, and a common challenge in most work is often being stressed, i.e. the shortage of accurate rate coefficients for a large number of elementary steps. Currently, many rate coefficients are calculated based on rough approximation methods, and their accuracy can be questioned when these approximations are used outside of the original application domain of the applied method. Furthermore, building these approximation methods requires a complete training and test set of reactions of which the rate coefficients are known from either experiments or theory. Manually obtaining all these rate coefficients is time consuming, and this work is aiming at automatically calculating a large number of rate coefficients using quantum chemistry, minimizing the manual work. The developed methods can furthermore be used after the model generation, i.e. when not all the kinetics are sufficiently accurate.

This work has been performed in the framework of Genesys[1], an automatic kinetic model generation tool developed at the Laboratory for Chemical Technology which employs both the rule-based and the rate-based termination criteria. The latter can in particular benefit from the automatic *ab initio* calculations in the search for more complete, more accurate and predictive kinetic models.

Methodology

Genesys offers a representation of chemical reactions, supported by the Chemistry Development Kit[2]. Molecules, radicals and transition states are treated as mathematical graphs, in which the atoms are nodes and the bonds are edges of the graph. This is a very powerful molecular representation to convert reactants into products, to find sub-molecular patterns, to search in databases for appropriate data, etc. However, it is insufficient to perform *ab initio* calculations, where the coordinates of the atoms in the 3D space are required. The conversion of connectivity information to 3D coordinates is already well established for molecules and radicals, for which rule-based heuristic methods and numerical methods exist. "Distance geometry" is a method belonging to the latter group in which the coordinates are estimated by minimizing an error function describing all the interatomic distances compared to expected distances, which are tabulated. Additional optimization steps, among others using the Merck Molecular Force Field (MMFF)[3], allow the generate a meaningful set of 3D coordinates, directly applicable for *ab initio* calculations. This has been implemented in Genesys.

To obtain accurate rate coefficients, quantum chemical calculations of transition states are also required. The 3D coordinates of atoms in transition states are more complex to deduce and cannot be summarized in the same simple rules used for molecules and radicals. Since Genesys has been programmed to offer the end-user flexibility in terms of which reaction families to use and how to constrain them, this philosophy is further employed for the generation of 3D coordinates for transition states. The end-user has to possibility to guess the interatomic distances of the atoms in the transition state that change in connectivity throughout the reaction, called the reactive atoms. This information, the so-called "template" is used in Genesys to obtain 3D coordinates of all the atoms via "distance geometry", i.e. similar to stable molecules and radicals. The template can be built from a small set of manually calculated transition states or from the vast literature data on quantum chemical calculations. Refining the "distance geometry" output to a more meaningful set of coordinates is also done. For this, the MMFF is used in which equilibrium distances for the reactive atoms originate from the template, and the force constants are taken from analogous stable molecules and radicals.

Once initial 3D coordinates are available for molecules, radicals and transition states, quantum chemical calculations are performed using the Gaussian09 suite of programs[4] in several steps. First, the geometry is energy-optimized using fast semi-empirical calculations such as PM3. Using the resulting geometry, the conformational space is scanned by rotating around single bonds in a systematic and exhaustive manner. Each of these conformers are optimized using DFT methods, and the geometries belonging to the lowest 5 kJ mol^{-1} are used for high-level optimization, i.e. CBS-QB3 calculations. Finally, hindered rotor potentials are obtained to correct for the harmonic oscillator approximation.

For transition states, it is necessary to verify if the imaginary frequency indeed corresponds to the expected reaction. This is done in two steps. First, Genesys controls if the final geometry is close to the user-provided template. Second, the normal mode corresponding to the imaginary frequency is analyzed and it is evaluated whether the bonds that are broken or formed have a large contribution to this normal mode. For this, the normal mode in Cartesian coordinates is transformed into a normal mode in internal coordinates.

Via statistical thermodynamics, standard enthalpies of formation, standard entropies and heat capacities of all the molecules, radicals and transition states are calculated. The frequencies that resemble internal rotations are automatically searched for via an analysis of the normal mode. Their contributions to the vibrational partition function are automatically replaced by the hindered rotor partition functions. Also, the total rotational symmetry number of the species is automatically calculated and used to obtain correct entropy values. In the case where stereocenters are present, several approaches are possible. In Genesys, it is possible to explicitly track each stereoisomer as a separate species and to calculate separate thermodynamic data. However, when this is not necessary, they can also be lumped together, for which the entropy is then corrected by the number of optical isomers.

The thermodynamic data of reactants and the transition state of a reactions are subsequently used to obtain high-pressure limit rate coefficients. Tunneling corrections are automatically calculated using Eckart potentials, and Arrhenius parameters are determined by regression of the rate coefficients.

Results

The new methods have been illustrated by calculating rate coefficients for 8 reactions, each belonging to a different reaction family, i.e. inter- and intramolecular hydrogen abstractions, inter- and intramolecular radical additions, a substitution, a dehydration, a dehydrogenation and a retro-ene reaction. For each reaction family, the template was created based on known geometries of transition states from the literature or from manual calculations. The resulting Arrhenius parameters and rate coefficients at 1000K are given in Table 1.

These rate coefficients have been compared to experimental and theoretical values in literature and an excellent agreement is found, which is the same agreement found when calculating rate coefficients from CBS-QB3 calculations "manually". Most of the rate coefficients are within a factor of 2 of experimental and theoretical literature rate coefficients.

Table 1: Arrhenius parameters and rate coefficients at 1000K of the eight reactions automatically calculated using

Genesys. The first line of each reaction corresponds to the forward reactions, the second line to the reverse reaction. The pre-exponential factor and rate coefficients are expressed in s^{-1} for monomolecular reactions and m^3 mol^{-1} s^{-1} for bimolecular reactions. The activation energies are expressed in kJ mol^{-1}.

Conclusions

New methods have been developed to automatically calculate rate coefficients from *ab initio* calculations. For this, Genesys is used, which provides a connectivity based representation of the species, transition states and reactions. This information is used to create 3D coordinates of the atoms in the species and transition states. These coordinates are inputted to quantum chemistry calculations, yielding high level thermodynamic and high pressure limit kinetic data for gas-phase reactions.

The new methods are of great value to calculate a large number of rate coefficients from *ab initio* calculations without the need for manual interventions or user knowledge. This is especially of great interest for (1) building approximation methods for rate coefficient calculations such as group additivity models and (2) increasing the accuracy of important rate coefficients in a kinetic model.

1. Vandewiele NM, Van Geem KM, Reyniers MF, Marin GB. Genesys: Kinetic model construction using chemo-informatics. *Chem Eng J.* 2012;207:526-538.

2. Steinbeck C, Han Y, Kuhn S, Horlacher O, Luttmann E, Willighagen E. The Chemistry Development Kit (CDK): An Open-Source Java Library for Chemo- and Bioinformatics. *Journal of Chemical Information and Computer Sciences.* 2003;43(2):493-500.

3. Halgren TA. Merck molecular force field .1. Basis, form, scope, parameterization, and performance of MMFF94. *Journal of Computational Chemistry.* 1996;17(5-6):490-519.

4. *Gaussian 09* [computer program]. Wallingford, CT, USA: Gaussian, Inc.; 2009.

492655 Quantifying the Catalytic Effect of Calcium on Initiation during Cellulose Pyrolysis

Abstract:

During high temperature biomass pyrolysis, biopolymers undergo rapid decomposition to form hundreds of organic compounds which condense to form bio-oil and can be subsequently upgraded to fuel and chemicals. Inorganic metals naturally present in lignocellulosic biomass such as calcium and magnesium have been shown to catalyze pyrolysis chemistry[1]. However, little information is available that quantifies the catalytic behavior of these metals on pyrolysis chemistry. This work focuses on identifying the fundamental chemical mechanisms and kinetics involved in this catalytic chemistry. Using Î±-cyclodextrin as a cellulose surrogate[2] and a novel reactor system PHASR (Pulsed Heated Analysis of Solid Reactions)[3-4], time-resolved conversion data is obtained between 370 and 430 ^{0}C. To study the effect of calcium catalysis, varying concentrations of $CaNO_3$ (0.4 to 2.0 wt%) were mixed with Î±-cyclodextrin to simulate calcium concentrations in lignocellulosic biomass[1]. Cellulose conversion exhibits first order like kinetic behavior among all temperatures and calcium concentrations. A catalyzed kinetic rate constant was extracted and exhibits similar energetics compared to a high-temperature non-catalyzed intra-chain scission regime determined in an earlier study[4].

References:

1. Zhu, C.; Maduskar, S.; Paulsen A. D.; Dauenhauer, P. J., *Alkaline-earth-metal-catalyzed thin film pyrolysis of cellulose*. ChemCatChem, **2016**, 8, 818-829.
2. Mettler, M. S.; Mushrif, S. H.; Paulsen, A. D.; Javadekar, A. D.; Vlachos, D. G.; Dauenhauer, P. J., Revealing pyrolysis chemistry for biofuels production: Conversion of cellulose to furans and small oxygenates. *Energy & Environmental Science* **2012,** 5, (1), 5414-5424.
3. Krumm, C.; Pfaendtner, J.; Dauenhauer, P.J., Millisecond pulsed films unify the mechanisms of cellulose fragmentation. *Chemistry of Materials* **2016**, 28, 3108-3114.
4. Zhu, C.; Krumm, C.; Facas, G. G.; Neurock, M.; Dauenhauer, P.J., Energetics of cellulose glycosidic bond cleavage. *Reaction Chemistry and Engineering* **2017**, 2, 201-214.

486729 Artificial Neural Networks for Flare Modeling and Set Point Determination

Abstract:

The refineries and chemical industries are mandated to meet the EPAâ€™s regulations for flare performance, i.e., smokeless flaring while maintaining a minimum combustion zone heating value of 270 BTU/scf and $NHV_{dil} > 22$ Btu/ft^2 for all steam/air/non-assisted flares. While field tests of the many flare operation parameters are expensive, it is highly efficient to use mathematical models to analyse data and develop the right tools to make decisions in the real plant scenario.Â Modelling flare data using artificial neural networks to achieve the desired combustion efficiency (CE) and eliminate smoke can be done robustly. Selected flare data obtained from various flare tests having a wide range of exit velocity, heating value, and fuel composition have been analysed and the most influencing independent variables are chosen. Both steam and air assisted flares were modelled using the neural network toolbox in MATLAB with high correlation coefficients of greater than 0.95. Determining the set point (amount of steam/air/make-up fuel required) at the Incipient Smoke Point (ISP) and for smokeless flaring has been performed as a part of this study. Desirable operating inputs can be set for the ISP ($NHV_{dil} > 22$ BTU/ft^2 & $Opacity < Opacity_{ISP}$) and for smokeless flaring with a high CE (>96.5).

490378 Unconventional OIL and Natural Gas - Science & Technology Advancement

Abstract:

Despite the significant growth in the production of Unconventional Oil and Gas (UOG) resources during the shale revolution, very little is known in detail about recovery efficiency in these hydraulically-fractured reservoirs. The limited information that is available suggests that recovery factors (the ratio of produced resource to total in-place resource) are typically quite low: approximately 20 percent in gas-rich shale reservoirs and less than 10 percent in liquid-rich plays. By increasing the recovery factors of hydrocarbons, the environmental impact of these activities would also be reduced.

Therefore, the primary research challenges include, but are not limited to: 1) understanding the nano-scale structure of UOG reservoirs in order to more effectively create and control stimulated rock volume; 2) improving recovery efficiency to reduce the incidence of stranded/unproduced resources and the related lost economic and energy security benefits; 3) understanding the anticipated continued growth in recoverable resource volumes with technology advance; and 4) accurately assessing the potential nature, scale, and rate (intensity) of future development in various areas of the country to guide assessment of region-specific and environmental impacts.

Hydraulic and natural fracture diagnostic technologies have not advanced beyond microseismic monitoring techniques, and there is a need to improve the characterization of the far-field fracture geometry for recovery factor calculations and improvement. Advanced diagnostic and monitoring tools, as well as subsurface models, can be capable of predicting induced seismic events and may provide insight and benefit to our understanding and modeling of the stimulated reservoir volume during hydraulic fracturing operations.

Once the geometries of natural and induced fractures are better understood, new physical and chemical treatments can be formulated for each specific well to achieve maximum recovery factors. Advanced stimulation technologies, such as energized foams or other novel fluids could work in tandem with advanced fracture diagnostic tools to improve recovery efficiency and reduce potential environmental impacts.

17AIChE
Annual Meeting, Minneapolis, MN

Alternatives to Decline-Curve Models for Unconventional Reservoirs: A Case for Data-Driven Discovery of Natural Laws

Bharat Thakur & Michael Nikolaou
Petroleum Engineering
Chemical & Biomolecular Engineering
University of Houston, TX

Advances in Unconventional Oil and Gas Modeling

University of Houston

UNIVERSITY of HOUSTON

Decline-Curve Analysis (DCA): Conventional Reservoirs

- Decline curve analysis:
  - Estimate future production
  - From limited initial production data
- DCA for conventional reservoirs (Arps, 1944):

$$q = \frac{q_0}{(1 + bD_i t)^{1/b}}$$

- Arps curves:

$$P = P_0 e^{-t/a}$$

$$P = P_0 \left(1 + \frac{1}{a_0} t\right)^{-1}$$

$$P = P_0 \left(1 + \frac{b}{a_0} t\right)^{-1/b}$$

- Respective cumulative production:

$$Q = a(P_0 - P)$$

$$Q = a_0 P_0 \log\left(\frac{P_0}{P}\right)$$

$$Q = \frac{a_0 P_0^b}{1-b}\left(P_0^{1-b} - P^{1-b}\right)$$

17AIChE
Annual Meeting, Minneapolis, MN

Thakur
Nikolaou

Data-Driven Decline Curves

UNIVERSITY of HOUSTON

Decline-Curve Analysis (DCA): Unconventional Reservoirs

Introduction

Data

Data Modeling

Conclusions

- DCA for unconventionals:
  - Arps not accurate
  - Alternatives cumbersome:
    - Different flow behaviors during:
      - ❑ Initial long transient,
      - ❑ Transition to boundary-dominated,
      - ❑ Boundary-dominated
    - Two models typically used
    - Unclear
      - ❑ when to switch models
      - ❑ which model to use in transition period

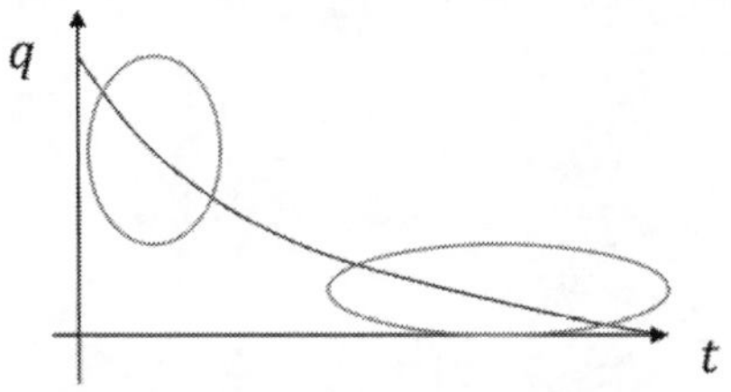

- Proliferation of Arps-like modified equations:
  - Fetcovich (1980)
  - Camacho-Velasquez (1987)
  - Robertson (1988)
  - Fulford & Blasingame (2013)
  - Duong (2011)
  - Valko (2009)

3

17AIChE
Annual Meeting, Minneapolis, MN

Thakur
Nikolaou

Data-Driven Decline Curves

UNIVERSITY of HOUSTON

Proposed Approach: Data-Driven DCA

Introduction

Data

Data Modeling

Conclusions

- Data-driven discovery of governing equation:
  - Collect long-time production data
  - Develop Arps counterpart
    - Parsimonious

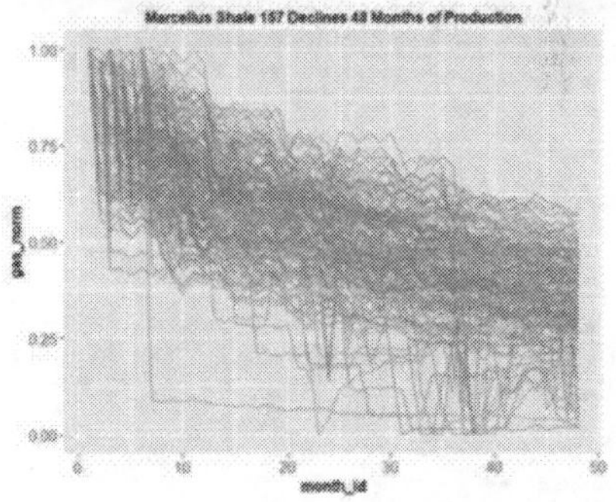

- Use Arps counterpart
  - Fit limited-time production with Arps counterpart
  - Extrapolate over time

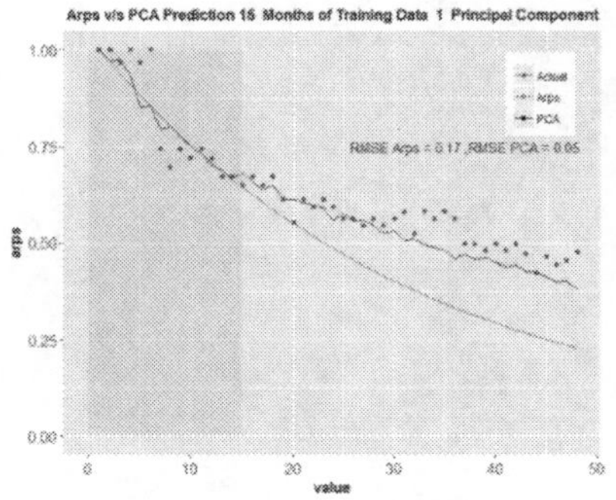

4

17AIChE
Annual Meeting, Minneapolis, MN

Thakur
Nikolaou

Data-Driven Decline Curves

UNIVERSITY of HOUSTON

Outline

- Introduction
- Data
- Data Modeling
- Conclusions

- Data of this study
- Modeling of data
- Conclusions

5

17 AIChE
Annual Meeting, Minneapolis, MN

Thakur
Nikolaou

Data-Driven Decline Curves

UNIVERSITY of HOUSTON

Data Analyzed

- Introduction
- Data
- Data Modeling
- Conclusions

- Lease Level Shale reservoir data analyzed:
  - Marcellus
  - Eagleford
  - Barnett

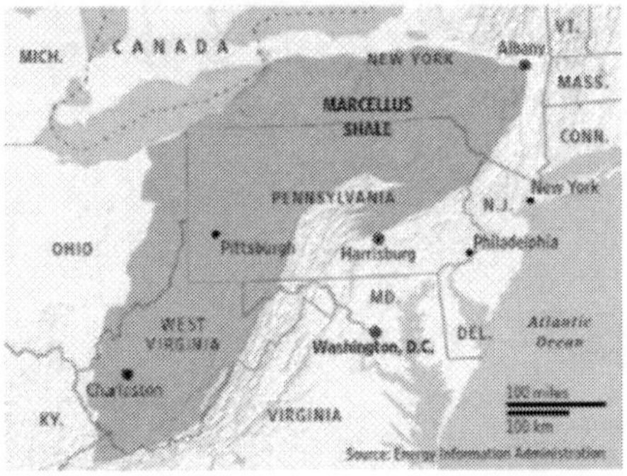

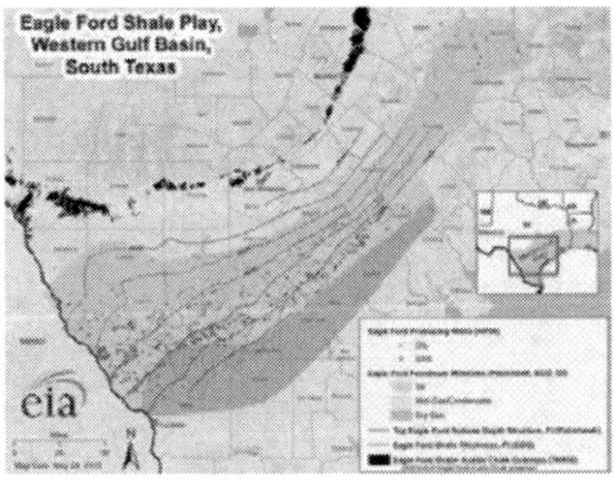

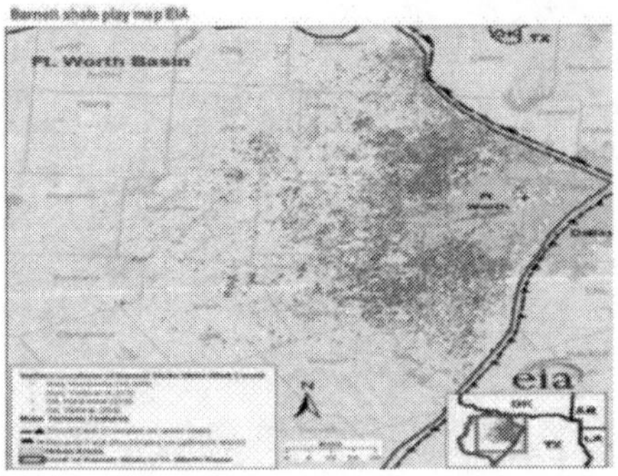

6

17 AIChE
Annual Meeting, Minneapolis, MN

Thakur
Nikolaou

Data-Driven Decline Curves

UNIVERSITY of HOUSTON

Data Analyzed

- Introduction
- Data
- Data Modeling
- Conclusions

- Shale reservoir data analyzed:
  - Marcellus
  - Eagle Ford
  - Barnett

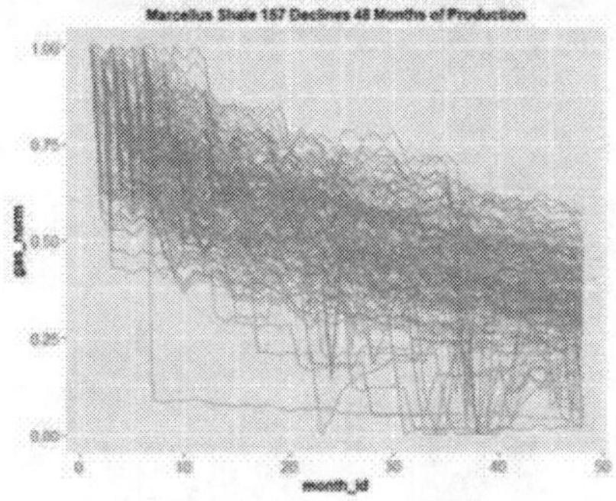

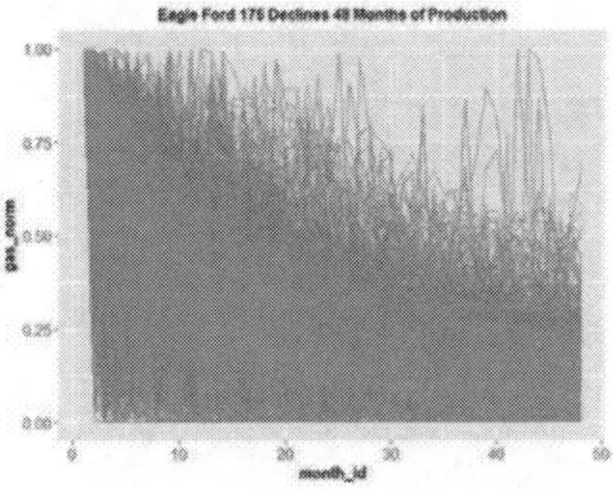

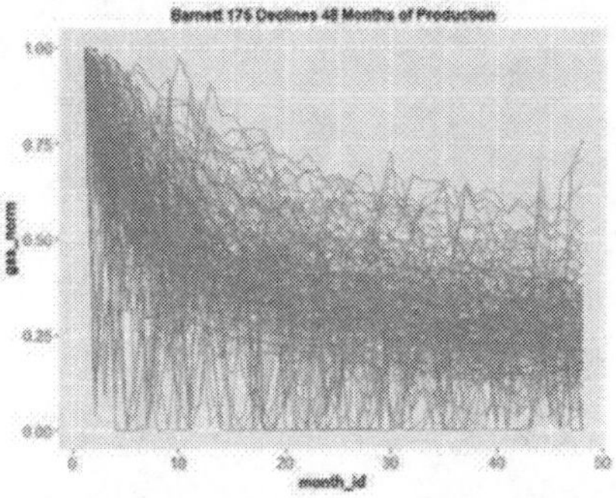

7

17 AIChE
Annual Meeting, Minneapolis, MN

Thakur
Nikolaou

Data-Driven Decline Curves

UNIVERSITY of HOUSTON

Data Pre-processing

- Introduction
- Data
- Data Modeling
- Conclusions

- Normalize the decline to 0-1 range with respect to Max Rate.
- Select decline curves with at least 48 months of production.
- Filter decline curves if average monthly CGR at least 1
- Remove if production increases for particular no of times and by particular percent
- From the Clean curves, subtract monthly average and perform PCA on deviation variable

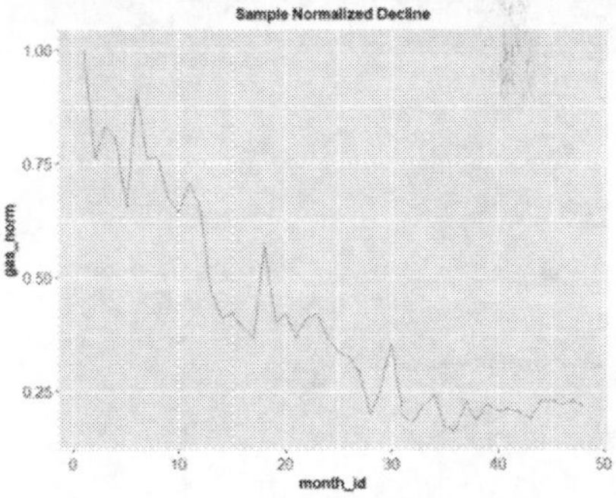

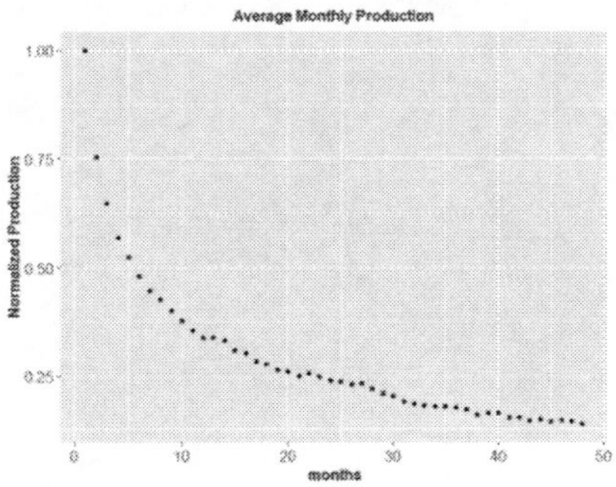

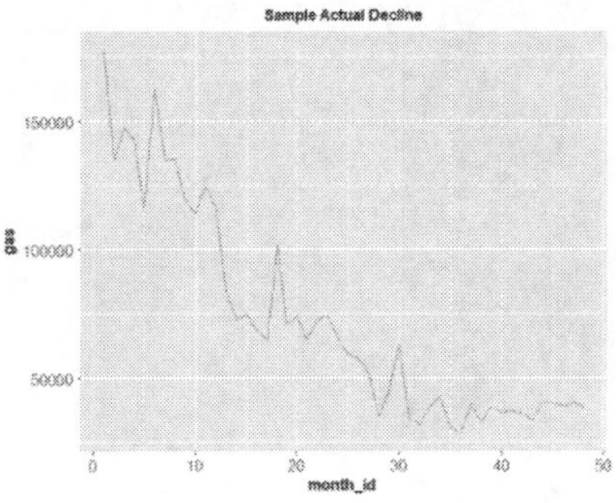

8

17 AIChE
Annual Meeting, Minneapolis, MN

Thakur
Nikolaou

Data-Driven Decline Curves

UNIVERSITY of HOUSTON

Marcellus

Introduction

Data

Data Modeling

Conclusions

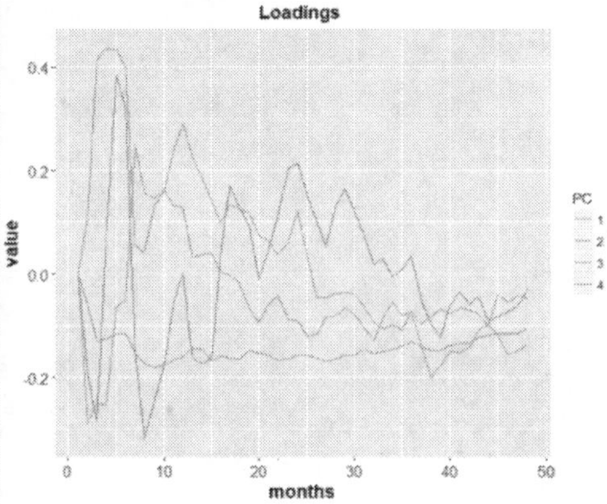

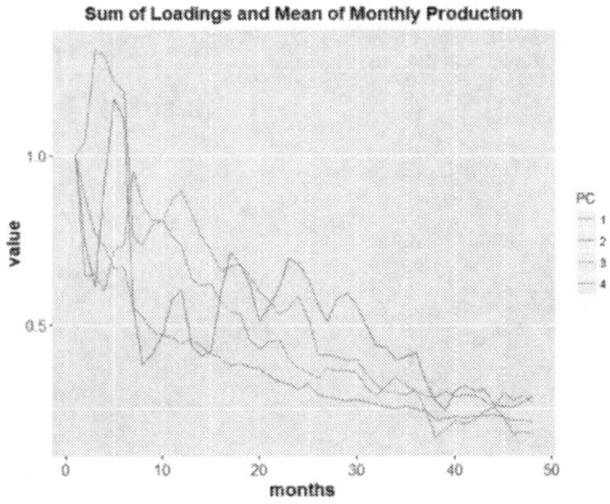

9

17 AIChE
Annual Meeting, Minneapolis, MN

Thakur
Nikolaou

Data-Driven Decline Curves

UNIVERSITY of HOUSTON

Eagle Ford

Introduction

Data

Data Modeling

Conclusions

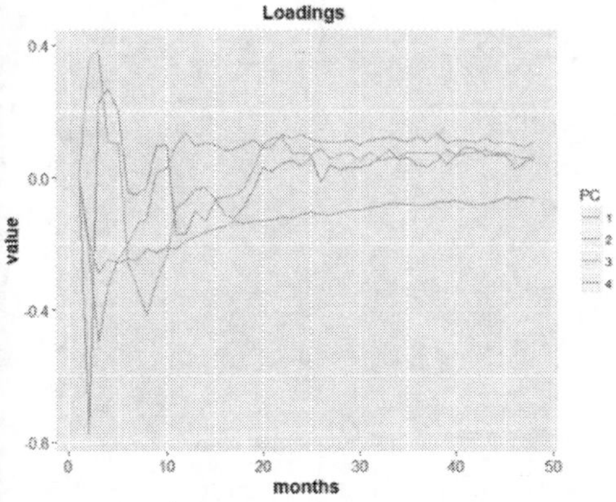

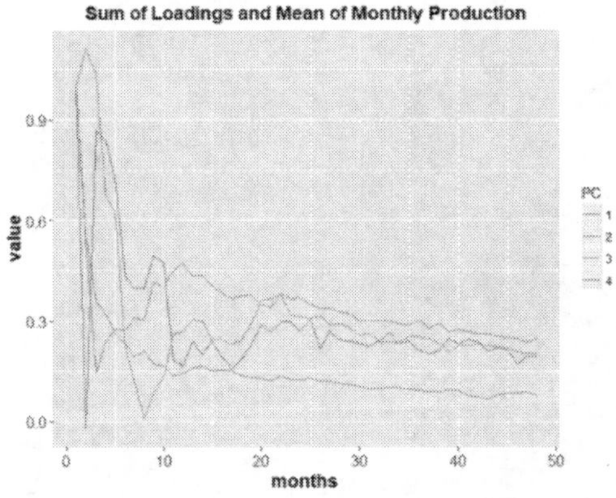

10

17 AIChE
Annual Meeting, Minneapolis, MN

Thakur
Nikolaou

Data-Driven Decline Curves

UNIVERSITY of HOUSTON

Barnett

Introduction

Data

Data Modeling

Conclusions

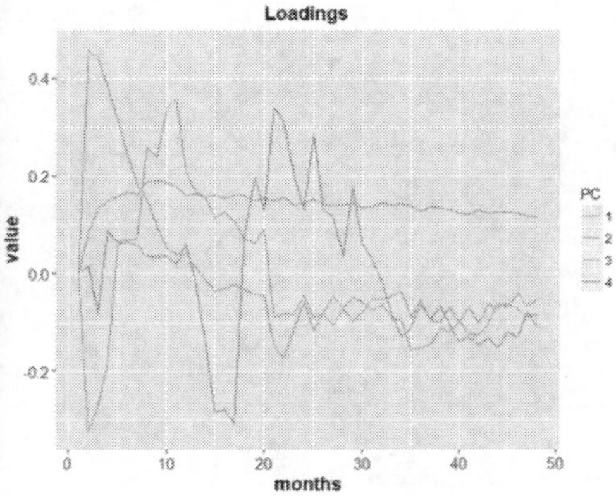

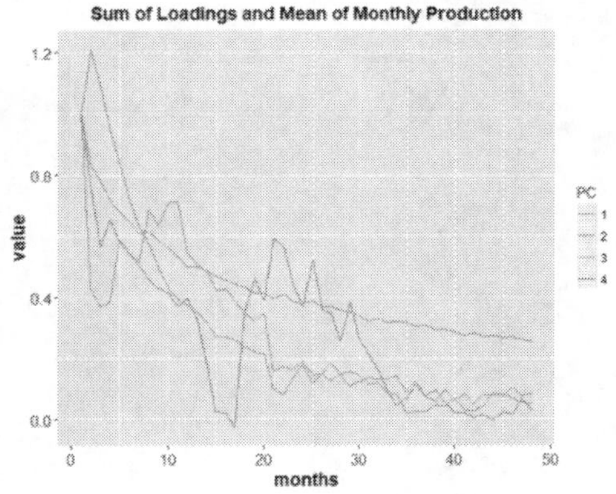

11

17 AIChE
Annual Meeting, Minneapolis, MN

Thakur
Nikolaou

Data-Driven Decline Curves

UNIVERSITY of HOUSTON

Marcellus Variance Captured By PCs

Introduction

Data

Data Modeling

Conclusions

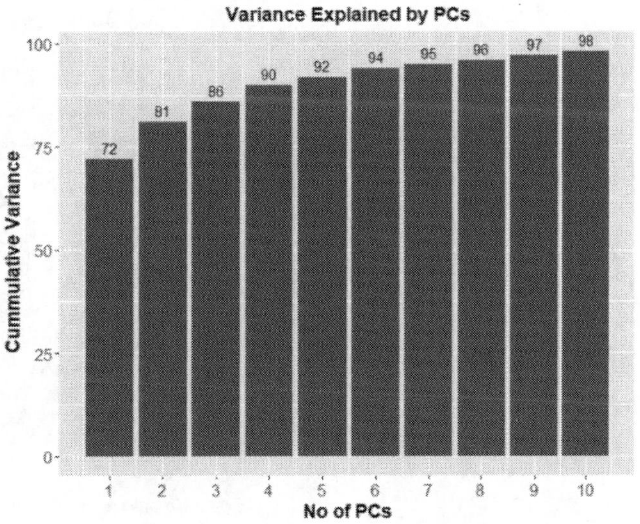

12

17 AIChE
Annual Meeting, Minneapolis, MN

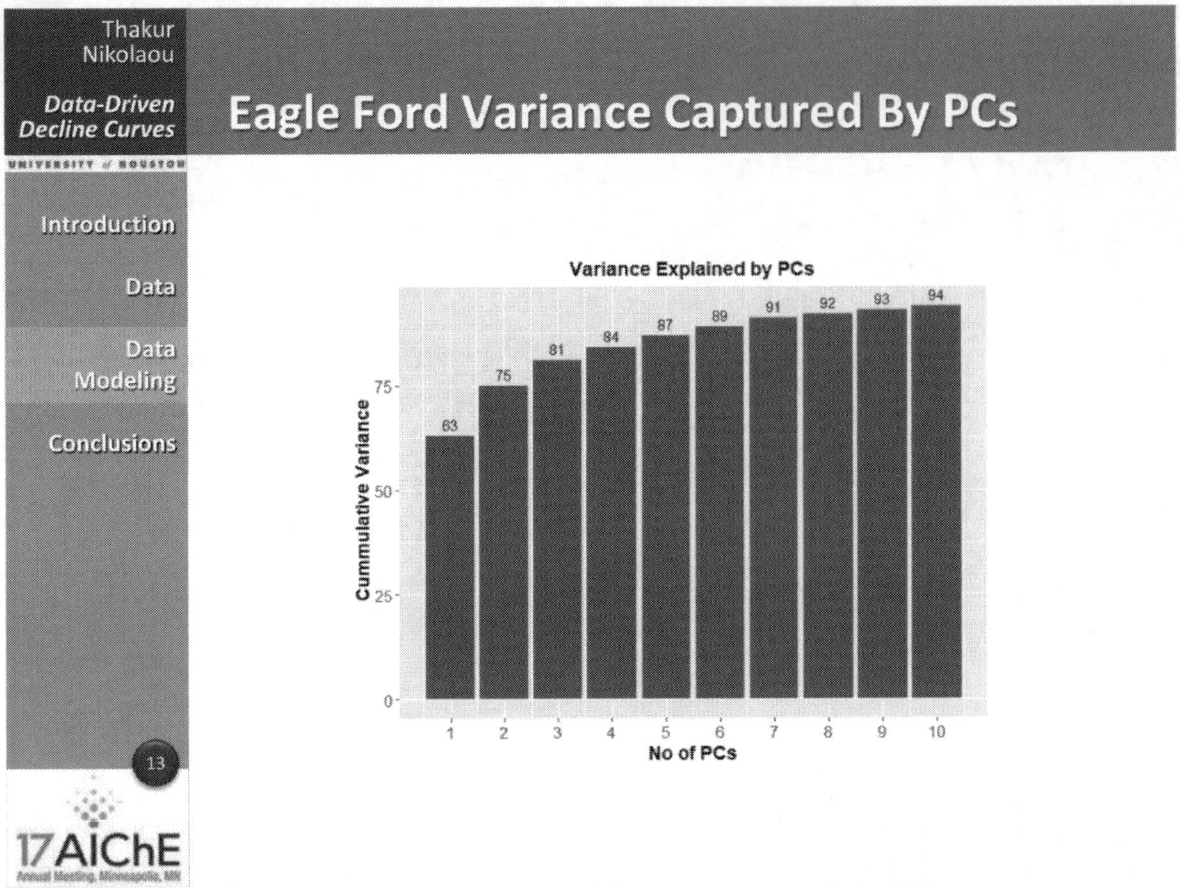
Thakur
Nikolaou
Data-Driven
Decline Curves
UNIVERSITY of HOUSTON
Eagle Ford Variance Captured By PCs
Introduction
Data
Data
Modeling
Conclusions
Variance Explained by PCs
63
75
81
84
87
89
91
92
93
94
Cummulative Variance
0
25
50
75
1
2
3
4
5
6
7
8
9
10
No of PCs
13
17AIChE
Annual Meeting, Minneapolis, MN

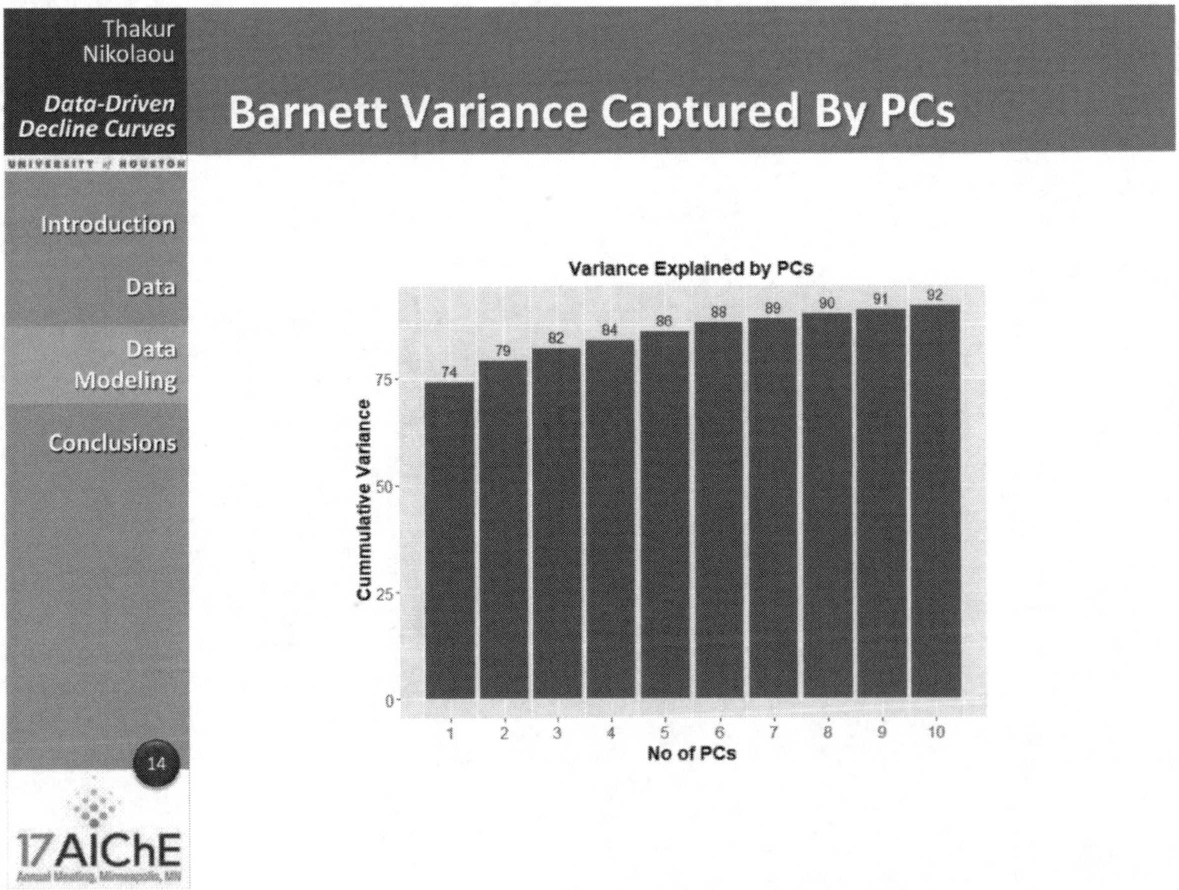
Thakur
Nikolaou
Data-Driven
Decline Curves
UNIVERSITY of HOUSTON
Barnett Variance Captured By PCs
Introduction
Data
Data
Modeling
Conclusions
Variance Explained by PCs
74
79
82
84
86
88
89
90
91
92
Cummulative Variance
0
25
50
75
1
2
3
4
5
6
7
8
9
10
No of PCs
14
17AIChE
Annual Meeting, Minneapolis, MN

Thakur
Nikolaou

Data-Driven Decline Curves

UNIVERSITY of HOUSTON

Deciding No of PC- Variance (Marcellus)

- Introduction
- Data
- Data Modeling
- Conclusions

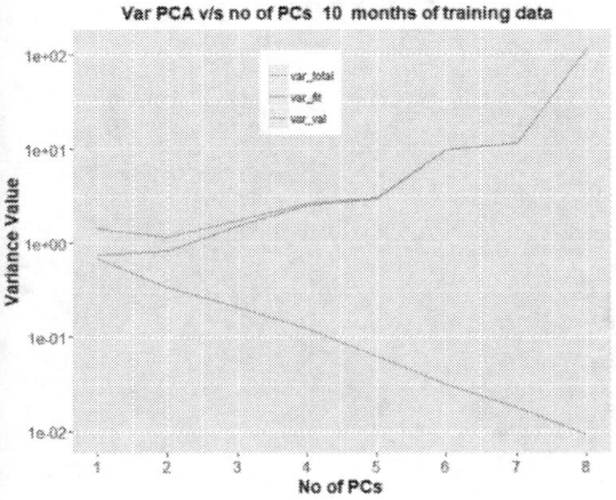

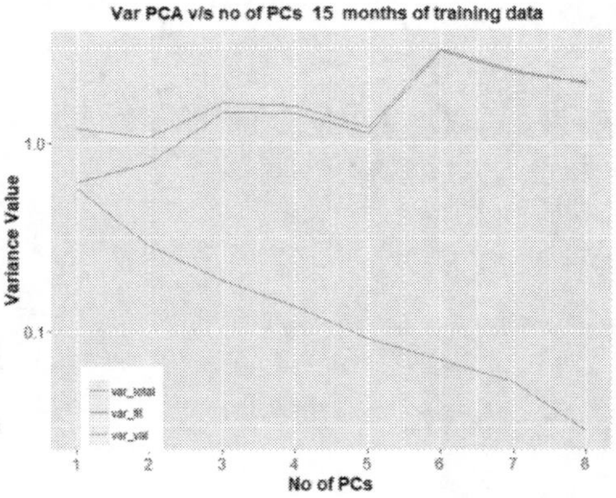

15

17AIChE
Annual Meeting, Minneapolis, MN

Thakur
Nikolaou

Data-Driven Decline Curves

UNIVERSITY of HOUSTON

Deciding No of PC-RMSE (Marcellus)

- Introduction
- Data
- Data Modeling
- Conclusions

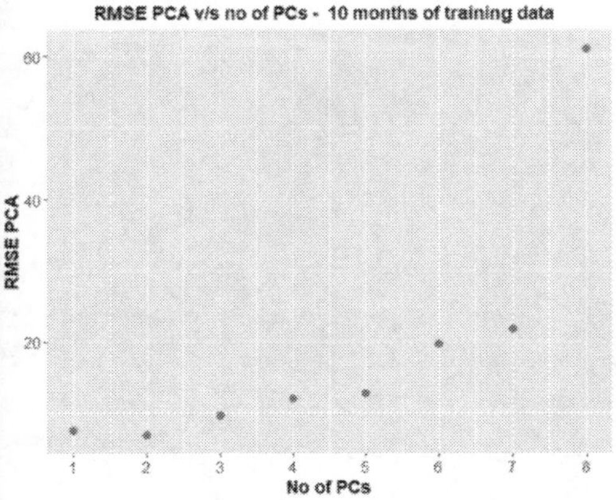

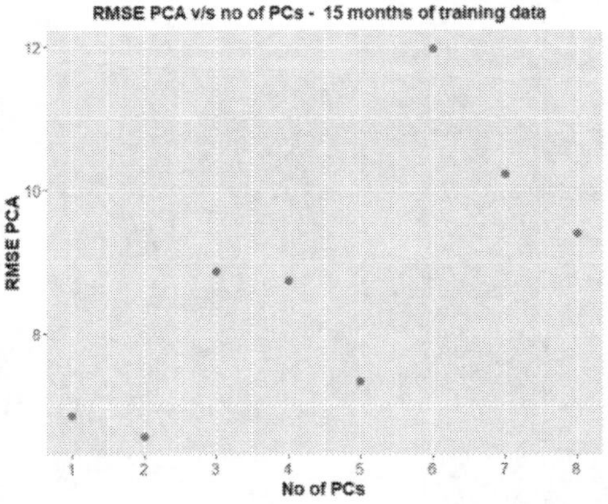

16

17AIChE
Annual Meeting, Minneapolis, MN

Thakur
Nikolaou

Data-Driven Decline Curves

UNIVERSITY of HOUSTON

Performance Overall - using 5 months of Training Data (Marcellus)

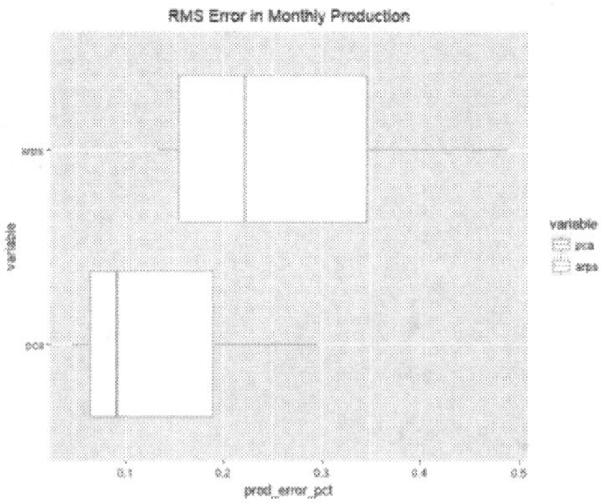

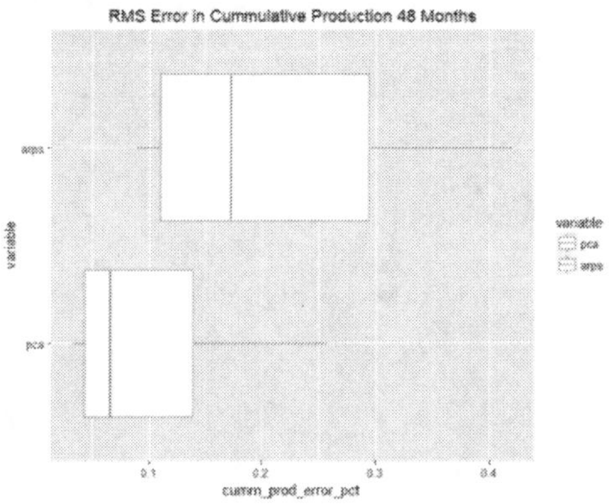

17

17AIChE
Annual Meeting, Minneapolis, MN

Thakur
Nikolaou

Data-Driven Decline Curves

UNIVERSITY of HOUSTON

Performance - using 10 months of Training Data (Marcellus)

Introduction
Data
Data Modeling
Conclusions

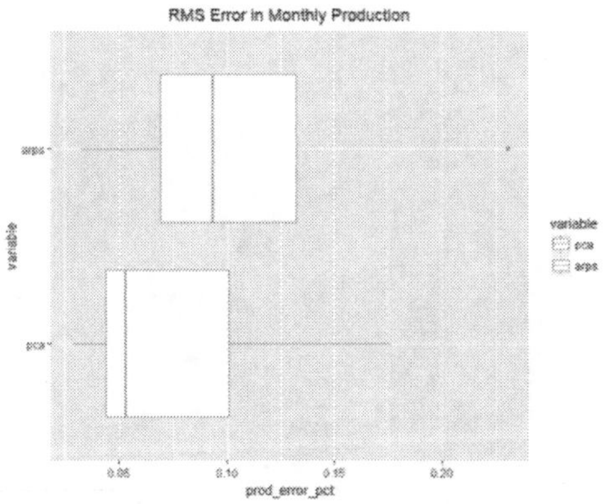

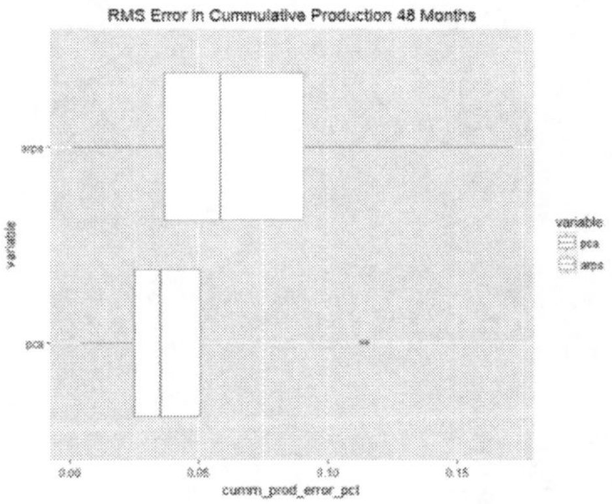

18

17AIChE
Annual Meeting, Minneapolis, MN

Thakur
Nikolaou

Data-Driven Decline Curves

UNIVERSITY of HOUSTON

Performance - using 15 months of Training Data (Marcellus)

Introduction

Data

Data Modeling

Conclusions

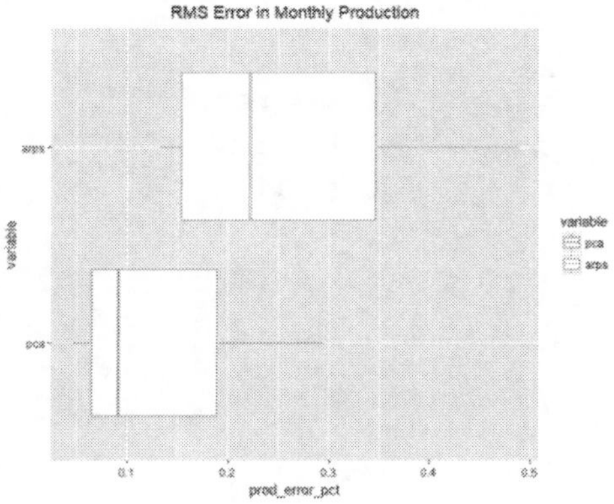

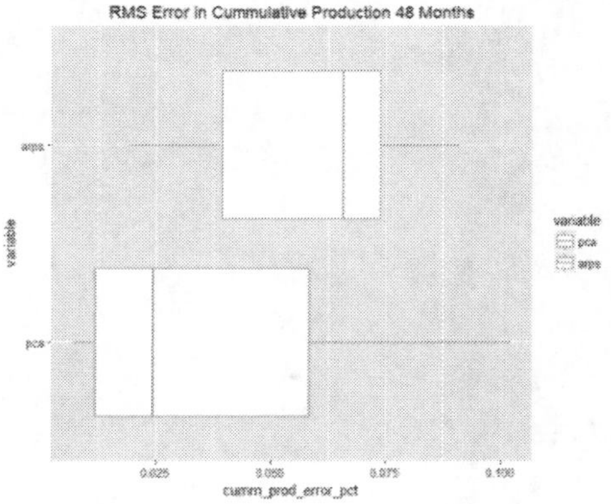

19

17 AIChE
Annual Meeting, Minneapolis, MN

Thakur
Nikolaou

Data-Driven Decline Curves

UNIVERSITY of HOUSTON

Effect of No of Training Months on Prediction (Marcellus)

Introduction

Data

Data Modeling

Conclusions

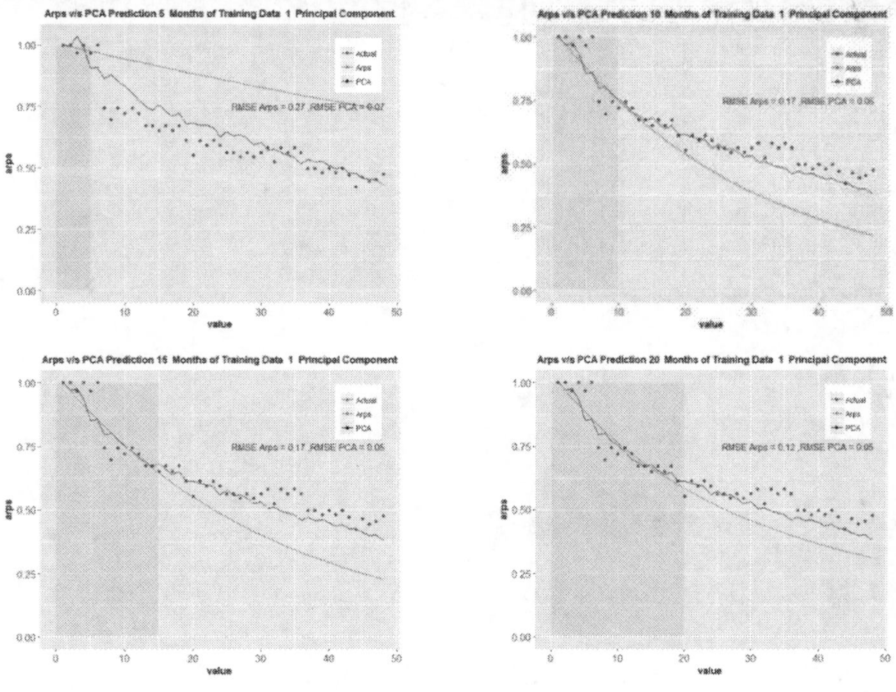

20

17 AIChE
Annual Meeting, Minneapolis, MN

Thakur
Nikolaou

Data-Driven Decline Curves

UNIVERSITY of HOUSTON

Effect of No of Cs on Prediction (Marcellus)

Introduction

Data

Data Modeling

Conclusions

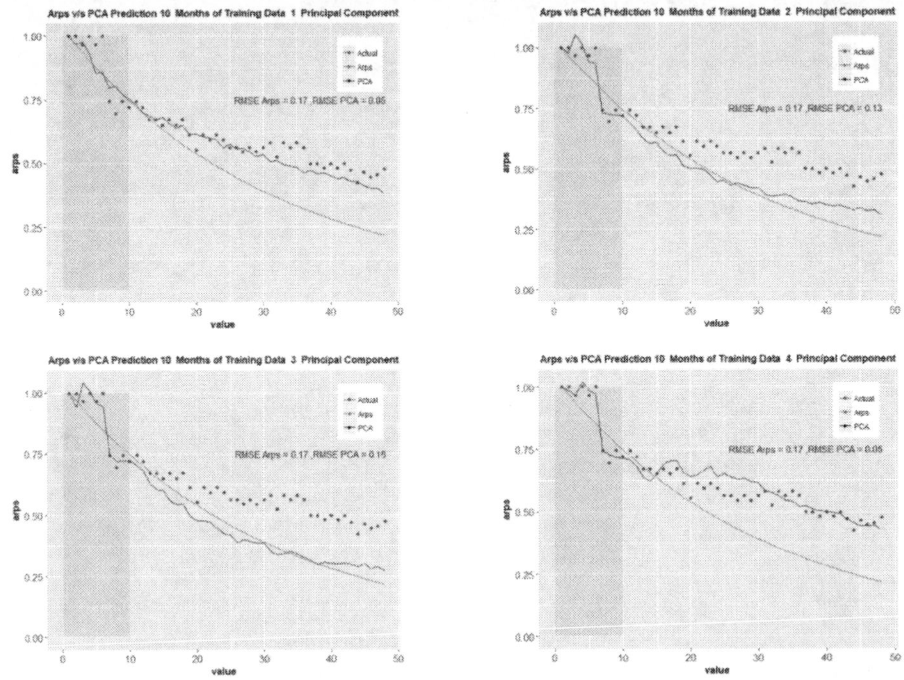

21

17 AIChE
Annual Meeting, Minneapolis, MN

Thakur
Nikolaou

Data-Driven Decline Curves

UNIVERSITY of HOUSTON

Confidence Intervals – PCA(using 1 PC) (Marcellus)

Introduction

Data

Data Modeling

Conclusions

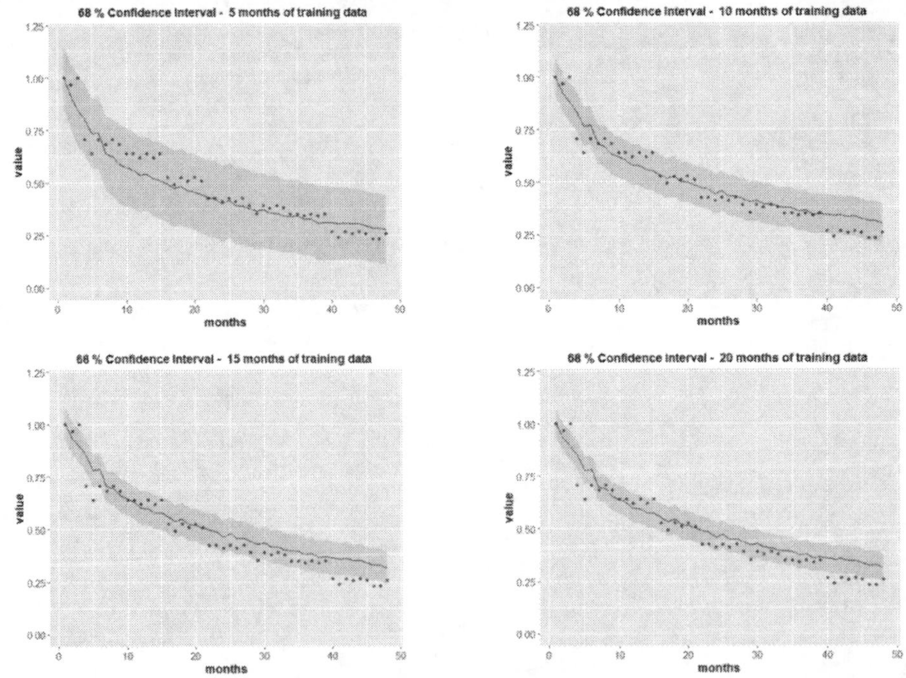

22

17 AIChE
Annual Meeting, Minneapolis, MN

Thakur
Nikolaou

Data-Driven Decline Curves

UNIVERSITY of HOUSTON

Introduction

Data

Data Modeling

Conclusions

23

17 AIChE
Annual Meeting, Minneapolis, MN

Conclusions

- Data are equations
  - …if data is representative of what the equations try to predict
- Non-standard functions usable in simple "equations"

Thakur
Nikolaou

Data-Driven Decline Curves

UNIVERSITY of HOUSTON

Introduction

Data

Data Modeling

Conclusions

24

17 AIChE
Annual Meeting, Minneapolis, MN

Acknowledgements

- Drilling Info
  - For access to data

Production Estimation and Well Classification for Hydraulically Fractured Horizontal Wells: A Data-Driven Model-Based Approach

Sunit Mathur and Michael Nikolaou
Petroleum Engineering
Chemical & Biomolecular Engineering
University of Houston, TX

Advances in Unconventional Oil and Gas Modeling

UNIVERSITY *of* HOUSTON

17 AIChE
Annual Meeting, Minneapolis, MN

Mathur-Nikolaou

Predicting Production for Unconventionals

UNIVERSITY *of* HOUSTON

Production from Shale Gas Wells

- Introduction
- Data processed
- Modeling
- Results
- Conclusions

2

17 AIChE
Annual Meeting, Minneapolis, MN

Mathur-Nikolaou

Predicting Production for Unconventionals

UNIVERSITY of HOUSTON

Introduction

Data processed

Modeling

Results

Conclusions

Predicting Shale Gas Well Production

- **Equations/simulator**:
  - Intense initial effort
  - Possibly inconclusive
- **Data-driven:**
  - Wealth of design and field data:
    - Exploration
    - Drilling job
    - Wellbore configuration
    - Fracture design
    - Historical production data

3

17AIChE Annual Meeting, Minneapolis, MN

Mathur-Nikolaou

Predicting Production for Unconventionals

UNIVERSITY of HOUSTON

Introduction

Data processed

Modeling

Results

Conclusions

Outline

- Data processed
- Modeling approach
  - PLS and variants
- Results
  - Accuracy of predictions
  - Ranking of variables in importance
  - Comparisons with other methods:
    - Combined ANN/PCA and variants
    - Random forests for ranking variables
- Conclusions

4

17AIChE Annual Meeting, Minneapolis, MN

Mathur-Nikolaou

Predicting Production for Unconventionals

UNIVERSITY of HOUSTON

Novelty of Work

Introduction
Data processed
Modeling
Results
Conclusions

- Previous works:

  Short-term production/completion data

  → predictions of future production

  → EUR

- This work:

$$\left\{\begin{array}{c}\text{Drilling job}\\ \text{Wellbore configuration}\\ \text{Fracture design \& construction}\\ +\\ \text{Historical production data}\end{array}\right\} \Rightarrow \left\{\begin{array}{c}\text{Production}\\ \text{in new wells}\end{array}\right\}$$

5

17AIChE Annual Meeting, Minneapolis, MN

Mathur-Nikolaou

Predicting Production for Unconventionals

UNIVERSITY of HOUSTON

Data Processed

IPPTC

Introduction
Data processed
Modeling
Results
Conclusions

- Southwestern Energy acreage
- Fayetteville shale

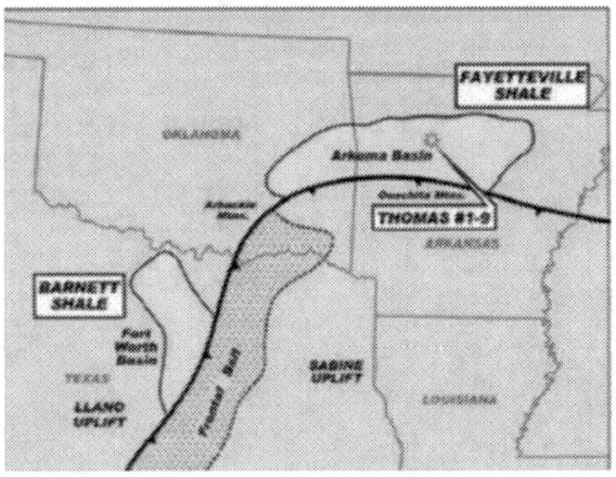

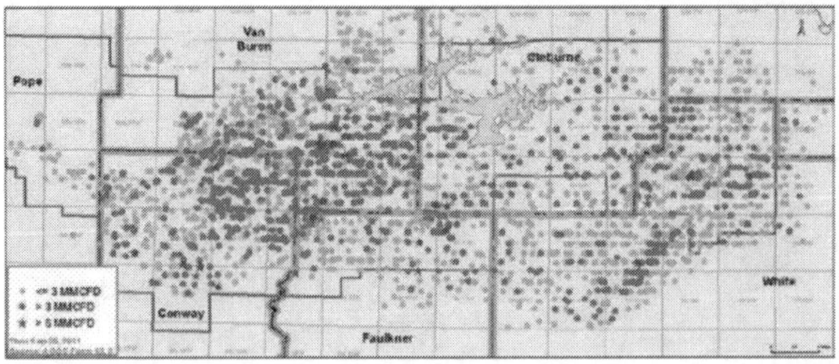

6

17AIChE Annual Meeting, Minneapolis, MN

Mathur-Nikolaou

Predicting Production for Unconventionals

UNIVERSITY of HOUSTON

Data Processed

Introduction
Data processed
Modeling
Results
Conclusions

- Southwestern Energy
  - Drilling reports
  - Completion reports
  - Overall wellbook reports
- Automated workflow adopted for data capture:
  - **250** wells
  - **9** wellbore/fracture configuration data points per well
  - **26** production data points (monthly) per well
  - **14** data points per stage (hydraulic fracturing job)
  - Total **2649** stages
  - Overall **45,836** data points to be captured

7

17AIChE
Annual Meeting, Minneapolis, MN

Mathur-Nikolaou

Predicting Production for Unconventionals

UNIVERSITY of HOUSTON

Snapshot of Hydraulic Fracturing Job Data

Introduction
Data processed
Modeling
Results
Conclusions

- Single stage fracture operation data
- Highlighted text important for this study

FRAC : Pressure test lines to 9,123 psi. Open well with 0,050 psi. Start down hole on stage 1 @ 5 bpm. Well broke @ 4,342 psi. Spearhead 1,500 gal of 15% HCl. Increase rate to 40 bpm, saw 1,386 psi drop from acid. Stepdown after acid hits as per procedure. Post-Acid ISIP was 2057 psi. Frac well as per procedure with SW fluid. Pumped a total of 9,596 bbls of clean fluid with 306,439 lbs of sand. Average rate/pressure was 66 bpm/6,426 psi. Max rate/pressure was 71 bpm/7,753 psi. ISIP: 2,415 psi. FG: 0.83 psi/ft

8

17AIChE
Annual Meeting, Minneapolis, MN

Mathur-Nikolaou

Predicting Production for Unconventionals

UNIVERSITY *of* HOUSTON

Modeling Overview

- Introduction
- Data processed
- Modeling
- Results
- Conclusions

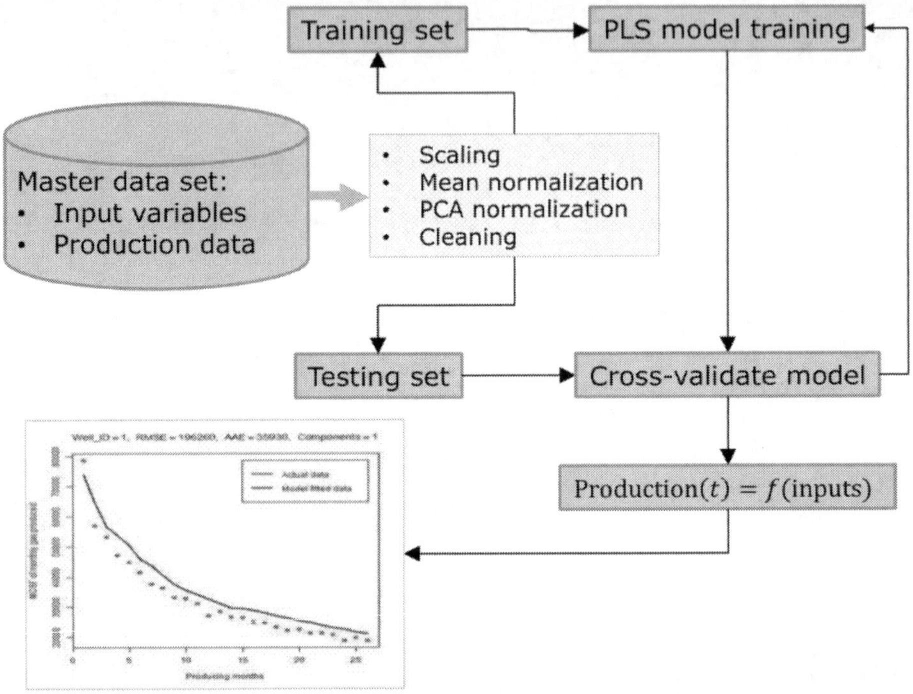

9

17AIChE
Annual Meeting, Minneapolis, MN

Mathur-Nikolaou

Predicting Production for Unconventionals

UNIVERSITY *of* HOUSTON

Input Variables

- Introduction
- Data processed
- Modeling
- Results
- Conclusions

- Pressure data (Hydraulic Fracturing)
  - Post Acid ISIP
  - Well Breaking Pressure
  - Opening Well Pressure
  - Pressure drop from acid
  - Pressure Test Line
  - Average Pressure during Injection
  - Maximum Pressure during Injection
  - ISIP
- Wellbore
  - Tubing Diameter
  - True vertical depth
  - Diameter of Horizontal section
  - Length of Horizontal section
- Fracture configuration
  - Number of cluster per stage
  - Number of perforations per cluster
  - Number of stages
  - Stage Length
  - Cluster spacing
  - Lbs of Sand added
  - Bbls of Clean fluid added
  - Bbls of HCl added
  - Increased Injection Rate
  - Average rate of injection
  - Maximum rate of injection

10

17AIChE
Annual Meeting, Minneapolis, MN

Mathur-Nikolaou

Predicting Production for Unconventionals

UNIVERSITY of HOUSTON

Output Variable

- Introduction
- Data processed
- Modeling
- Results
- Conclusions

- Production over 26 months
- Derived output: EUR

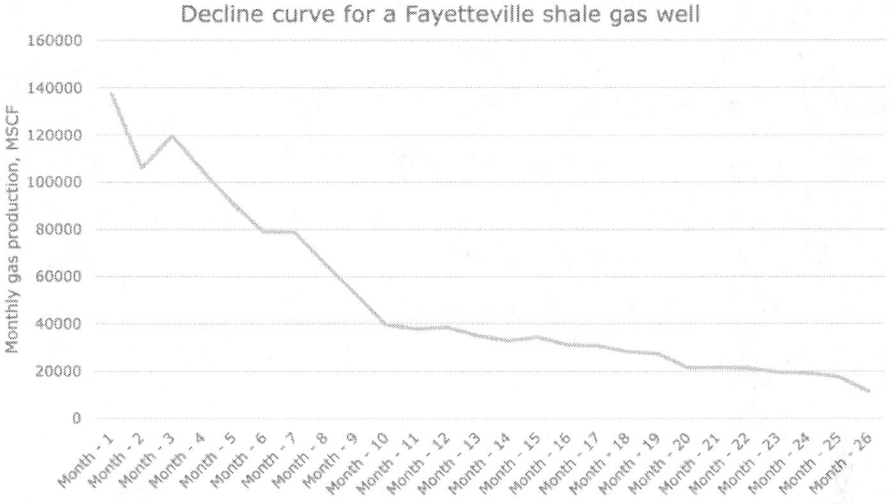

11

17AIChE Annual Meeting, Minneapolis, MN

Mathur-Nikolaou

Predicting Production for Unconventionals

UNIVERSITY of HOUSTON

Testing Model Quality

- Introduction
- Data processed
- Modeling
- Results
- Conclusions

- Cross-validation:
  - 80/20 split for training/validation

1	2	3	4	5
Train	Train	Validation	Train	Train

12

17AIChE Annual Meeting, Minneapolis, MN

Data Pre-processing

Introduction
Data processed
Modeling
Results
Conclusions

- Scaled:
  - 0 to 1
- Mean normalized
  - Deviation from average
- For ANN: PCA normalized
  - Principal components for 26 months of production
- Removal of outliers based on manual judgement:
  - Refractured wells
  - Wells temporarily closed

Results

Introduction
Data processed
Modeling
Results
Conclusions

- PLS-based production curves for
  - Outlier points included or removed
  - Input/output scaled or unscaled
  - Input data with averages removed or not
- PLS-model comparisons with
  - ANN

17AIChE
Annual Meeting, Minneapolis, MN

Results

Decline curves PLS

UNIVERSITY *of* HOUSTON

Mathur-Nikolaou

Predicting Production for Unconventionals

UNIVERSITY *of* HOUSTON

PLS Uncleaned Unscaled 2 Latent Variables

- Introduction
- Data processed
- Modeling
- Results
- Conclusions

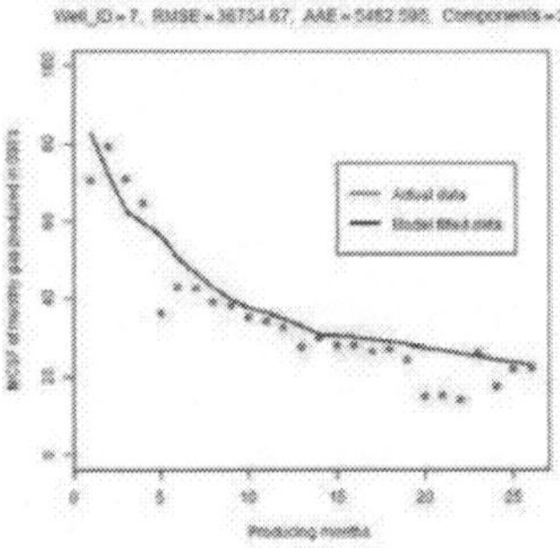

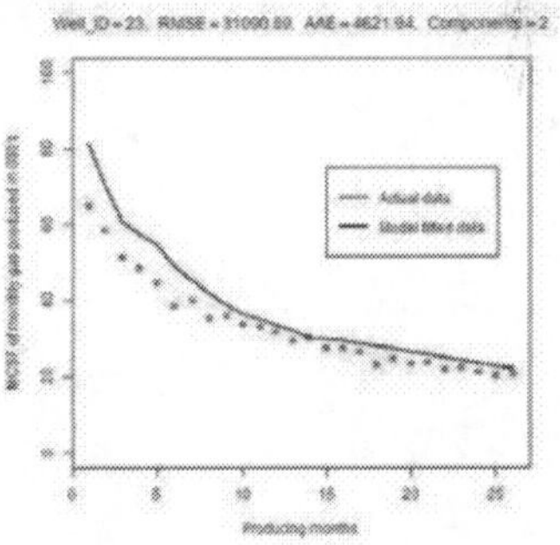

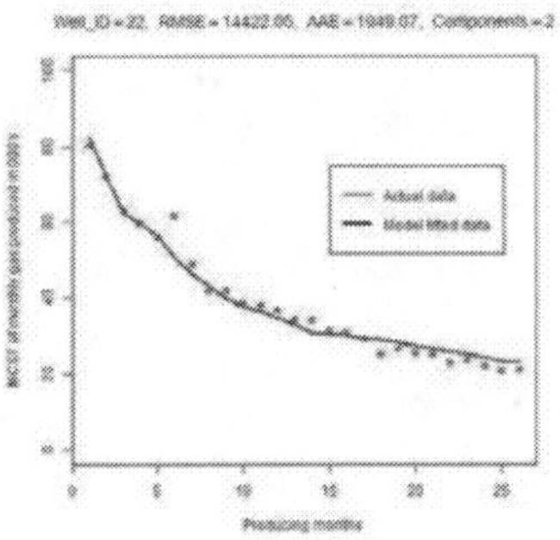

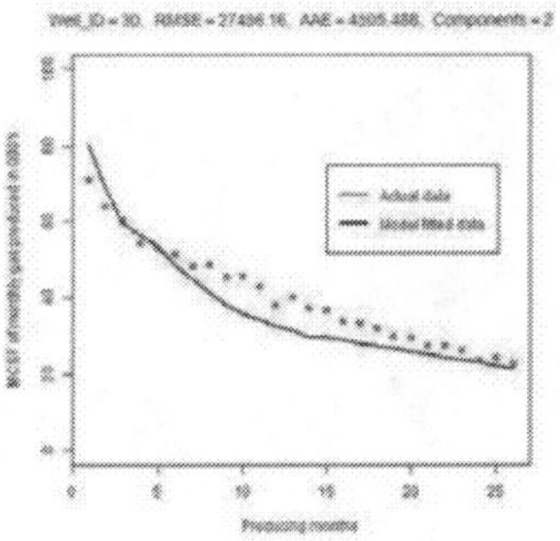

16

17AIChE
Annual Meeting, Minneapolis, MN

Mathur-Nikolaou

Predicting Production for Unconventionals

UNIVERSITY *of* HOUSTON

PLS Uncleaned Scaled 2 Latent Variables

- Introduction
- Data processed
- Modeling
- Results
- Conclusions

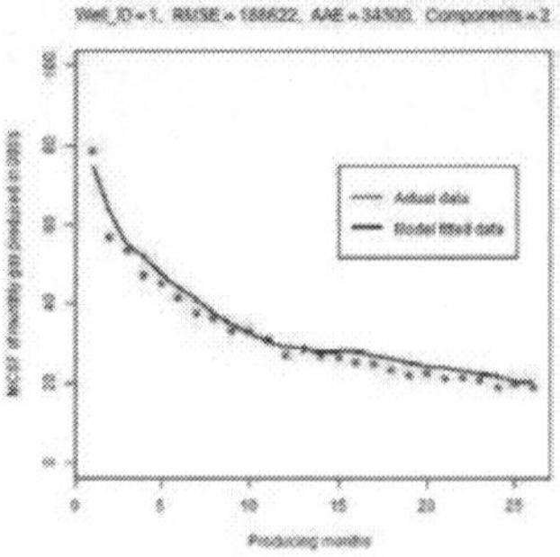

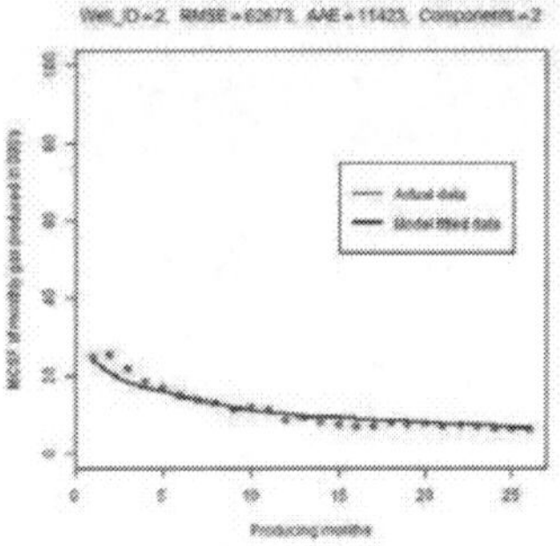

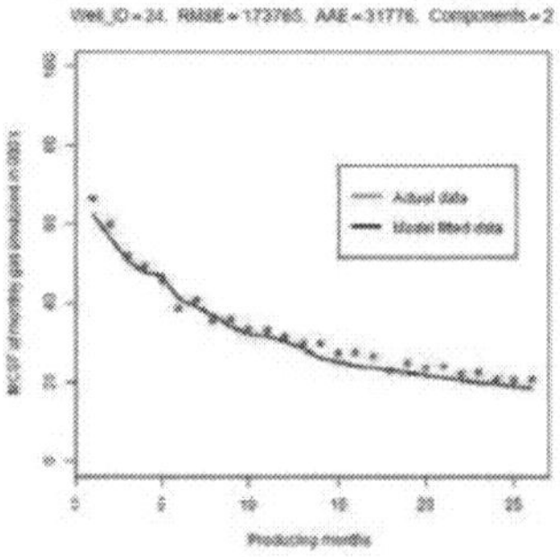

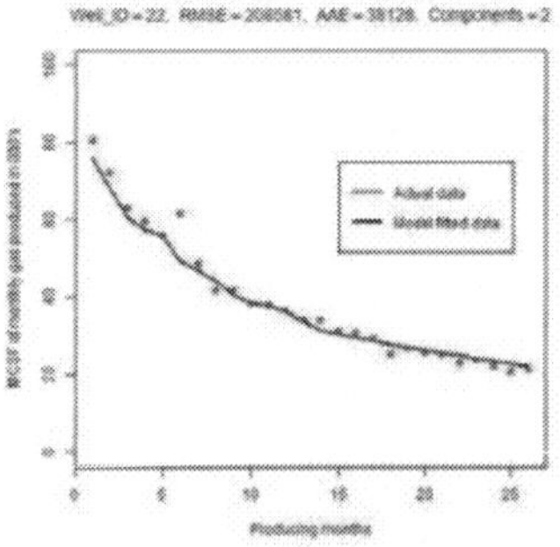

17

17AIChE
Annual Meeting, Minneapolis, MN

Mathur-Nikolaou

Predicting Production for Unconventionals

UNIVERSITY *of* HOUSTON

PLS Cleaned Average Corrected 2 Latent Variables

- Introduction
- Data processed
- Modeling
- Results
- Conclusions

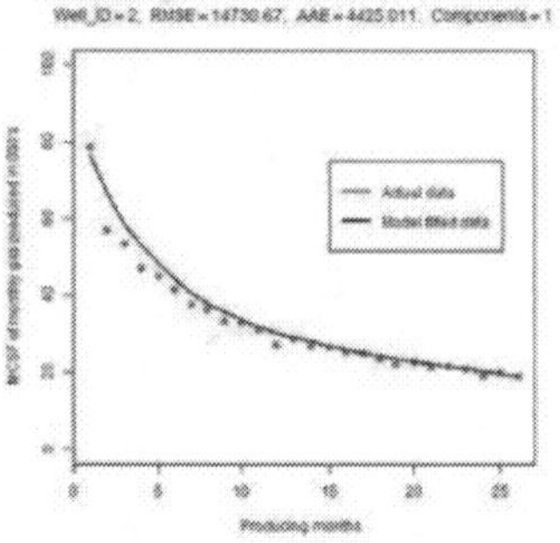

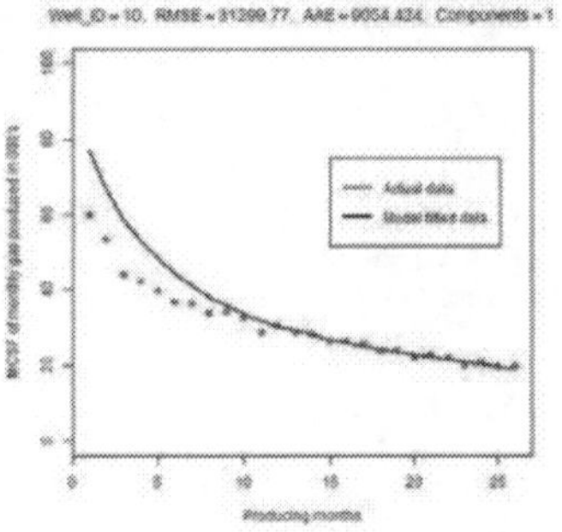

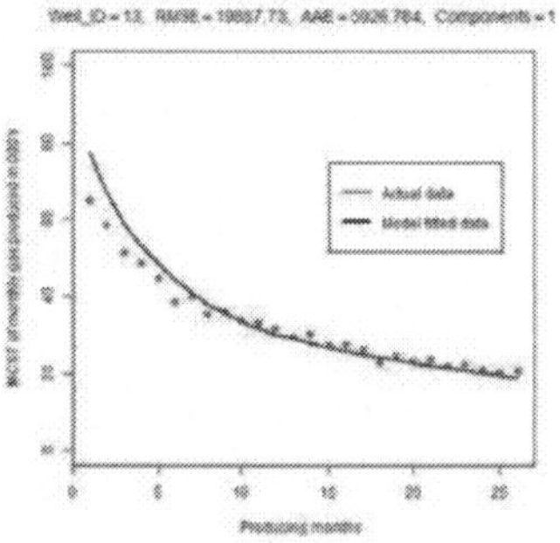

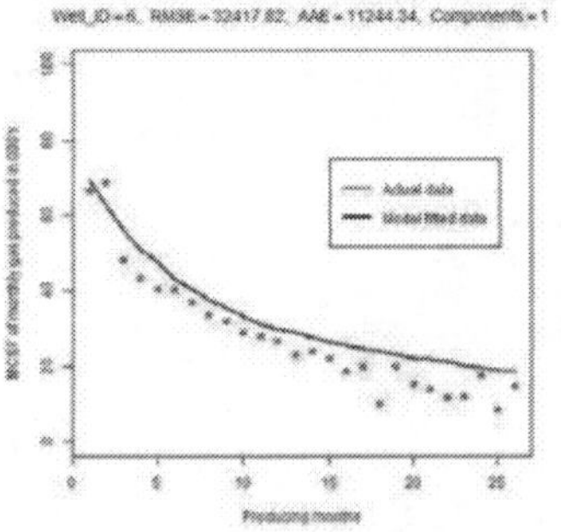

18

17AIChE
Annual Meeting, Minneapolis, MN

Mathur-Nikolaou

Predicting Production for Unconventionals

UNIVERSITY of HOUSTON

PLS Cleaned Unscaled 2 Latent Variables

- Introduction
- Data processed
- Modeling
- Results
- Conclusions

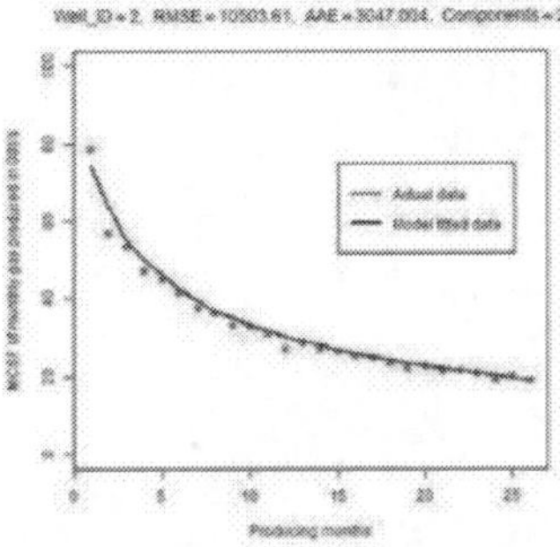

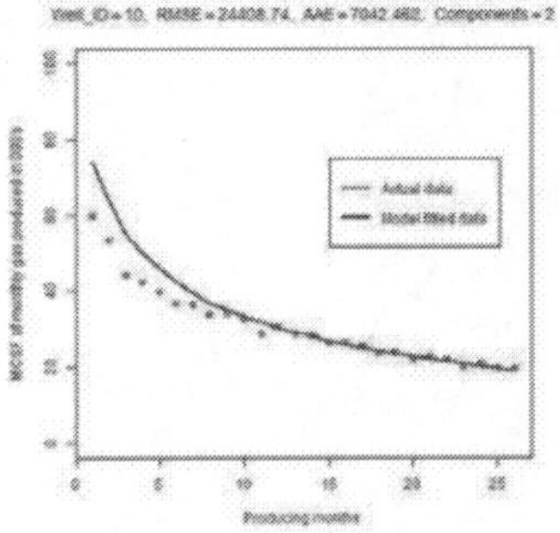

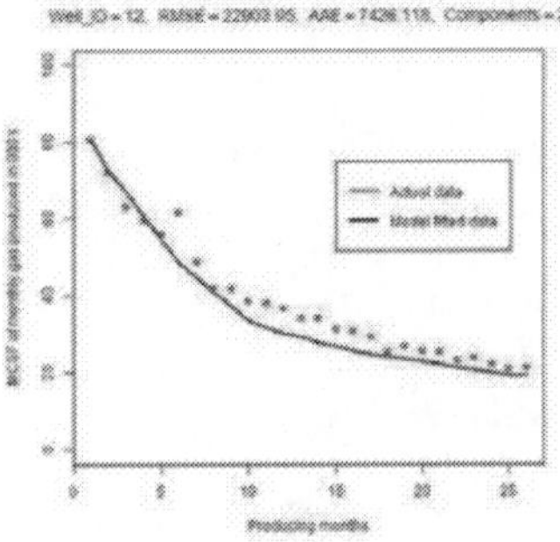

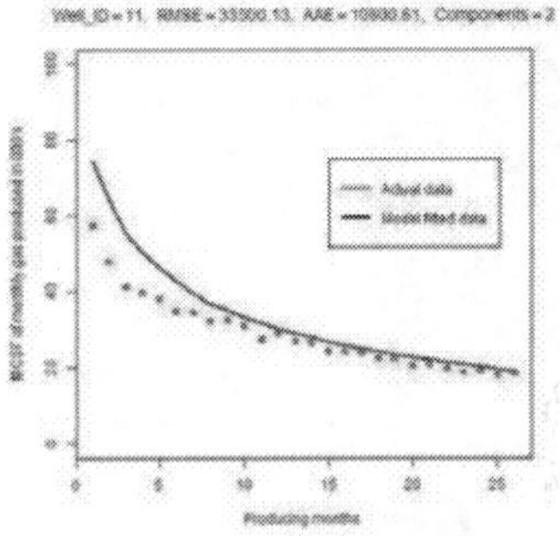

19

17AIChE
Annual Meeting, Minneapolis, MN

Mathur-Nikolaou

Predicting Production for Unconventionals

UNIVERSITY of HOUSTON

PLS Cleaned Scaled 2 Latent Variables

- Introduction
- Data processed
- Modeling
- Results
- Conclusions

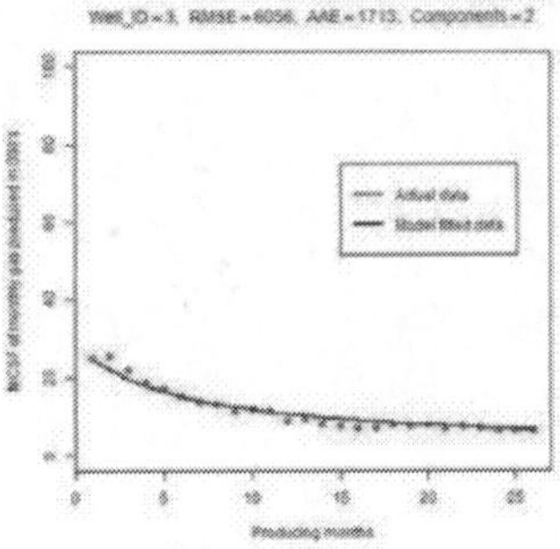

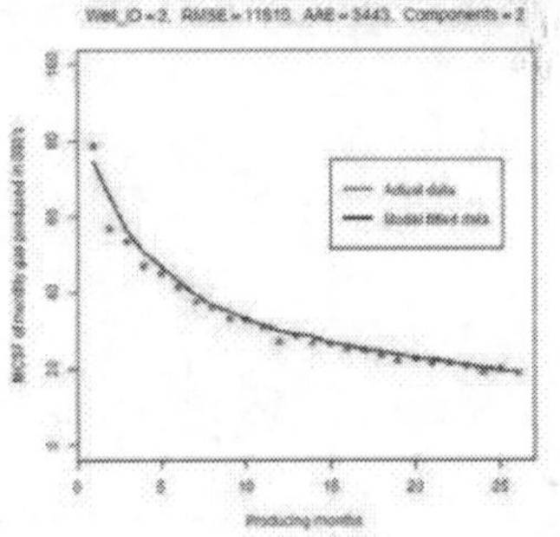

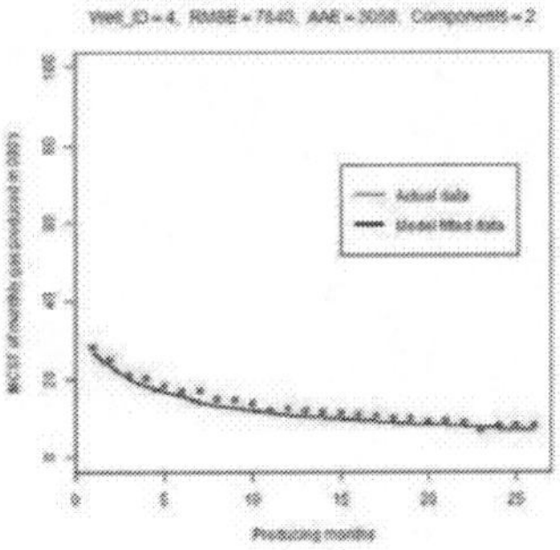

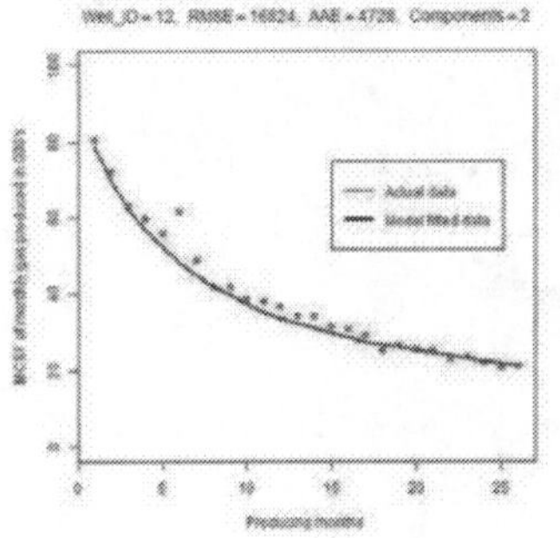

20

17AIChE
Annual Meeting, Minneapolis, MN

Mathur-Nikolaou

Predicting Production for Unconventionals

UNIVERSITY of HOUSTON

Loading Weights for Uncleaned Scaled PLS

Introduction

Data processed

Modeling

Results

Conclusions

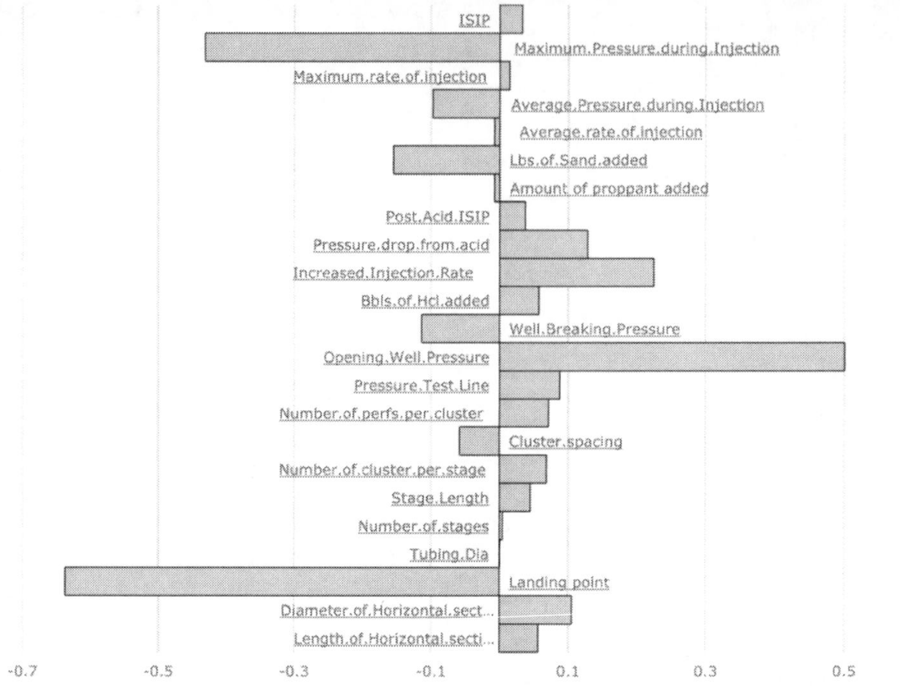

21

17AIChE Annual Meeting, Minneapolis, MN

17AIChE Annual Meeting, Minneapolis, MN

Importance of Inputs

Variable Importance in Projection (VIP) statistic

UNIVERSITY of HOUSTON

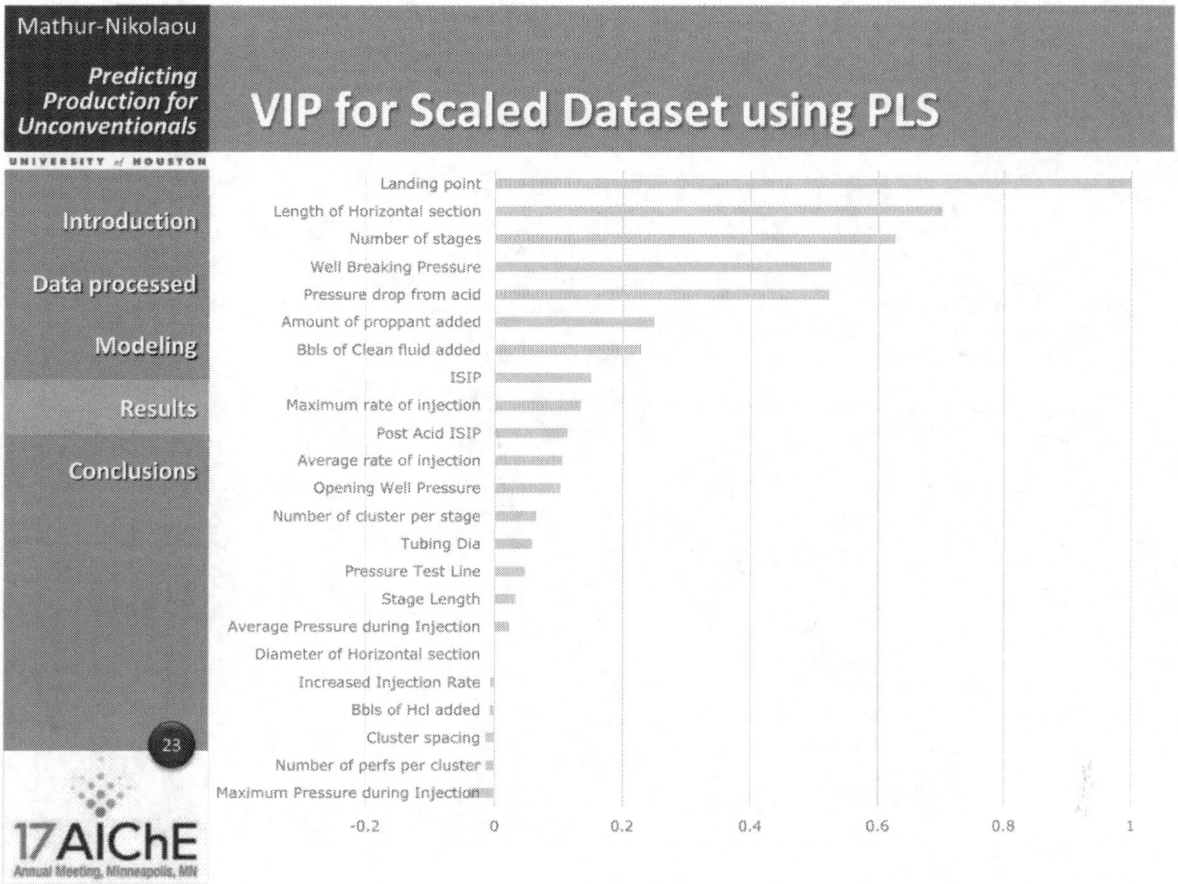
Mathur-Nikolaou
Predicting Production for Unconventionals
UNIVERSITY of HOUSTON
VIP for Scaled Dataset using PLS
Introduction
Data processed
Modeling
Results
Conclusions
23
17AIChE
Annual Meeting, Minneapolis, MN
Landing point
Length of Horizontal section
Number of stages
Well Breaking Pressure
Pressure drop from acid
Amount of proppant added
Bbls of Clean fluid added
ISIP
Maximum rate of injection
Post Acid ISIP
Average rate of injection
Opening Well Pressure
Number of cluster per stage
Tubing Dia
Pressure Test Line
Stage Length
Average Pressure during Injection
Diameter of Horizontal section
Increased Injection Rate
Bbls of Hcl added
Cluster spacing
Number of perfs per cluster
Maximum Pressure during Injection
-0.2
0
0.2
0.4
0.6
0.8
1

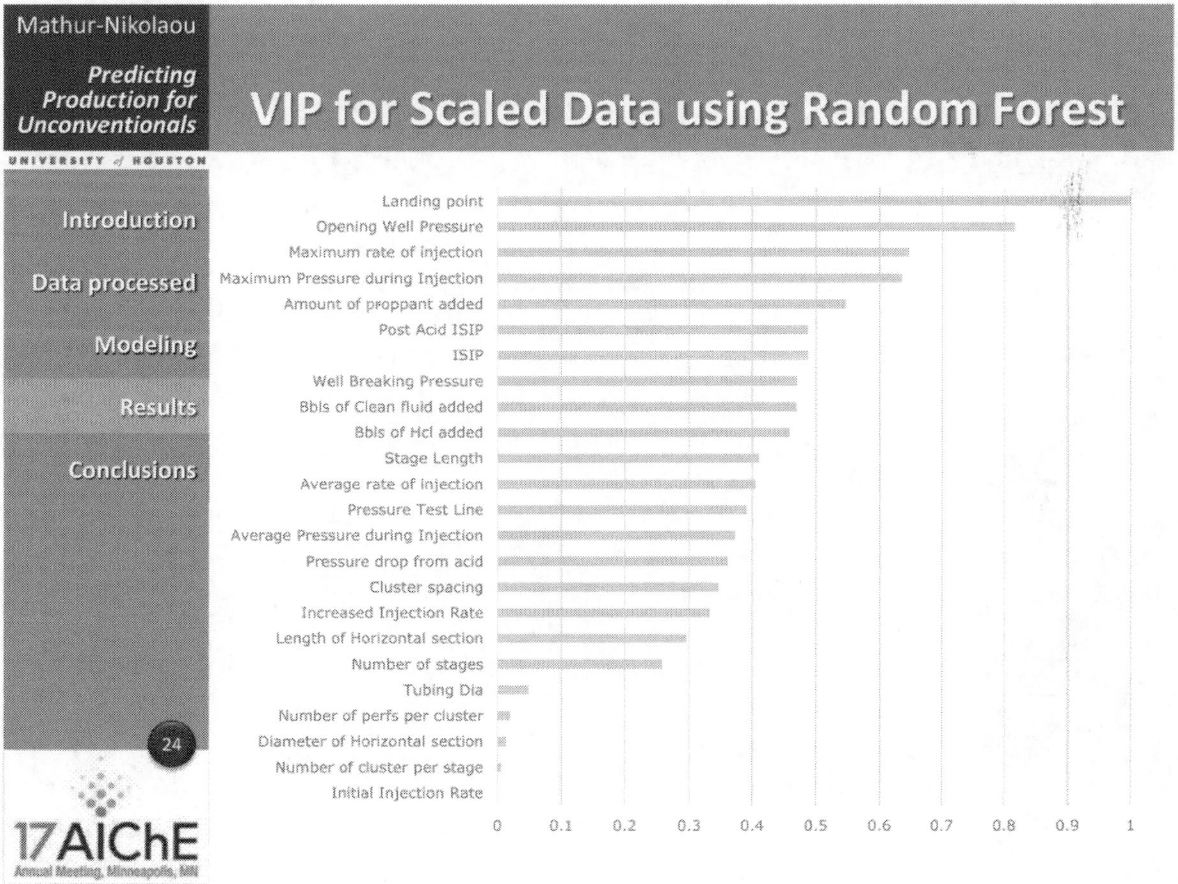
Mathur-Nikolaou
Predicting Production for Unconventionals
UNIVERSITY of HOUSTON
VIP for Scaled Data using Random Forest
Introduction
Data processed
Modeling
Results
Conclusions
24
17AIChE
Annual Meeting, Minneapolis, MN
Landing point
Opening Well Pressure
Maximum rate of injection
Maximum Pressure during Injection
Amount of proppant added
Post Acid ISIP
ISIP
Well Breaking Pressure
Bbls of Clean fluid added
Bbls of Hcl added
Stage Length
Average rate of injection
Pressure Test Line
Average Pressure during Injection
Pressure drop from acid
Cluster spacing
Increased Injection Rate
Length of Horizontal section
Number of stages
Tubing Dia
Number of perfs per cluster
Diameter of Horizontal section
Number of cluster per stage
Initial Injection Rate
0
0.1
0.2
0.3
0.4
0.5
0.6
0.7
0.8
0.9
1

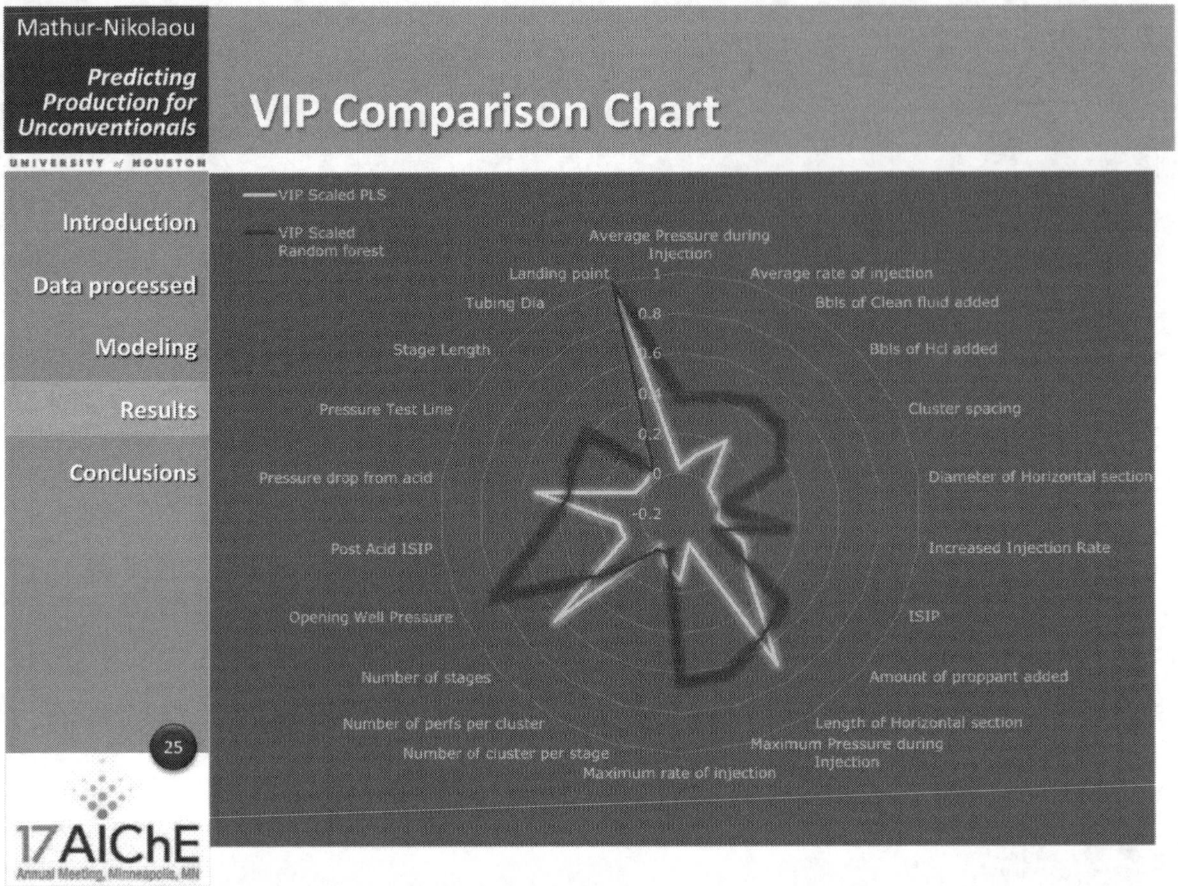

Mathur-Nikolaou
Predicting Production for Unconventionals
UNIVERSITY of HOUSTON
VIP Comparison Chart
Introduction
Data processed
Modeling
Results
Conclusions
25
17AIChE
Annual Meeting, Minneapolis, MN
VIP Scaled PLS
VIP Scaled Random forest
Average Pressure during Injection
Average rate of injection
Bbls of Clean fluid added
Bbls of Hcl added
Cluster spacing
Diameter of Horizontal section
Increased Injection Rate
ISIP
Amount of proppant added
Length of Horizontal section
Maximum Pressure during Injection
Maximum rate of injection
Number of cluster per stage
Number of perfs per cluster
Number of stages
Opening Well Pressure
Post Acid ISIP
Pressure drop from acid
Pressure Test Line
Stage Length
Tubing Dia
Landing point
1
0.8
0.6
0.4
0.2
0
-0.2

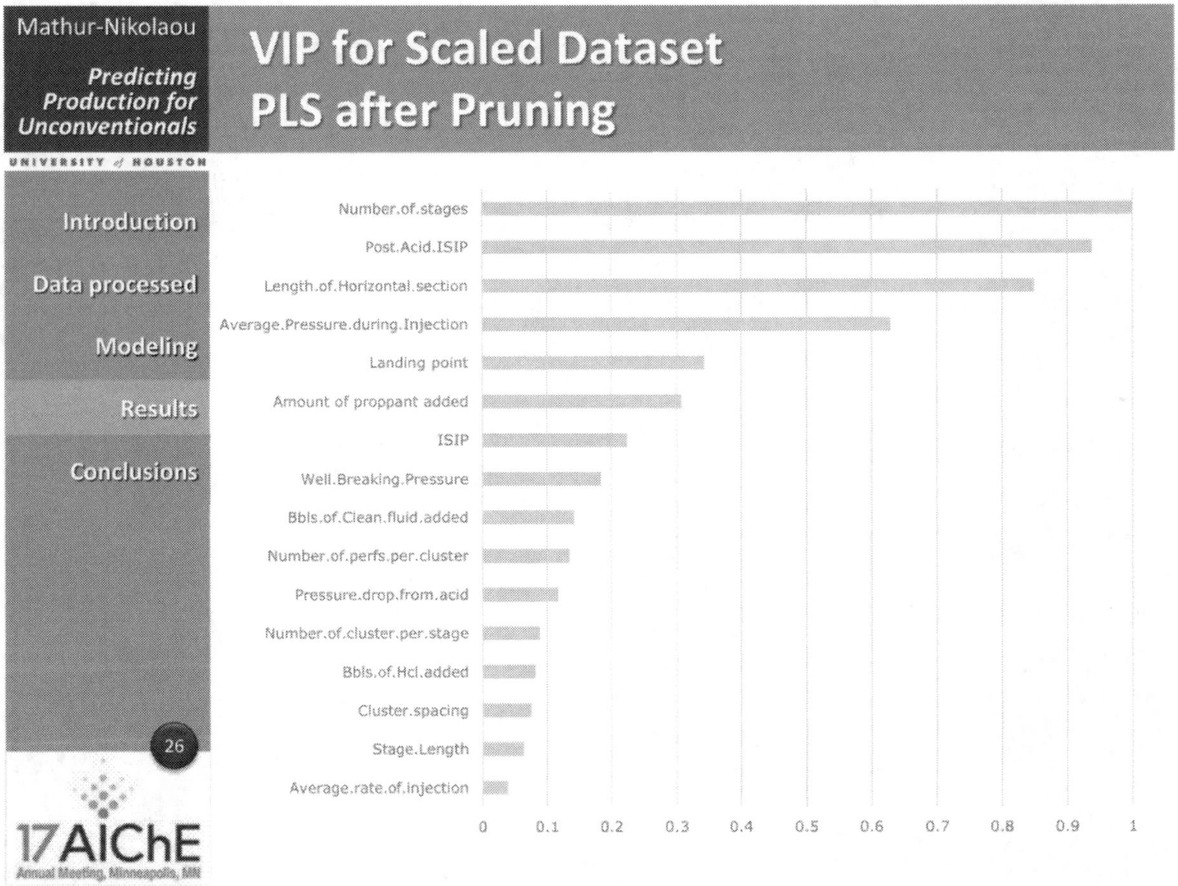

Mathur-Nikolaou
Predicting Production for Unconventionals
UNIVERSITY of HOUSTON
VIP for Scaled Dataset PLS after Pruning
Introduction
Data processed
Modeling
Results
Conclusions
26
17AIChE
Annual Meeting, Minneapolis, MN
Number.of.stages
Post.Acid.ISIP
Length.of.Horizontal.section
Average.Pressure.during.Injection
Landing point
Amount of proppant added
ISIP
Well.Breaking.Pressure
Bbls.of.Clean.fluid.added
Number.of.perfs.per.cluster
Pressure.drop.from.acid
Number.of.cluster.per.stage
Bbls.of.Hcl.added
Cluster.spacing
Stage.Length
Average.rate.of.injection
0
0.1
0.2
0.3
0.4
0.5
0.6
0.7
0.8
0.9
1

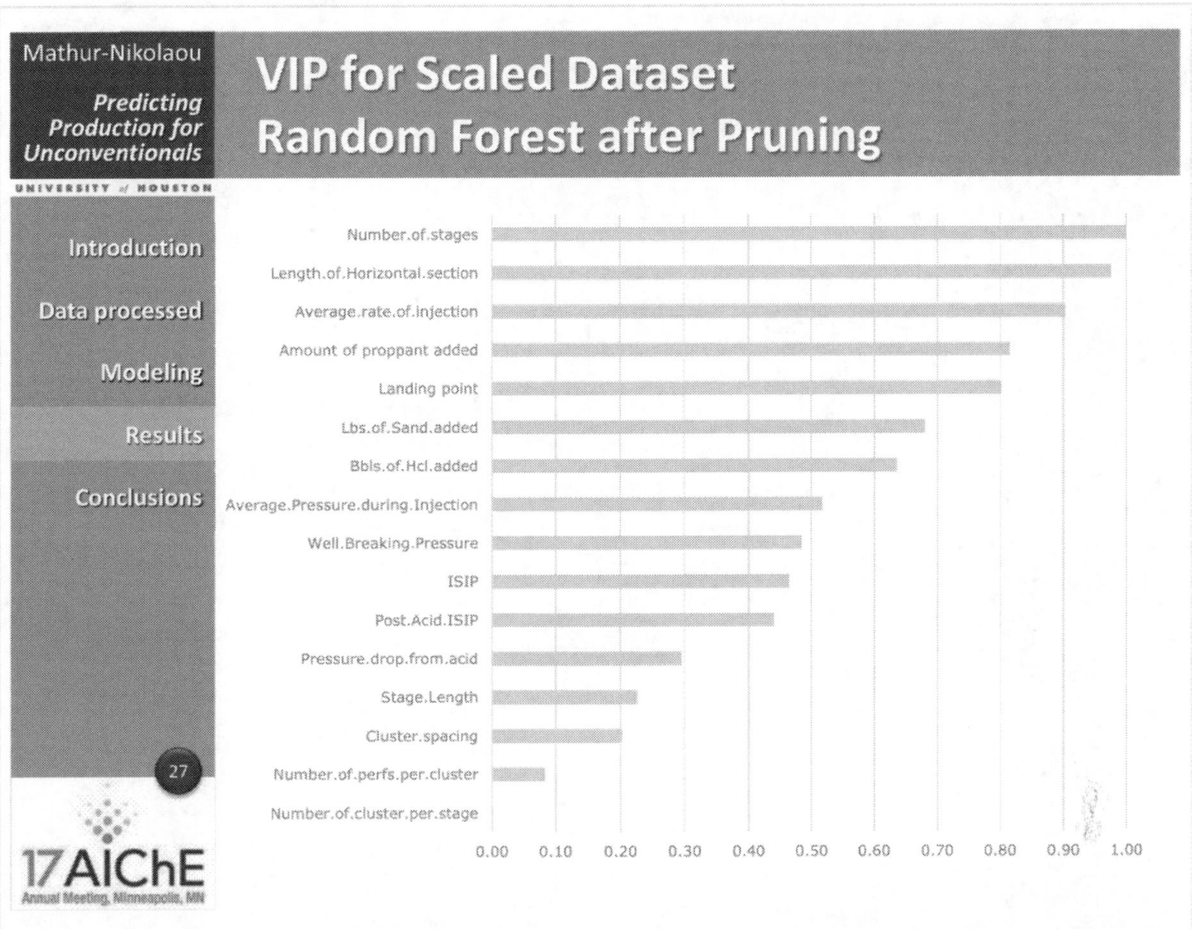
Mathur-Nikolaou
Predicting Production for Unconventionals
UNIVERSITY of HOUSTON
VIP for Scaled Dataset
Random Forest after Pruning
Introduction
Data processed
Modeling
Results
Conclusions
27
17 AIChE
Annual Meeting, Minneapolis, MN
Number.of.stages
Length.of.Horizontal.section
Average.rate.of.injection
Amount of proppant added
Landing point
Lbs.of.Sand.added
Bbls.of.Hcl.added
Average.Pressure.during.Injection
Well.Breaking.Pressure
ISIP
Post.Acid.ISIP
Pressure.drop.from.acid
Stage.Length
Cluster.spacing
Number.of.perfs.per.cluster
Number.of.cluster.per.stage
0.00 0.10 0.20 0.30 0.40 0.50 0.60 0.70 0.80 0.90 1.00

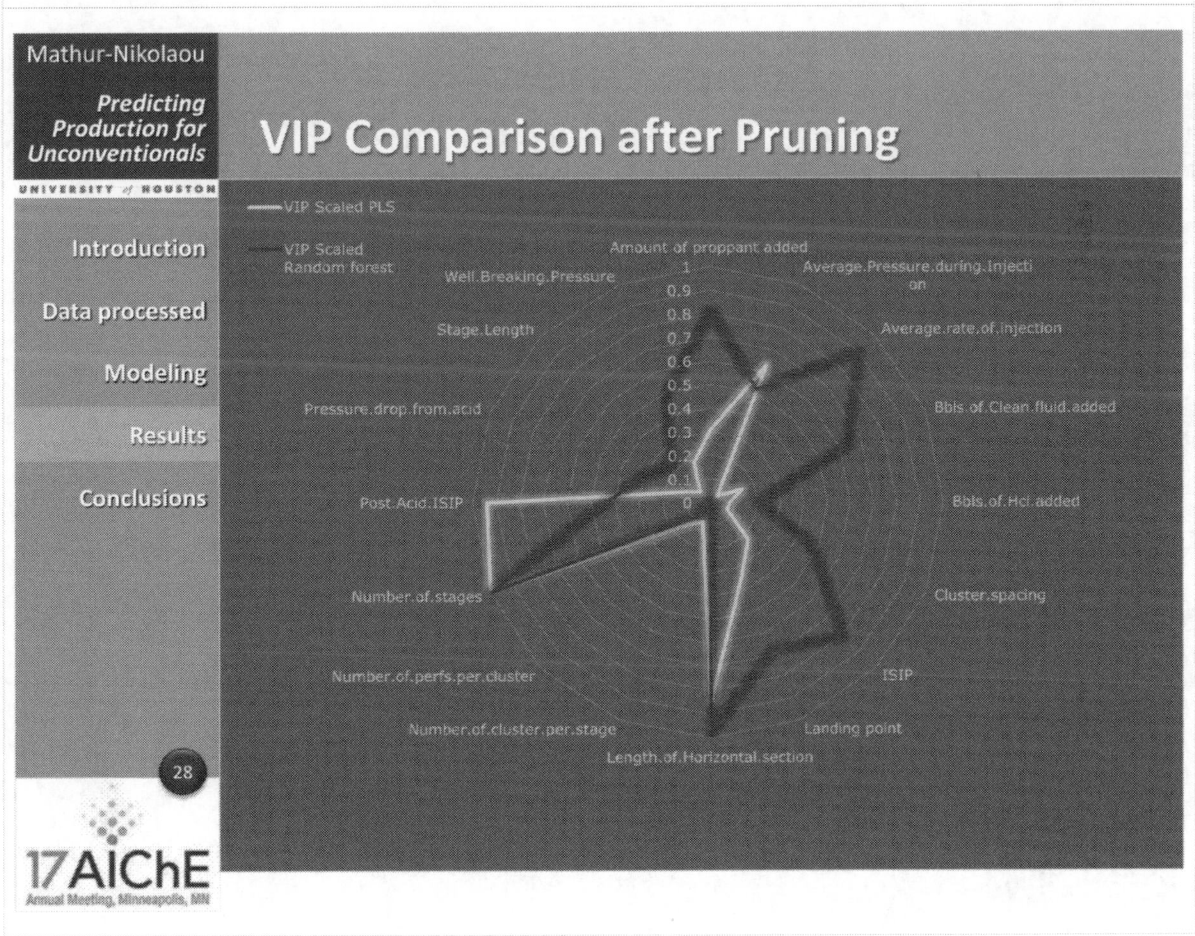
Mathur-Nikolaou
Predicting Production for Unconventionals
UNIVERSITY of HOUSTON
VIP Comparison after Pruning
Introduction
Data processed
Modeling
Results
Conclusions
28
17 AIChE
Annual Meeting, Minneapolis, MN
VIP Scaled PLS
VIP Scaled Random forest
Amount of proppant added
Average.Pressure.during.Injection
Average.rate.of.injection
Bbls.of.Clean.fluid.added
Bbls.of.Hcl.added
Cluster.spacing
ISIP
Landing point
Length.of.Horizontal.section
Number.of.cluster.per.stage
Number.of.perfs.per.cluster
Number.of.stages
Post.Acid.ISIP
Pressure.drop.from.acid
Stage.Length
Well.Breaking.Pressure
1 0.9 0.8 0.7 0.6 0.5 0.4 0.3 0.2 0.1 0

Mathur-Nikolaou

Predicting Production for Unconventionals

UNIVERSITY of HOUSTON

Introduction

Data processed

Modeling

Results

Conclusions

29

17 AIChE Annual Meeting, Minneapolis, MN

Comparison of Modeling Methods

- PLS
  - Scaled
  - Unscaled
  - Mean removed
- ANN
  - Scaled
  - Unscaled
  - PCA reduced

Mathur-Nikolaou

Predicting Production for Unconventionals

UNIVERSITY of HOUSTON

Introduction

Data processed

Modeling

Results

Conclusions

30

17 AIChE Annual Meeting, Minneapolis, MN

RMSE comparison for PLS models

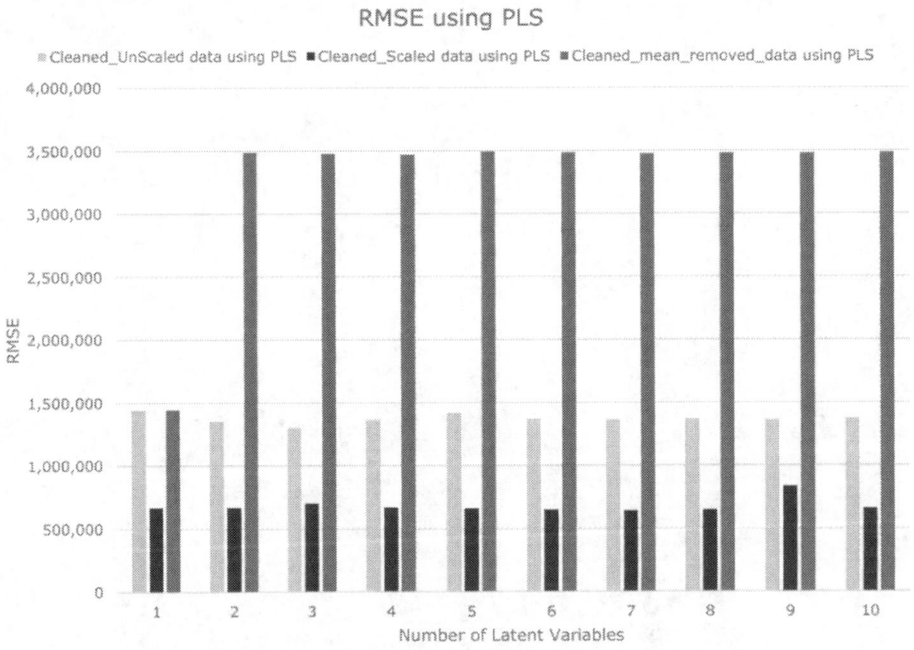

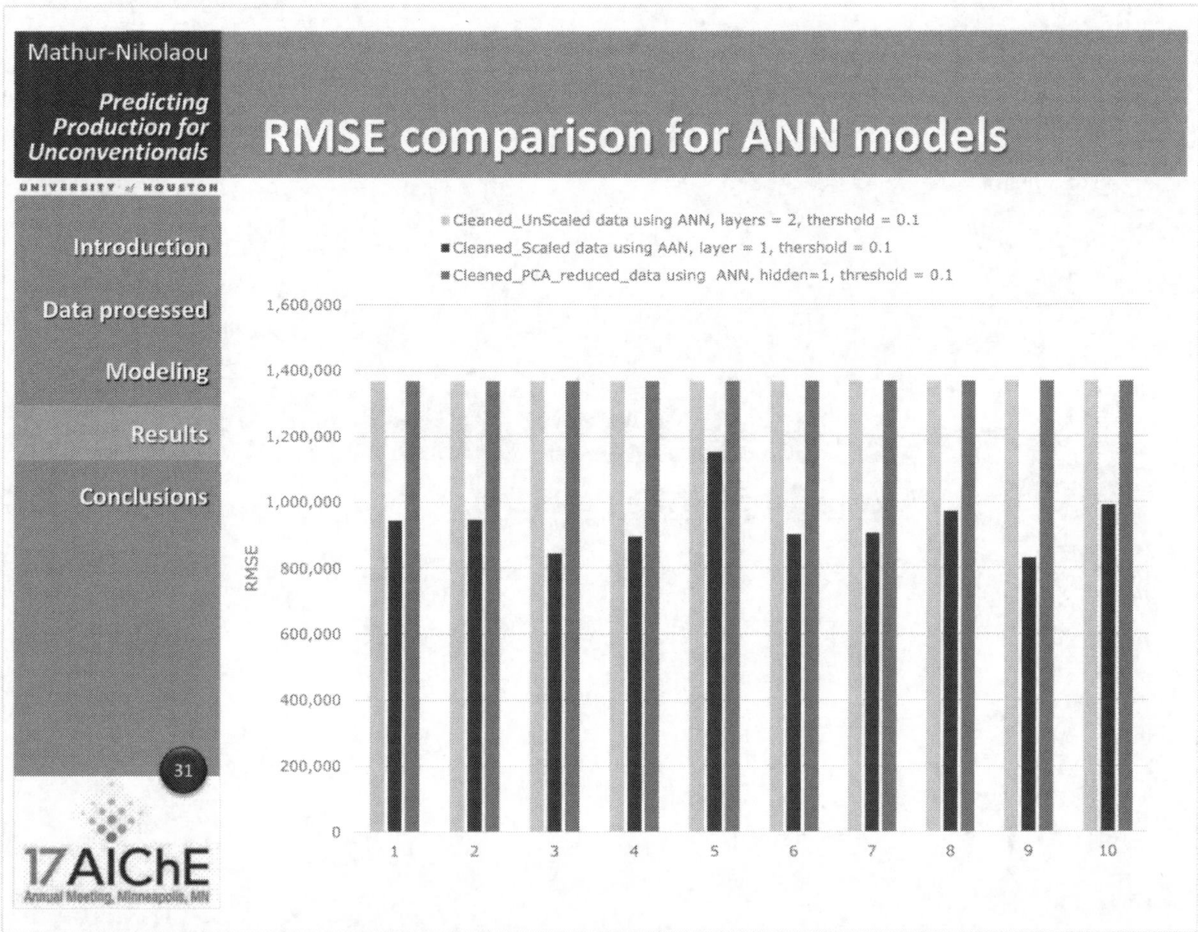
Mathur-Nikolaou
Predicting Production for Unconventionals
UNIVERSITY of HOUSTON
RMSE comparison for ANN models
Introduction
Data processed
Modeling
Results
Conclusions
31
17AIChE
Annual Meeting, Minneapolis, MN
Cleaned_UnScaled data using ANN, layers = 2, thershold = 0.1
Cleaned_Scaled data using AAN, layer = 1, thershold = 0.1
Cleaned_PCA_reduced_data using ANN, hidden=1, threshold = 0.1
RMSE
1,600,000
1,400,000
1,200,000
1,000,000
800,000
600,000
400,000
200,000
0
1
2
3
4
5
6
7
8
9
10

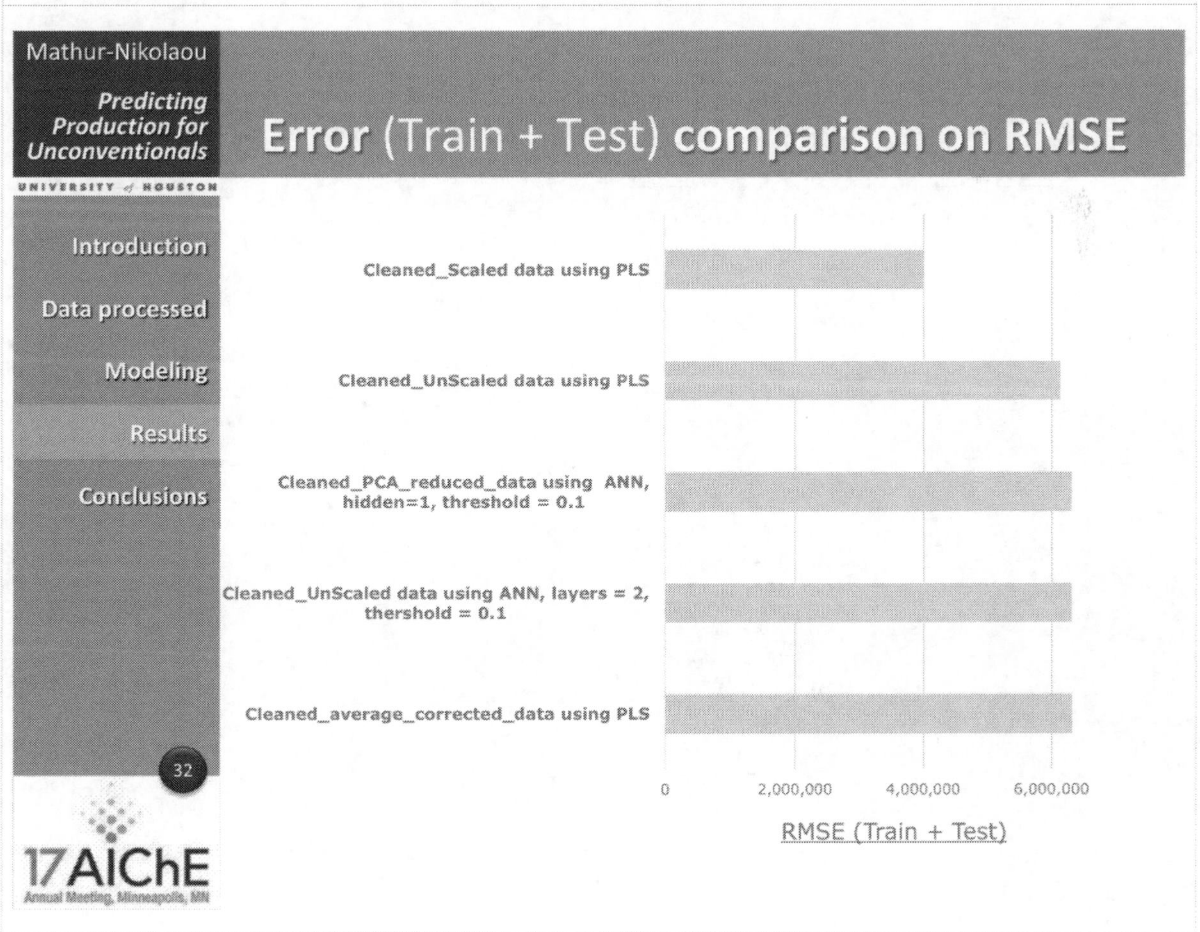
Mathur-Nikolaou
Predicting Production for Unconventionals
UNIVERSITY of HOUSTON
Error (Train + Test) comparison on RMSE
Introduction
Data processed
Modeling
Results
Conclusions
32
17AIChE
Annual Meeting, Minneapolis, MN
Cleaned_Scaled data using PLS
Cleaned_UnScaled data using PLS
Cleaned_PCA_reduced_data using ANN, hidden=1, threshold = 0.1
Cleaned_UnScaled data using ANN, layers = 2, thershold = 0.1
Cleaned_average_corrected_data using PLS
0
2,000,000
4,000,000
6,000,000
RMSE (Train + Test)

17AIChE
Annual Meeting, Minneapolis, MN

Results

Accuracy of Future Predictions

UNIVERSITY *of* HOUSTON

Mathur-Nikolaou

Predicting Production for Unconventionals

UNIVERSITY of HOUSTON

Error propagation with time on training data for uncleaned unscaled PLS

Introduction

Data processed

Modeling

Results

Conclusions

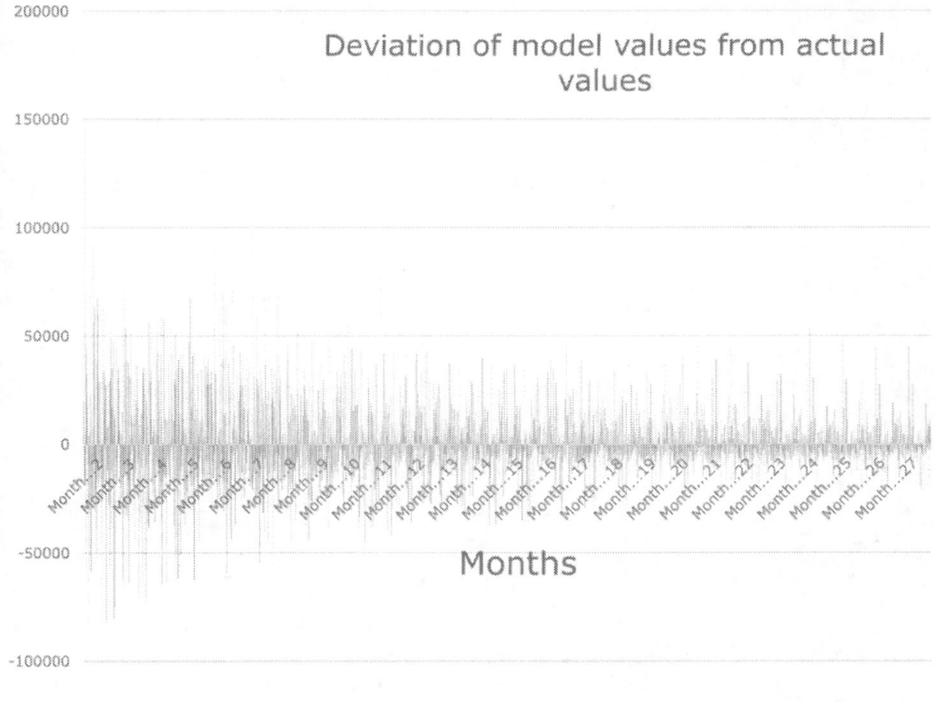

34

17AIChE
Annual Meeting, Minneapolis, MN

Mathur-Nikolaou

Predicting Production for Unconventionals

UNIVERSITY of HOUSTON

Error propagation with time on testing data for uncleaned unscaled PLS

Introduction

Data processed

Modeling

Results

Conclusions

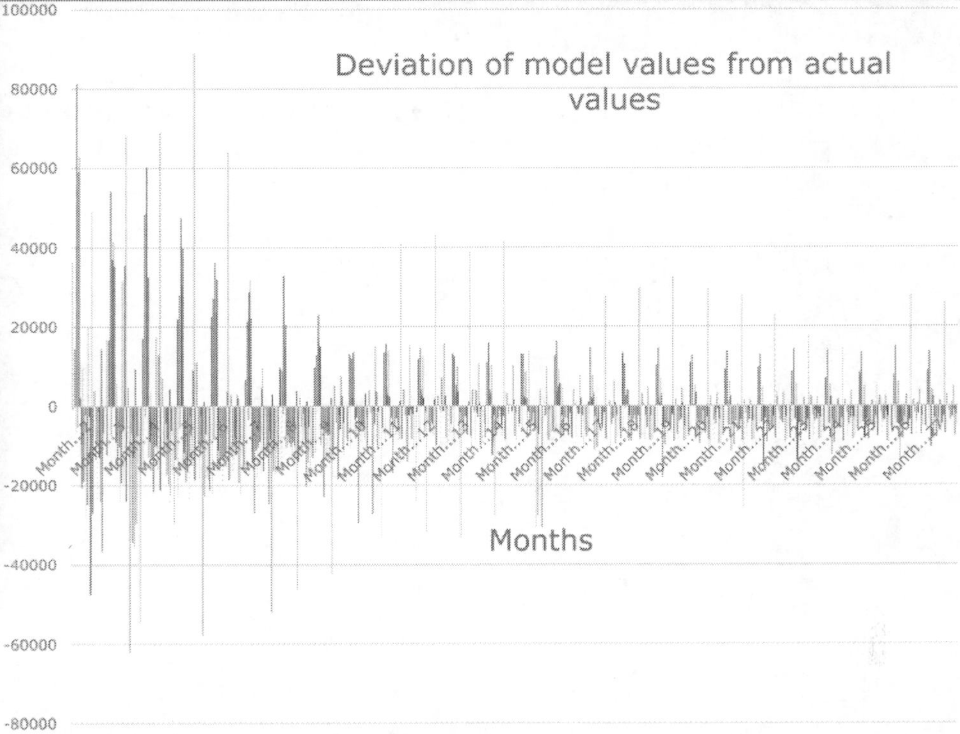

35

17AIChE

Annual Meeting, Minneapolis, MN

Mathur-Nikolaou

Predicting Production for Unconventionals

UNIVERSITY of HOUSTON

Conclusions

Introduction

Data processed

Modeling

Results

Conclusions

- Data-driven modeling of shale gas production
  - Capitalizes on wealth of design and field data collected
    - **Drilling job**
    - **Wellbore configuration**
    - **Fracture design**
    - **Historical production data**
- Data-driven PLS model
  - Predicts future production in new wells remarkably well
  - Can rank variables in order of importance

36

17AIChE

Annual Meeting, Minneapolis, MN

Mathur-Nikolaou

Predicting Production for Unconventionals

UNIVERSITY of HOUSTON

Acknowledgements

- Southwestern Energy
- Dr. Matteo Marongiu-Porcu, Schlumberger

Introduction

Data processed

Modeling

Results

Conclusions

37

17 AIChE Annual Meeting, Minneapolis, MN

Mathur-Nikolaou

Predicting Production for Unconventionals

UNIVERSITY of HOUSTON

Introduction

Data processed

Modeling

Results

Conclusions

THANKS

38

17 AIChE Annual Meeting, Minneapolis, MN

493977 Measurement and Modeling of Competitive Sorption of Methane/Ethane Mixtures on Marcellus Shale â€“ Isotherms and Kinetics

Abstract:

As the primary mechanism of gas storage in shale, sorption phenomena of CH_4 and other hydrocarbons in the micropores and mesopores are critical to estimates of gas-in-place and of the long-term productivity from a given shale play. Since C_2H_6 is another important component of shale gas, besides CH_4, knowledge of CH_4-C_2H_6 binary mixture sorption on shale is of fundamental significance and plays a central role in understanding the physical mechanisms that control fluid storage, transport, and subsequent shale-gas production.

In this work, measurements of pure component sorption isotherms for CH_4 and C_2H_6 for pressures up to 114 bar and 35 bar, respectively, have been performed using a thermogravimetric method in the temperature range (40-60 ^oC). Sorption experiments of binary (CH_4-C_2H_6) gas mixtures containing up to 10% (mole fraction) of C_2H_6, typical of shale-gas compositions, for pressures up to 125 bar under the aforementioned temperature conditions have also been conducted.

To facilitate investigation of depletion dynamics (e.g. via application of the Extended Langmuir Model), excess sorption measurements must be converted to absolute sorption. This conversion requires accurate estimates of the adsorbate density. Here we investigate the performance of several of such methods proposed for evaluating the adsorbate density. We demonstrate that the equilibrium constant and hence the adsorption and desorption kinetic constants are sensitive to the choice of the adsorbate density model and conclude by providing recommendations for selection of adsorbate density modeling. Lastly, we apply the best-fit adsorbate density method to model the sorption dynamics of pure and binary gas mixtures on to the shale.

To the best of our knowledge, this is the first time that systematic measurements of CH_4, C_2H_6 pure and binary mixture sorption on the Marcellus shale have been conducted, thus providing a comprehensive set of CH_4-C_2H_6 competitive sorption data, which can help to improve the fundamental understanding of shale-gas storage mechanisms and its subsequent production.

SENSITIVITY STUDY OF FRACTURE PROPAGATION BY FOAMED FLUIDS AND SLICKWATER IN UNCONVENTIONAL RESERVOIRS

Fatick Nath, Chongwei Xiao, Texas A&M University-Kingsville, Kingsville, TX, USA

Abstract

Low water-content frac fluids are demanding for hydraulic fracturing application in order to improve oil and gas production from unconventional reservoirs. Non-Newtonian CO_2 foams stabilized by nanoparticles has been recently studied as a promising frac fluid. The proppant placement and fracture propagation are significantly affected by frac fluid's properties. High quality CO_2 foams can provide significant environmental-friendly benefits, such as low water usage and treatment, minimal effect to the formations, high-conductivity channels, deeper proppant placement. The viscosity of CO_2 foams, which is critical for the process, changes significantly along with gas quality and pumping rate. It is important to understand fracture propagation in unconventional reservoirs by using CO_2 foams under different foam quality and pumping rate.

This study investigated fracture propagation during hydraulic fracturing process by a 3D numerical model in an unconventional reservoir using a commercial fracturing simulator. The heterogeneous reservoir properties of a horizontal well perforated in Eagle Ford Shale formation of Zavala County, Texas were applied for the study. Effects of injection rate (20 ~ 40 BPM), CO_2 foam quality (50 ~ 80 %), incremental proppant distribution (0 ~ 5 PPA) on fracture propagation were investigated in multistage fracturing process. Fluid leakoff, which is important for the fracturing process, was investigated by pressure dependent leakoff method for different CO_2 foam qualities. The results showed that the viscosity of fracture fluid significantly influences fracture propagation. Along with the foam quality increased from 50% to 80%, the fracture width increased from 0.29 inches to 0.35 inches, whereas fracture half-length decreased from 780 ft. to 660 ft. The fluid leakoff rates decreased from 13 to 9 ml/cm^2 at 10 $\sqrt{\text{min}}$. Fracture propagation using CO_2 foams with varied injection rates compared to slick water fracturing. The results showed that CO_2 fracturing develop less fracture length compare to slick water fracturing but superior fracture width and height growth than slick water ranging from 15% to 20%. The fracture propagation by 2D and 3D numerical models were compared as well.

This study provides an insight and improved fracture treatments design by non-Newtonian frac fluids - CO_2 foams application with increased fracture conductivity and efficiency compare to slick water, which is vital for fossil fuel exploitation from unconventional reservoirs with reduced environmental impacts.

Keywords: Fracture propagation, CO_2 foam, Eagle Ford Shale, Fossil fuel, Slick water.

Introduction

Exploration and production from unconventional formations have increased in recent twenty years and the production from shale formation will last to rise due to two major technologies: horizontal drilling (Fisher et al., 2004) and hydraulic fracturing with pressurized frac fluids (Britt et al., 2006). Most of the fracturing operations used slick water or oil-based fluid as a base frac fluid to transport proppant (Barati and Liang, 2014; Schein, 2005). Carbon dioxide (CO_2) foams stabilized by nanoparticles have been recently studied as a low water content frac fluid to increase fracture conductivity (Fang et al., 2014; Xiao et al., 2017). The high-quality foams with water content less than 60% are eco-friendly as they reduce the potential damage for water sensitive formations and the volume of produced water. CO_2 foams have good mobility that result in better sweeping efficiency into formations and fast clean effect. The higher viscosity of CO_2 foams leads to better proppant transport than slickwater (Fang et al., 2014; Liu et al., 2014). Frac fluids play a key role in hydraulic fracturing treatment design (Barati and Liang, 2014) by extending fractures and transporting proppant within fractures. Some viability cases of supercritical CO_2 foams were studied from foam fracturing in high temperature and high pressure formations to control mobility (Harris, 1987; McAndrew et al., 2014; Alotaibi et al., 2016).

The mechanics of fracturing using CO_2 foams as a frac fluid needs more investigation, but precise prediction methods are still lacking. Typical 2D models are KGD (Geertsma and De Klerk 1969), PKN (Nordgren 1972), and radial model as shown in Figure 1. Both KGD and PKN models, deformation is assumed by increasing plane strain with fixed fracture height to the net pay and calculate width of the fracture by analytical solutions. The PKN models assumes the plain strain in vertical direction, which leads to an elliptical shape in cross section; whereas KGD model assumes the plain strain in horizontal direction, which leads to a rectangle shape in cross section. The radial model assumes propagation on a symmetrical plane, which leads to a circular shape in cross section for radial development. These models are found to be useful, but accuracy and simplification is the problem in most of the cases. KGD model is more appropriate for propagation of early time injection when height is larger than the length of the fracture ($x_f << h_f$), whereas PKN model for the late time of injection when length exceeds fracture height ($x_f >> h_f$). Now these classic 2D models are substituted by pseudo-3D models, but still aid as a validation tools for numerical models and providing insight of the fracturing process.

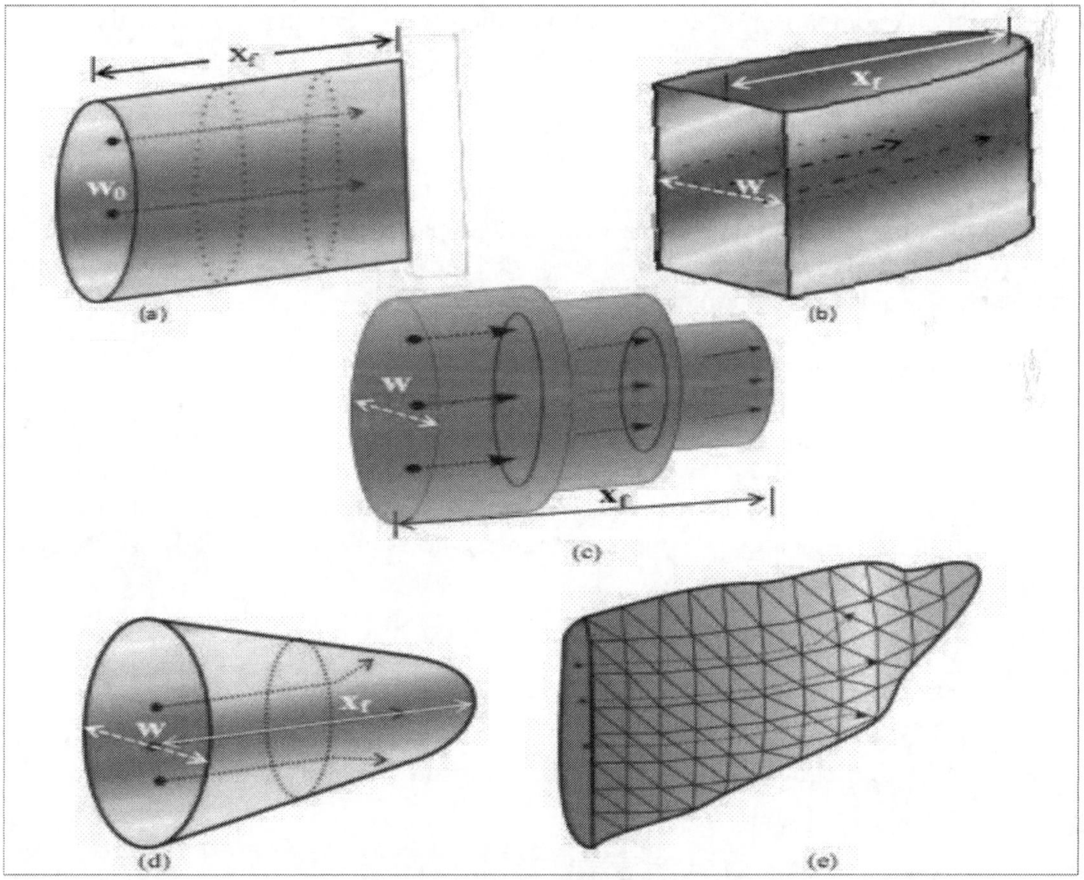

Figure 1. Fracture geometry

(a) PKN Model (b) KGD Model (c) P-3D Model (d) 3D-parametrised (e) Full 3D, gridded

The analytical models of hydraulic fracturing have greatly advanced from simple 2D to 3D models. Typical 2D models are focusing on the fracture width and length assuming constant height while the P-3D, full 3D and planar 3D models predict the height along with width and length variation. Based on the flexibility and advantages, different numerical models are used to simulate hydraulic fracturing process. The pseudo-three-dimensional (P-3D) and true 3D models were eventually established for height of fracture growth and to describe the insight of hydraulic fracturing process with enhanced accuracy. P-3D models are inexact but proficient to characterize the behavior of height growth of the fracture. A P-3D model demonstrates the height enlargements in a symmetric 3-layer formation investigating the material properties, in-situ stresses and pay zone pressure gradients and the fault of the formation (Simonson, Abou-Sayed et al. 1987). The P-3D model by Settari (Settari 1980) described a fluid flow in lateral direction and crack initiation for fracture and coupled with the development of fracture vertically. The typical methodology for true 3D models is based on 3D deformation of rock layers and 2D fluid flow within the fractures (Abou-Sayed, Sinha et al. 1984).

As CO_2 foams is emerging as a new frac fluid for unconventional resources development, it is important to study its effect on the hydraulic fracturing process. This research developed a 3D numerical model with mechanical heterogeneity in Eagle Ford shale formation for the fracturing process using CO_2 foams as the frac fluid. The combination of rock mechanics and two-phase fluid flow both in reservoir and inside the fractures will provide a guidance for future fracture treatment design. The sensitivity studies of fracture propagations affected by CO_2 foam quality, injection rate, and proppant concentration were analyzed accordingly. A comparison of fracture propagation from 3D model to 2D analytical models is performed to increase the understanding of the mechanism of hydraulic fracturing.

Fracture Propagation Model

The sensitivity studies is performed using finite element based numerical model and the model is based on a corner point gridding system on Cartesian plane and all parameters are defined at each corner point of grid blocks (GOHFER Manual). Each grid block is defined by all state variables - fluid composition, viscosity, pressure, rock mechanics, proppant concentration, etc.

The formation mechanical properties were derived from four different types of logs, including Gamma ray, Resistivity, Density Porosity, and Density Neutron, were acquired for the reservoir properties. Log-derived input data were analyzed by a built-in log analysis package and accurate lithologic description was generated, including lithology, in-situ stress and poroelastic properties of rock, etc. The in-situ stress profile was calculated from the pore pressure and poroelastic constant. The mechanical properties of the formation, including Young's modulus, Poisson's ratio, and Biot's constant, were determined by empirical correlations.
The complete stress equation is given by the closure pressure,

$$P_c = \sigma_h = \frac{\nu}{(1-\nu)}[\sigma_v - \alpha_v P_p] + \alpha_h P_p + \varepsilon_x E + \sigma_t \quad (1)$$

The Carreau rheology model (Elise, 1995) is used to describe the time-dependent fluid behavior in a different geometry including sphere motion, pipe flow, and plates.

$$\eta(\dot{\gamma}) = \eta_\infty + (\eta_o - \eta_\infty)\left[1 + (\dot{\gamma}\lambda)^2\right]^{\frac{n-1}{2}} \quad (2)$$

Where, $\eta(\dot{\gamma})$, Viscosity as a function of shear rate; η_o, zero shear viscosity, cp; η_∞: infinite shear viscosity, cp and λ : characteristic time in sec.

The fracture propagation was simulated using the fully 3D model based on finite difference scheme (Barree, 1983). Each grid node can be assigned an individual value of net stress, pore pressure,

permeability, porosity, wall-building coefficient, rock strength, Poisson's ratio and Young's modulus and variables describing fracture-wall roughness and tortuosity. The fracture width equation used is the general formula given by Boussinesq (Sneddon, 1946) in terms of displacement for a semi-infinite pace acting by the distributed load.
In a dynamic flowing system, the fracture fluid pressure is associated with the apparent fluid viscosity, flow rate and fracture width at each point along the fracture –

$$Q = \frac{h}{12\mu_a}\frac{d}{dl}(P + \gamma h)w^3 \quad (3)$$

Where, h is the length of a grid block normal to the direction of flow, γ is the fracture gradient is the fracture width. For flow between two nodes harmonic average of the cube is used.

PKN model is applied when $x_f \gg h_f$. The maximum fracture width for non-Newtonian fluid is calculated by:

$$w_{max} = 12\left[\left(\frac{128}{3\pi}\right)(n'+1)\left(\frac{2n'+1}{n'}\right)^{n'}\left(\frac{0.9775}{144}\right)\left(\frac{5.61}{60}\right)^{n'}\right]^{1/(2n'+2)} \cdot \left(\frac{q_i^{n'} K' x_f h_f^{1-n'}}{E}\right)^{1/(2n'+2)} \quad (4)$$

$$\bar{w} = 0.625 w_{max} \quad (5)$$

Where, x_f : fracture half-length, ft
w_f: fracture width, in
h_f : fracture height, ft

KGD model is applicable where $h_f \gg x_f$. The average fracture width, for Newtonian fluid is:

$$\bar{w} = 0.34\left[\frac{q_i \mu x_f{}^2}{E' h_f}\right]^{1/4}\left(\frac{\pi}{4}\right) \quad (6)$$

For radial fracture geometry, the maximum fracture width is calculated by

$$w_{max} = 2\left[\frac{q_i \mu (1-v) R}{G}\right]^{1/4} \quad (7)$$

, Where, R is the fracture radius.

Study Area: Eagle Ford Shale Formation

The hydrocarbon-rich Eagle Ford formation is located in South Texas. The productive zones vary from subsea 4,500 to 18,000 ft. (Shelley et al., 2012). with the thickness among 50 to 300 ft. The formation is thermogenically kerogen-rich and poroelastic properties are in attractive range including medium-ranged Young moduli, low Poisson's ratios, and least stress anisotropy. To represent formation heterogeneity, the rock mechanical properties were derived from well logs including Gamma Ray, Resistivity, Neutron Porosity, and Density Porosity logs, characterized by Young's modulus ($3\times10^6 \sim 6\times 10^6$ psi), Biot constant (0.6 ~ 0.8), and Poisson's ratio (0.2 ~ 0.4). The horizontal oil well (API: 42-507-32744) in Zavala County, TX was completed in Eagle Ford Shale formation in 15 March, 2010. The well was perforated in the interval of measured depth, 5567-10771 ft. The reservoir temperature is 150°F and formation pressure is 2300 psia with reservoir pressure gradient of 0.41 psi/ft.

Simulation Model Input

The simulation inputs of different properties were given in the Table 1 and fluid modeling procedure with determined parameters are given Table 2.

Table 1. Simulation Model Input

Rock Properties	
Young's modulus E, MMpsi	3.7-4.2
Poisson's ratio v,	0.27-0.32
Matrix Density, lb/ft3	165
Fracture Toughness, psi.(in) $^{0.5}$	1000
In-situ Stress data	
Vertical Stress, σ_v ,psi	~5200
Max. horizontal Stress, σ_H psi	~4200
Min. horizontal Stress, σ_H psi	~3200

Reservoir data	
Formation depth D, ft.	5300
Formation thickness h, ft.	300
Porosity ϕ,	0.08~0.1
Formation compressibility, C_t psi^{-1}	1E-6
Permeability k, md	~0.001
Water Saturation, Sw	0.32
Pore Pressure,psi	2286
Formation Tem., F	150
Fluid gradient, psi/ft.	0.41

Treatment Schedule

Each Stage	Pump Rate, BPM	Fluid Name	Stage Fluid Volume, Gal	Proppant Concentration, PPA per gal	Proppant Type + Mesh
Pad	20~40	CO_2 Foams	1000	0.00	-
Slurry	20~40	CO_2 Foams	1000	0.50	100 mesh sand
Slurry	20~40	CO_2 Foams	1000	1.00	100 mesh sand
Slurry	20~40	CO_2 Foams	1000	2.00	100 mesh sand
Slurry	20~40	CO_2 Foams	1000	3.00	100 mesh sand
Slurry	20~40	CO_2 Foams	1000	4.00	100 mesh sand
Flash	20~40	Slick H_2O	125	0.00	100 mesh sand

Fracturing fluid Properties	
Fluid Name	CO_2 Foam
Foam Quality	0.5-0.8
Frac fluid density $\rho_L, lb/ft^3$	~65
Temperature, F	104
Power law exponent n',	0.5~0.7
Consistency index $K', lb_f sec^{n'}/ft^2$	0.0023-0.0286
Spurt loss coefficient	0
Proppant Data	
Proppant	100 mesh sand
Proppant porosity ϕ_p,	0.35
Proppant density $\rho_p, lb/ft^3$	165
EOJ slurry concentration c_t, ppg	0~5

Table 2: Frac-fluid modeling using experimental data and parameters determination

Shear Rate, 1/s	Viscosity, cp			
	Foam quality 50%	Foam quality 60%	Foam quality 70%	Foam quality 80%
5950	1.45	2.02	2.78	3.49
8925	1.21	1.80	2.47	3.04
11900	1.06	1.65	2.28	2.76
14875	0.96	1.55	2.14	2.55
17850	0.88	1.46	2.03	2.40

Parameter	flowrate, ml/min	Foam Quality			
		50%	60%	70%	80%
n'	6	0.5506	0.7063	0.7156	0.6574
k', Pa.s^n	9	0.072	0.026	0.033	0.069
$\eta_{0, cp}$	12	1.80	2.50	3.35	4.25
$\eta_{\infty, cp}$	15	0.95	1.5	2.1	2.5
λ, s	18	0.0011	0.0012	0.001	0.001

Sensitivity Analysis of Fracture Propagation

During the process of CO_2 foams injection and flow in the unconventional reservoir, the effects of shear rates, elastic response of porous media, and leak-off on the fracture propagation were investigated. The effects of the fracture geometry, and non-Newtonian fluid flow were numerically analyzed.

Increasing injection flow rate augments the fracture length and width of the fracture. The viscosity of CO_2 foam is a dominant factor for fracture propagation. As the viscosity increases with varying foam quality (50%-80%), slurry fluid proppant-carrying capacity increases and average fracture width increases but the fracture extent reduces. Figure 1-4 showed the effects of CO_2 foam quality and injection rate on reservoir pressure and fracture propagation.

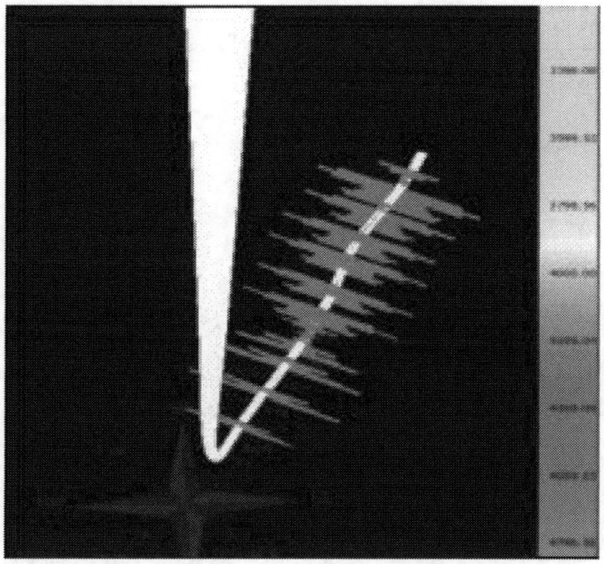

Figure 1. The profile of reservoir pressure during the process

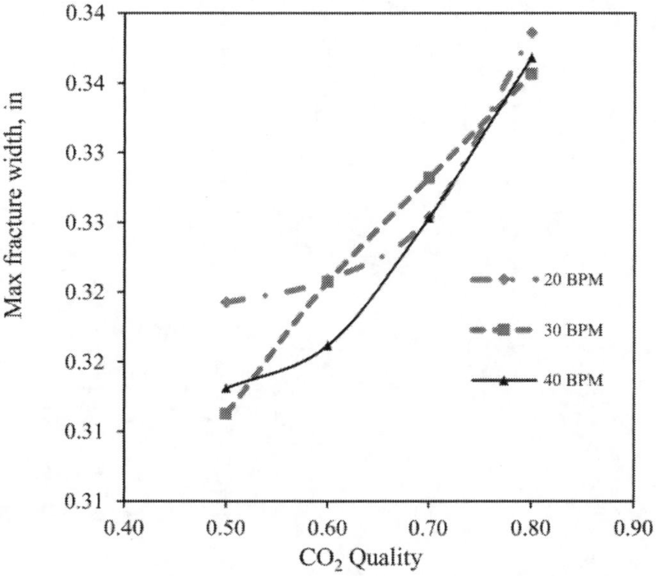

Figure 2. Effects of CO_2 foam quality and pumping rate on fracture width

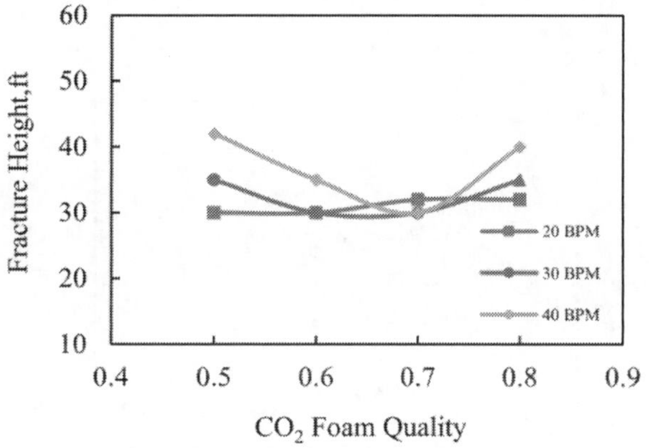

Figure 3. Effects of CO_2 foam quality and injection rate on fracture height

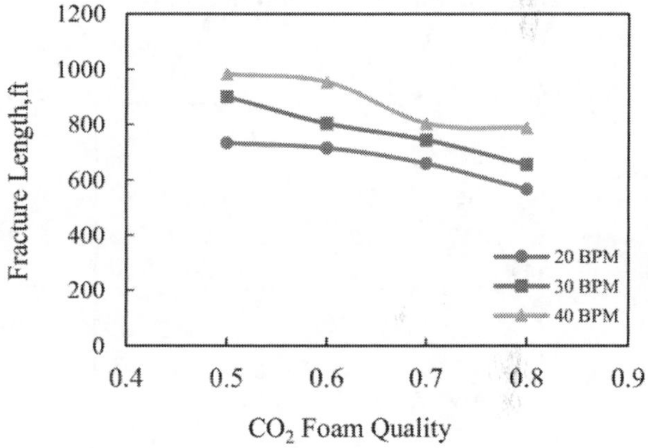

Figure 4. Effects of CO_2 foam quality and injection rate on fracture length

Fluid Leakoff

Fluid leakoff associated for different CO_2 foams quality (50% - 80%) was investigated. Pressure dependent leakoff method is used y to quantify the fluid loss during fracturing. The fluid leakoff rates along with resident time under different foam quality was shown Figure 5. The results showed that the fluid leakoff decreased with increasing CO_2 foams quality that leakoff rate reached the maximal 13 ml/cm^2 at 50% CO_2

foams quality and the minimal 9 ml/cm^2 at 80% of foam quality.

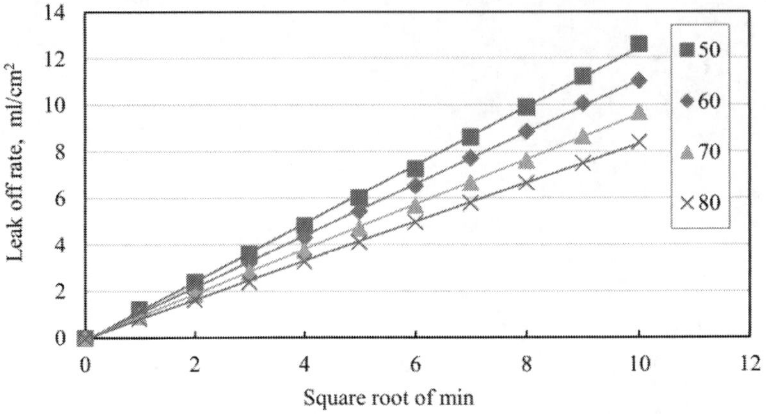

Figure 5. The fluid leakoff under different CO_2 foams quality

Comparison of fracture propagation with slick water

Fracture propagation using 60% quality CO_2 foams was compared (in 1-10 stages) to slick water fracturing and observed maximum 15% greater fracture width and height growth than slick water. The comparison of fracture width and height of CO_2 foams. Overall fracture propagation result is given in Table 4.

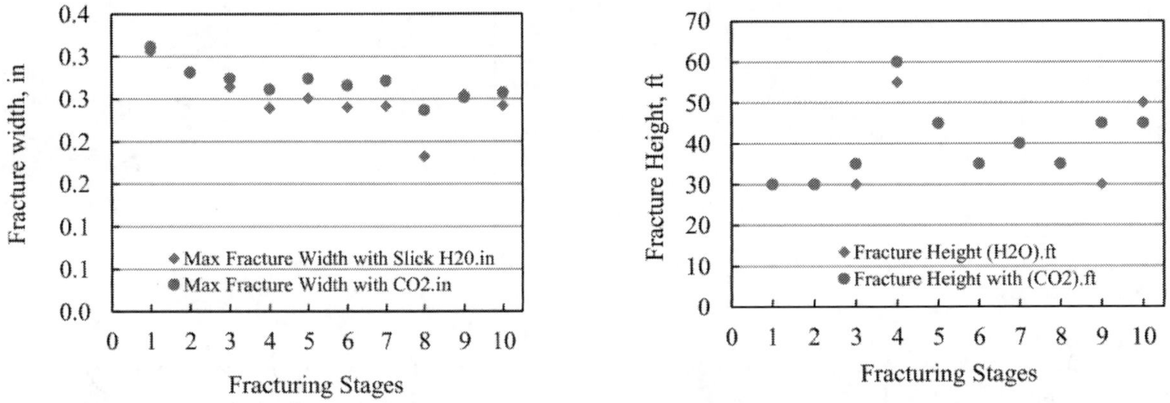

Figure 6. The comparison of fracture width and height fractured by CO_2 foams and slick water

Table 4. The comprehensive results of different treatment

Parameters	Slick H_2O	50% CO_2	60% CO_2	70% CO_2	80% CO_2
Half-length, ft.	852	743	612	560	524
Frac. height, ft.	35	30	35	35	40
Max. width, in	0.284	0.314	0.322	0.332	0.356
Avg.width, in	0.15	0.14	0.155	0.167	0.178

Comparison of fracture propagation among 3D and 2D numerical models

The fracture propagation was also studied with analytical 2D models. Fracture width from 3D model is compared for three different injection rates with 2D analytical model-PKN, KGD and radial model. The comparison shows a good agreement with fracture width variation. 3D model gives the highest width compared to the other models (Figure 7).

Fracture propagation by width is increased with increasing injection rate of CO_2 foams. For a given injection rate, 3D model considers the fundamental theories of linear elastic fracture mechanics with complex fluid flow behaviors inside the fractures whereas 2D models are not. In 3D model, fracture growth is simultaneously apply mixed-mode fracture propagation criterion which resulted the vertically and laterally fracture with highest extend. 2D models are good for quick checking the usual growth which gives a rough idea about the fracture propagation. 3D model is widely used in industry for commercial purposes as it incorporates complex behaviors of formation representing rock mechanical properties, fluid rheology, and complex stress distribution around the bore hole.

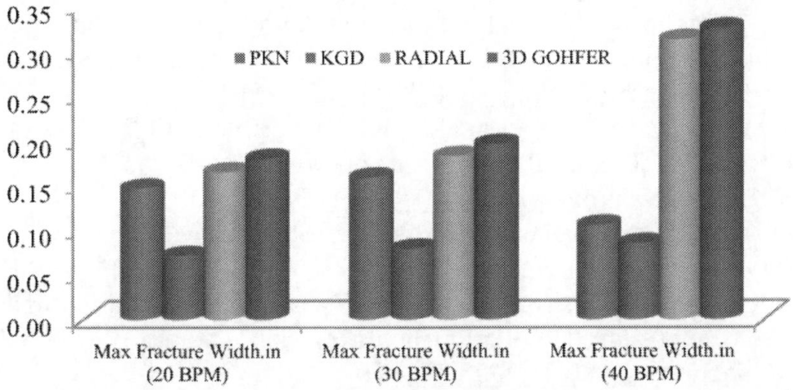

Figure 7. 3D model is compared with 2D analytical models

Conclusions

This study numerically simulated the process of hydraulic fracture propagation with the frac fluid - CO_2 foams in the Eagle Ford Shale formation with a 3D heterogeneous reservoir model.

The results showed that fracture propagation is obviously affected by the properties of frac fluid and injection rate. Along with increasing fluid viscosity and injection rate, both fracture length and width of the fracture augments increase; the proppant-carrying capacity of the slurry increases as well and proppant were placed deeper in fractures. However, the fracture extent reduces along with increasing fluid viscosity. Along with CO_2 foam quality increased from 50% to 80%, the fluid leakoff rates decreased from 13 to 9 ml/cm^2 and fluid efficiency increased from 85% to 96%. The fracture width and height growth using CO_2 foams is 15% to 20% greater than those of slick water fracturing. Finally, the comparison with 2D analytical model (PKN, KGD and radial) to 3D model showed a good agreement in fracture propagation.

Acknowledgement

The authors would like to acknowledge Houston Advanced Research Center (HARC) and the Department of Chemical and Natural Gas Engineering at Texas A&M University, Kingsville for their intellectual and financial support to conduct this research study. Authors also thankful to Barree & Associates, LLC to support this study by providing the simulator GOHFER.

References

1. Al-Dhamen, M., and Soriano, E., 2015. Increased Well Productivity from the Use of Carbon Dioxide to Foam Fracturing Fluids During a Refracturing Treatment in Saudi Arabia. Presented at the SPE Latin American and Caribbean Petroleum Engineering Conference, Quito, Ecuador, 18-20 November. SPE-177112-MS. http://dx.doi:10.2118/177112-MS.
2. Abou-Sayed, A. S., Sinha, K. P., and Clifton, R. J.,1984. Evaluation of the Influence of In-Situ Reservoir Conditions on the Geometry of Hydraulic Fractures Using a 3-D Simulator: Part 1-Technical Approach. Presented at the SPE Unconventional Gas Recovery Symposium, Pittsburgh, Pennsylvania, 13-15 May, . SPE-12877-MS. http://dx.doi.org/10.2118/12877-MS
3. Alotaibi, F., Zhou, X., Habah, A. et al 2016. EOR by CO_2 Emulsions Better Sweep Efficiency and Improved Recovery. Presented at the Abu Dhabi International Petroleum Exhibition and Conference, Abu Dhabi, UAE, 7-10 November.SPE-183014-MS. http://dx.doi:10.2118/183014-MS
4. Al-Qasim, A., AlOtaibi, F., Kokal, S. et al 2017. CO_2-Foam Rheology Behavior under Reservoir Conditions.Presented at the SPE KSA Annual Technical Symposium and Exhibition held in Dammam, Saudi Arabia, 24-27 April. SPE-188059-MS. http://dx.doi:10.2118/188059-MS
5. Barati, R. and Liang, J.T., 2014, A review of fracturing fluid systems used for hydraulic fracturing of oil and gas wells. J. Appl. Polym. Sci., 131, 40735, doi: 10.1002/app.40735
6. Barree, R. D., 1983. A Practical Numerical Simulator for Three-Dimensional Fracture Propagation in Heterogeneous Media. Presented at the Reservoir Simulation Symposium, San Francisco, CA, 15-18 November.SPE-12273-MS. http://dx.doi.org/10.2118/12273-MS.
7. Britt, L. K., Smith, M. B., Haddad, Z. A., et al., 2006. Waterfracs: We Do Need Proppant After All. Presented at the SPE Annual Technical Conference and Exhibition, San Antonio, TX, U.S.A. 24-27 Septermber. SPE-102227-MS. http://dx.doi.org/10.2118/102227-MS.
8. Economides, M., Oligney,R., and Valko, P., 2002. Unified fracture design: bridging the gap between theory and practice Orsa Press, TX,USA
9. Elise, A., 1995. An Improved Viscosity Equation to Characterize Shear-Thinning Fluids. Presented at the SPE International Symposium on Oilfield Chemistry, San Antonio, TX, U.S.A. 14-17 February. SPE-28973-MS. http://dx.doi.org/10.2118/28973-MS
10. Fang, C., Chen, W. and Amro, M., 2014. Simulation Study of Hydraulic Fracturing Using Super Critical CO_2 in Shale. Presented at the Abu Dhabi International Petroleum Exhibition and Conference, Abu Dhabi, UAE, 10-13 November. SPE-28973-MS. http://dx.doi.org/10.2118/172110-MS
11. Fisher, M. K., Heinze, J. R., Harris, C. D., et al., 2004. Optimizing Horizontal Completion Techniques in the Barnett Shale Using Microseismic Fracture Mapping. Presented at the SPE Annual Technical Conference and Exhibition, Houston, 26–29 September. SPE-90051-MS. http://dx.doi.org/10.2118/90051-MS
12. Fisher, M. K. and Warpinski, N. R., 2012. Hydraulic-Fracture-Height Growth: Real Data. SPE Production & Operations, February. SPE-145949-PA. http://dx.doi.org/10.2118/145949-PA
13. Geertsma, J. and De Klerk, F., 1969. A Rapid Method of Predicting Width and Extent of Hydraulically Induced Fractures. 10.2118/2458-PA
14. GOHFER-Manual, v8.3.0.184. GOHFER Manual.
15. Gupta, D. V. S., 2009. Unconventional Fracturing Fluids for Tight Gas Reservoirs. Presented at the SPE Hydraulic Fracturing Technology Conference, The Woodlands, Texas, USA, 19-21 January. SPE-119424-MS. http://dx.doi.org/10.2118/119424-MS
16. Gupta, A. P., Gupta, A., and Langlinais, J., 2005. Feasibility of Supercritical Carbon Dioxide as a Drilling Fluid for Deep Underbalanced Drilling Operation. Presented at the SPE Anuual Technical Conference and Exhibition,Dallas,Texas,USA, 9-12 October. SPE-96992-MS.http://dx.doi.org/10.2118/96992-MS

17. Gupta, S., Hlidek, B. T., Hill, E.S. W., and Dinsa, H.S., 2007. Fracturing Fluid for Low-Permeability Gas Reservoirs: Emulsion of Carbon Dioxide With Aqueous Methanol Base Fluid: Chemistry and Applications. Presented at the SPE Hydraulic Fracturing Technology Conference. College Station, Texas, U.S.A., 29-31 January.SPE-106304-MS. http://dx.doi.org/10.2118/106304-MS
18. Gupta, D. V. S., Leshchyshyn, T. T., and Hlidek, B. T., 2005. Surfactant Gel Foam/Emulsion: History and Field Application in Western Canadian Sedimentary Basin. Presented at the SPE Anuual Technical Conference and Exhibition,Dallas,Texas,USA, 9-12 October. SPE-97211-MS. http://dx.doi.org/10.2118/97211-MS
19. Harris, P. C., 1987. Dynamic Fluid-Loss Characteristics of CO2-Foam Fracturing Fluids. Production Engineering Journal, Issue. May, p. 89-94. SPE-13180-PA.. http://dx.doi.org/10.2118/13180-PA
20. McAndrew, J. J., Fan, R., Sharma, M., and Ribeiro, L., 2014. Extending the Application of Foam Hydraulic Fracturing Fluids. Presented at the Unconventional Resources Technology Conference,Denver,Colorado,USA,25-27 August. SPE-1926561-MS URTEC. http://dx.doi.org/10.15530/URTEC-2014-1926561
21. Geertsma, J., and De Klerk, F.,1969. A Rapid Method of Predicting Width and Extent of Hydraulically Induced Fractures. Journal of Petroleum Technology, Vol. 246 (01),Pg 1571-81. December. SPE-2458-PA. http://dx.doi.org/10.2118/2458-PA
22. Schein, G., 2005. The Application and Technology of Slickwater Fracturing. SPE-108807-DL SPE General. http://dx.doi.org/10.2118/108807-DL
23. Settari, A.,1980. Simulation of Hydraulic Fracturing Processes. Society of Petroleum Engineers Journal,Vol. 20,Issue 06,Dec 1980. SPE-7693-PA. https://doi.org/10.2118/7693-PA
24. Shelley, R. F., Saugier, L. D., Al-Tailji, W., Guliyev, N. and Shah, K., 2012. Understanding Hydraulic Fracture Stimulated Horizontal Eagle Ford Completions,Presented at the SPE/EAGE European Unconventional Resources Conference and Exhibition,Vienna,Austria, 20-22 March. SPE-152333-MS. http://dx.doi.org/10.2118/152533-MS
25. Simonson, E. R., Abou-Sayed, A. S., and Clifton, R. J.,1978. Containment of Massive Hydraulic Fractures. Society of Petroleum Engineers Journal,Vol. 18,Issue 01,Feb 1978. SPE-6089-PA. http://dx.doi.org/10.2118/6089-PA
26. Sneddon, I. N., 1946. The opening of Griffith Crack Under Internal Pressure. Quart. Appl. Math **4**: 262-267.
27. U.S. Energy Information Administration, A. E. O. E. R. (2015). U.S. Energy Information Administration, Annual Energy Outlook 2015 Early Release. http://www.eia.gov/todayinenergy/detail.cfm?id=21072 Date of accessed: 2/25/2016
28. Warnock, W. E., Jr., Harris, P. C., et al., 1985. Successful Field Applications of CO_2-Foam Fracturing Fluids in the Arkansas-Louisiana-Texas Region. Journal of Petroleum Technology, 37 (01), January. SPE-11932-PA. http://dx.doi.org/10.2118/11932-PA
29. Warpinski, N. R., Lorenz, J. C., Branagan, P. T., et al.,1991. Examination of a Cored Hydraulic Fracture in a Deep Gas Well (includes associated papers 26302 and 26946). SPE Production & Facilities, 8 (03),August. http://dx.doi.org/10.2118/22876-PA
30. Xiang, J., 2011. A PKN Hydraulic Fracture Model Study and Formation Permeability Determination, MS Thesis,Petroleum Engineering, Texaa A&M University, Dec 2011.
31. Xiao, C., Balasubramanian, S. N. and Clapp, L. W., 2017. Rheology of Viscous CO2 Foams Stabilized by Nanoparticles under High Pressure, Industrial & Engineering Chemistry Research 2017 56 (29), 8340-8348 DOI: 10.1021/acs.iecr.7b01404.

SCREENING IMPROVED RECOVERY METHODS IN TIGHT-OIL FORMATIONS BY INJECTING AND PRODUCING THROUGH FRACTURES

Harpreet Singh[1]

[1]National Energy Technology Laboratory, Morgantown, WV, USA

Abstract

Tight-oil reservoirs exhibit two characteristic behaviors that limit potential for improved recovery: i) limited fluid movement from the unfractured matrix limits the potential to sweep additional oil towards the production wells with injection of a displacing fluid, ii) the wettability of these reservoirs tend to be oil-wet which holds oil in relatively smaller pores, as compared to gas or water, making it difficult to mobilize that trapped oil. Primary oil production from these formations can decline to half of initial rates in the first year due to low permeability in the unfractured reservoir matrix, and a large amount of unrecovered oil remains in smaller pores that is not able to move out. Declining oil rates and low ultimate recoveries from oil-rich shale formations could benefit from application of secondary fluid injection schemes.

To address these challenges, this study investigates the potential to improve oil recovery from tight-oil reservoirs by a method that includes injecting a fluid into the fractured reservoir to produce hydrocarbons from adjacent fractures intersecting the same wellbore. The injection and production through fractures has a potential economic advantage over huff-n-puff scheme such that there is no lag-period between injection and production. We compare the estimates of incremental oil recovery obtained by injecting water, immiscible CO_2, and surfactant over the oil recovered by primary depletion (with no injection). The results suggest that injection of fluids may not always improve recovery from tight-oil reservoirs, especially in a heterogeneous reservoir (having different rock types) whose average matrix permeability is lower than 0.01 mD. In cases where recovery is improved over primary depletion, water flood and surfactant flood perform almost equally well with no noticeable difference between the two. Although, surfactant results in favorable fluid properties to mobilize oil, the reason surfactant did not perform well in tight-oil formation studied here is because the unswept oil is inaccessible due to small permeability, and not because of high capillary pressure.

Keywords: tight-oil; shale; EOR; water; surfactant; immiscible CO_2;

1. Introduction

Oil and gas production from unconventional tight-oil resources has become an important source of energy security in the United States (US) and it is emerging to be as important in other countries, especially Russia, Argentina, and Canada will see significant tight-oil production by 2040 (Aloulou, 2016). In the U.S., the contribution of oil produced from unconventional shale resources has jumped from 0.5 million barrels per day (BPD) in 2005 to ~5 million BPD in 2015, i.e. ~52% of the total US crude oil production (US EIA, 2016). This increase in production from unconventional shale resources has been mainly due to advancements in drilling efficiency and well configuration (Rystad Energy, 2014). However, similar improvements have not been achieved for well performance; it has been observed (Rystad Energy, 2014) that year after year wells in mature plays (e.g. Eagle Ford and Bakken) have shown similar initial rates and decline trends. Production from tight-oil resources has two drawbacks with a potential for improvement: i) low ultimate recovery factor (Gaswirth and Marra, 2015; Gaswirth et al., 2013), and ii) decline in daily production rate by more than 50% within the first year of production from an average well in the Bakken (Lund, 2014; Wachtmeister et al., 2017) and Eagle Ford (Delaihdem, 2013; Gong, 2013) plays. For instance, estimated oil resources in the Bakken formation is estimated to be 300 billion barrels (Flannery, 2006), out of which 7.38 billion barrels is estimated to be technically recoverable using current best-available technology (Gaswirth and Marra, 2015; Gaswirth et al., 2013), i.e. just ~2% of the original oil in place (OOIP). In addition to low recovery factor, the production rate of a new well from tight-oil reservoirs have been reported to decline by more than half within a year (Delaihdem, 2013; Gong, 2013; Lund, 2014; Wachtmeister et al., 2017), primarily because of fast depletion from fractures and slow fluid movement from low matrix permeability.

To arrest the declining production rates and recover additional oil from tight-oil resources, traditional methods of improved oil recovery (IOR) can be considered (e.g. water/gas/surfactant flooding etc.). However, unlike IOR in conventional reservoirs with relatively high permeability, unconventional tight-oil resources present two major challenges to IOR: i) very limited fluid movement outside the fractures limits the possibility of an efficient sweep between the wells, ii) the wettability of these reservoirs tend to be oil-wet that keeps the oil in the smallest pores compared to gas or water, making it difficult to drive oil out.

We investigate the potential of IOR from tight-oil reservoirs by a method that includes numerical reservoir simulation-based evaluation of the performance of fluid injection through a set of fractures to stimulate additional hydrocarbon recovery from a different set of fractures intersecting the same wellbore (Dombrowski et al., 2015) as shown in Figure 1. This injection-production scheme is less expensive compared to IOR using several injection and production wells in tight-oil reservoir (e.g. Ghaderi et al., 2017; Hustak et al., 2017), and unlike huff-n-puff, does not require shutting-in the well.

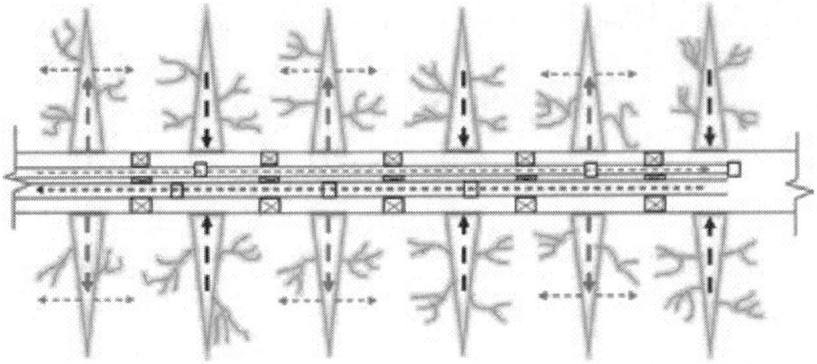

Figure 1: Schematic for fluid injection and production through hydraulic fractures in tight-oil reservoir. Dashed red and black colored arrows represent the direction of the injected and produced fluids, respectively.

To investigate the potential of improving recovery in tight-oil reservoirs, we study the injection of three different fluids (water, immiscible CO_2, and surfactant) in a single wellbore that is intersected by three injection-fractures and three recovery-fractures. This representation used for simulation of what is very likely a network of engineered fractures and pre-existing natural fractures. We compare the estimates of oil recovery obtained by injecting water, immiscible CO_2, and surfactant with the oil recovered by primary depletion (from all six fractures) with no injection. The impact of fluid injection on oil recovery is assessed for three different average permeabilities (very low, low, and relatively high), three different injection rates (low, high, and very high), and when injection is constrained by constant bottom-hole pressure in a homogeneous and heterogeneous medium with rock characteristics representative of a typical tight-oil reservoir.

2. Method

Below, we discuss the development of the reservoir model and the fluid model used in simulations. The reservoir model is composed of discrete hydraulic fractures that are connected with the matrix grid by a nearest neighborhood connection approach (Lie, 2016). The flow simulations are performed using Matlab Reservoir Simulation Toolbox (MRST), a full-physics reservoir simulator (Lie, 2016), with fully implicit solutions of blackoil model.

2.1. Reservoir Model

2.1.1. Injection and Production in Tight-Oil Reservoirs

A schematic of a single horizontal well (henceforth referred to as *lateral*) intersecting six hydraulic fractures in a tight-oil reservoir is depicted in Figure 2. The conventional approach of recovering hydrocarbons from tight reservoirs would lead to low injection rates and very limited fluid movement due to extremely low permeability, therefore this drawback can be addressed by either having extremely long fractures between injection and production wells (which is not technically possible at present) or drilling additional infill wells that are closely spaced. The former option is difficult to achieve with current state of fracturing technology where fracture half-length is on the order of a few hundred feet (Zuloaga-Molero et al., 2016), while the option of drilling closely spaced additional wells may prove uneconomical for reservoirs with low rates of production. Dombrowski et al. (2015) proposed an economical approach that addresses these concerns by a method that includes injecting a fluid through a set of fractures and producing hydrocarbons from a different set of fractures within the same wellbore. As the gravity forces have relatively small impact on light oil in tight reservoirs compared to viscous forces, the hydrocarbons can be recovered from both above and below the fractures.

In this study, numerical simulation of a single horizontal wellbore intersecting six hydraulic fractures is used to investigate the injection of fluids as a measure to improve recovery after production rate drops significantly. Initially, all six fractures are used for production by primary depletion (without any injection) for ~1800 days when the production rate is significantly lower (< 50 %) than its initial value.

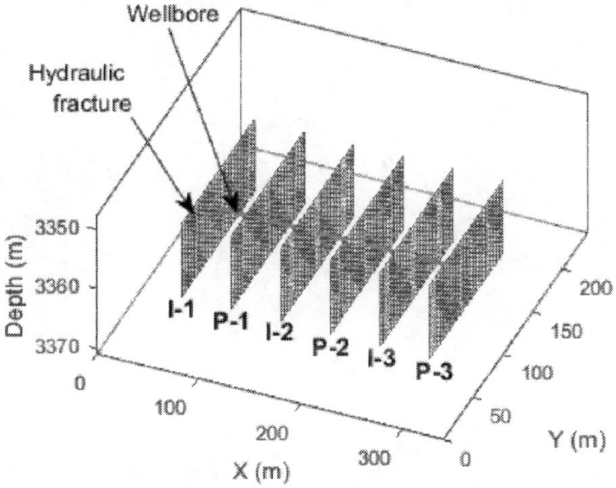

Figure 2: Computational domain for fractured tight-oil reservoir with six hydraulic fractures intersecting the lateral. After 1800 days (~ 5 years) of production by primary depletion, three existing hydraulic fractures are used to inject fluid (I-1, I-2, I-3) and the remaining three fractures are used for production (P-1, P-2, P-3).

Although complex fracture geometries (including interaction of hydraulic and natural fractures) are common in tight-oil reservoirs, such geometries are not considered here in order to control the number of variables and to avoid the complexity beyond necessary.

2.1.2. Reservoir Properties

The reservoir and fracture properties are taken from other studies (Sanchez-Rivera et al., 2015; Zhu et al., 2017; Zuloaga-Molero et al., 2016) on Bakken tight-oil. The parameters for reservoir including well, and hydraulic fractures (shown in Figure 2) are given in Table I. The hydraulic fractures in Figure 2 are modeled using embedded discrete fracture and are linked to the matrix grid through nearest neighborhood connection approach (Lie, 2016).

Table I: Parameters for (a) reservoir including well, and (b) hydraulic fractures.

(a)

Reservoir parameters	Value	Unit
Initial oil saturation (S_{oi})	80	%
Initial reservoir pressure	8000	psi
Production pressure (BHP)	1800	psi
Reservoir temperature	240	F
Reservoir depth	11000	ft
Formation compressibility	5e-06	psi^{-1}
Time when injection begins	1800	$days$
Time when injection ends	20	$years$
Mean permeability (matrix)	0.1, 0.01, 0.005	mD
Mean porosity (matrix)	8	%

(b)

Hydraulic Fracture Parameters	Value	Unit
Aperture	5e-03	ft
Permeability	10	D
Porosity	50	%
Spacing	160	ft
Half-length	200	ft
Height	40	ft

2.2. FLUID MODEL

2.2.1. Fluid Properties

In-situ fluid properties

The PVT analysis conducted on in-situ fluids from different tight-oil reservoirs (Ghaderi et al., 2017; Hawthorne et al., 2017a; Li and Luo, 2017; Mavor, 2013) reveal the crude oil from these reservoirs to be light, with low density and viscosity. The density, viscosity, compressibility, and formation volume factor of light oil and brine in a typical tight-oil formation like Bakken at initial reservoir conditions (Zhu et al., 2017; Zuloaga-Molero et al., 2016) are shown in Table II. Only brine and oil are considered as the in-situ fluid present at initial conditions in the reservoir. The bubble point pressure of Bakken oil is ~1480 psi (Ghaderi et al., 2017), therefore, it is safe to assume that no free gas exists at initial reservoir condition (8000 psi, 240 ^oF), with all the gas dissolved in oil. Another reason to assume no initial free gas in the reservoir is due to limited capability of surfactant flooding module in MRST that does not provide an option for the gas phase; therefore, having no initial gas in the reservoir would allow for consistent comparison from all four types of recovery processes, i.e. primary depletion, water flooding, gas flooding, and surfactant flooding.

Table II: Properties of formation (a) light oil, and (b) brine, with fluid properties reported at in situ conditions.

(a)

Oil Properties	Value	Unit
Density ($\rho_{o,i}$)	40.47	lb/ft^3
Viscosity ($\mu_{o,i}$)	0.142	cp
Compressibility (c_o)	6.65e-05	bar^{-1}
Formation volume factor	$0.94e^{c_o(p-p_{ini})}$	$-$

(b)

Brine Properties	Value	Unit
Density ($\rho_{w,i}$)	1000	kg/m^3
Viscosity ($\mu_{w,i}$)	0.61	cp
Compressibility (c_w)	4.28e-05	bar^{-1}
Formation volume factor	$0.99e^{c_w(p-p_{ini})}$	$-$

Injection fluid properties

Three different fluids are injected to study their impact on improving recovery from tight-oil formations, which are water, immiscible CO_2, and surfactant. The properties of CO_2 (Andersen, 2017) at initial reservoir conditions (8000 psi, 240 F) are given in Table III. We assume CO_2 is immiscible with reservoir oil, i.e. both CO_2 and oil exist as separate phases.

Table III: Properties of CO_2.

Gas Properties	Value	Unit
Density ($\rho_{g,i}$)	803.5	kg/m^3
Viscosity ($\mu_{g,i}$)	0.075	cp
Compressibility (c_g)	6.25e-04	bar^{-1}
Formation volume factor	$e^{c_g(p-p_{ini})}$	$-$

Although surfactant flooding has been studied extensively in conventional reservoir rocks (Sheng, 2015) to improve recovery, only few studies (He et al., 2014; Mirchi et al., 2015; Shuler et al., 2016; Wang et al., 2011, 2012, 2016; Xu et al., 2011) exist about their application in tight reservoir rocks like shale, especially on real reservoir rock. Surfactant's ability to reduce the interfacial tension between oil and water (σ_{ow}) results in changing the relative permeability and decreasing the residual oil saturation. The decrease in interfacial tension (IFT) by adding surfactant results in increasing the capillary number (N_c), which is a measure of viscous force relative to the capillary force, i.e.:

$$N_c = \frac{|k\nabla p|}{\sigma_{ow}} \tag{1}$$

Therefore, the consequential effect of an increased capillary number is the reduction in residual saturation of oil. The reduction in residual saturation is a function of miscibility condition, which is defined in terms of miscibility factor ($m \in [0, 1]$) that is calculated by interpolating the logarithmic value of the capillary number (Jørgensen, 2013).

$$m(N_c) = \frac{\log_{10}(N_c) - \log_{10}(N_c^{misc})}{\log_{10}(N_c^{immisc}) - \log_{10}(N_c^{misc})} \quad (2)$$

The surfactant adsorption is assumed to be negligible such that it does not result in any change in porosity or permeability of the rock. Relevant details on the surfactant model implementation can be found in the Appendix or in Jørgensen (2013) for specific details. The parameters used for surfactant flooding are taken from Xu et al. (2011) who conducted surfactant flooding experiments for a tight-oil formation (Yanchang) in China with various concentrations of surfactant. The IFT and viscosity as a function of surfactant concentration, and the magnitude of miscibility for different capillary numbers (function of σ_{ow}) as reported in Xu et al. (2011) are given in Table IV:

Table IV: (a) interfacial tension and viscosity as functions of surfactant concentration, and (b) miscibility factor as a function of capillary number.

(a)

$C_s\ (kg/m^3)$	$\sigma_{ow}\ (N/m)$	$\mu_w\ (cp)$
0	0.05	0.61
0.1	5e-04	-
0.5	1e-05	-
1	1e-06	-
30	1e-06	0.8
100	1e-06	1

(b)

$log_{10}(N_c)$	$m(N_c)$
-10	0
-5.5	0
-4	0.5
-3	1
2	1

2.2.2. Relative Permeabilities

Relative permeability data govern the flow movement and relative ease of flow for multi-fluids in tight-oil reservoirs. Unfortunately, there are not many experimental data for relative permeability in tight-oil reservoirs, however, there is one notable study by Honarpour et al. (2012) that published relative permeability data for tight-oil Eagle Ford samples categorized as calcite-rich, organic-rich, and fractured. Besides this study, Liu et al. (2010) published the relative permeability data for a tight-oil reservoir, Yanchang oilfield, in China. The data from these two studies suggest that end-point relative permeability values in tight-oil reservoirs are lower than generally higher values in conventional reservoir rocks (Moghadasi et al., 2015) and the range of saturation within which fluids are mobile is tighter than conventional rocks. To the authors' knowledge there are no experimental data available tight-oil reservoir in Bakken, but many simulation studies for Bakken (Sanchez-Rivera et al., 2015; Zhu et al., 2017; Zuloaga-Molero et al., 2016) have used data from Yu et al. (2014) who assumed Brooks-Corey model for relative permeability and estimated its parameters by history-matching pressure and production data from the Middle Bakken. However, the relative permeability parameters estimated by Yu et al. (2014) for the Middle Bakken show two contradictions:

i) their estimated end-point relative permeability values are highly optimistic ($k_{rw}^o = k_{row}^o = k_{rog}^o = 1; k_{rg}^o = 0.85$) when compared to experimental observations (Honarpour et al., 2012; Liu et al., 2010) from other tight-oil formations, e.g. Eagle Ford in Texas (Honarpour et al., 2012) and Yanchang oilfield in China (Liu et al., 2010; Xu et al., 2011).

ii) their estimated parameters (end-point and residual saturations) suggest that Middle Bakken is water-wet ($S_{wr} > S_{or}$; $k_{rw}^o = k_{row}^o$), however, experimental survey for Bakken (Wang et al., 2012) reveals that it is slightly oil-wet. A review on wettability of various shale formations around the world by Singh (2016) revealed that most shales tend to be oil-wet.

Therefore, we use the relative permeability data that are representative of a slightly oil-wet medium ($k_{row}^o \leq k_{rw}^o$; $S_{wr} = S_{or}$) and a tight-oil formation ($k_{row}^o, k_{rw}^o < 1$). The relative permeability data for oil-water flow are taken from an experimental surfactant flooding study (Xu et al., 2011) on tight-oil cores from Yanchang oilfield in China, while for oil-gas flow the relative permeability parameters from Yu et al. (2014) are modified slightly to account for oil-wet and tight-oil medium properties as discussed in Zhang et al. (2016). The Brooks-Corey relative permeability model for oil-water and gas-liquid flow can be defined as:

$$k_{rw}(S_w) = k_{rw}^0.(S_{wn})^{n_w} \quad (3)$$

$$k_{row}(S_w) = k_{row}^0.(1 - S_{wn})^{n_{ow}} \quad (4)$$

$$k_{rg}(S_g) = k_{rw}^0.(S_{gn})^{n_g} \quad (5)$$

$$k_{rog}(S_g) = k_{rog}^0.(1 - S_{gn})^{n_{og}} \quad (6)$$

Where, k_{ri} is the relative permeability of phase $i(= o, w, g)$, k_{ri}^o is the end-point relative permeability of phase i, n_i is the Corey curve exponent for k_{ri}, and S_{in} is the normalized saturation of phase i. The parameters in above relative permeability model for different fluids are defined in Table V. The

Table V: Relative permeability model parameters.

Water-Oil					
S_{wr}	S_{or}	k_{rw}^0	k_{row}^0	n_w	n_{ow}
0.2	0.2	0.6	0.5	3	3
Gas-Oil					
S_{gr}	S_{or}	k_{rg}^0	k_{rog}^0	n_g	n_{og}
0.0	0.2	0.85	0.7	3.15	5.75
Water-Surfactant-Oil					
S_{wr}	S_{or}	k_{rw}^0	k_{row}^0	n_w	n_{ow}
0.05	0.05	1	1	1.5	1.5

2.3. HETEROGENEITY

Based on laboratory (Dong et al., 2017; Hawthorne et al., 2017b; Tuero et al., 2017) and field (Kenomore et al., 2017; Steiner et al., 2016; Witkowsky et al., 2012) observations of shale rocks, it can be broadly divided into three rock types, i.e. inorganic (poor quality), organic (good quality), and intermediate (fair quality) that lies between the inorganic and organic. A spatially heterogeneous reservoir model based on this concept is developed using sequential indicator simulation algorithm (Deutsch and Journel, 1997) by considering marginal probabilities of 0.6, 0.2, and 0.2 for the inorganic, intermediate, and organic rock types, respectively. These marginal probabilities are chosen to reflect the volumetric percentages of organic pores found in many shales, which is about 20% by volume (Witkowsky et al., 2012) based on 1.1 g/cc kerogen grain density. The spatial correlation of each rock type is defined using three nested spherical variograms along 60 degrees to 120 degrees.

The rock properties of inorganic, organic, and intermediate type are considered based on laboratory-scale and field-scale observations of shale rock (Hawthorne et al., 2017b; Kenomore et al., 2017; Tuero et al., 2017; Witkowsky et al., 2012) that show organic rock type, which are also associated with micro-fractures (Hawthorne et al., 2017b), have larger pore sizes (Dong et al., 2017) in comparison to intermediate and inorganic rock types that occupy larger pore volume of the rock (Tuero et al., 2017) with significant saturation of hydrocarbons (Tuero et al., 2017). Therefore, these qualitative observations are used to assign rock properties (porosity, permeability, and initial oil saturation) to each rock type as shown in Figure 3 .

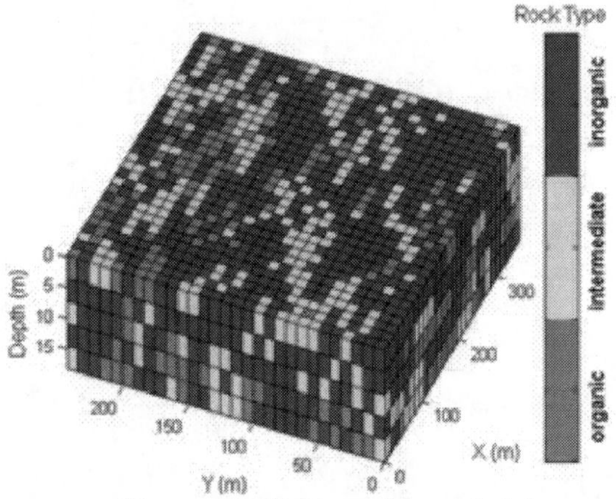

Geology		Φ (%)	k (mD): Scenario			S_{oi} (%)
			I	II	III	
Heterogeneous	organic	3	0.4	0.04	0.02	60
	intermediate	6	0.1	0.01	0.005	70
	inorganic	10	0.01	0.001	0.0005	90
	Mean	8	0.1	0.01	0.005	80
Homogeneous		8	0.1	0.01	0.001	80

Figure 3: Reservoir model is composed of three different rock types (shown by colors) for a spatially heterogeneous model, where each rock type has a distinct porosity, permeability, and initial oil saturation value. The average values of porosity, permeability, and initial oil saturation for the heterogeneous model are equal to their corresponding values in a homogeneous model. Three matrix permeability scenarios considered are very low (mean 0.005 mD), low (0.01 mD), and relatively high (0.1 mD).

2.4 Injection Rates

Three injection rates are chosen and denoted as low (~7 rb/d), high (32 rb/d), and very high (48 rb/d). The magnitude of each injection rate is denoted as low, high, or very high based on range of injection rates adopted in other studies (Sanchez-Rivera et al., 2015; Zhu et al., 2017) for tight-oil reservoirs with different permeability and fracture spacing.

3. Results

The reservoir is depleted naturally without any fluid injection for ~1800 days when a fluid (water, surfactant, or immiscible CO_2) is injected for the remaining life of the reservoir (~20 years) to arrest the declining production rates and recover additional oil.

The impact of injected fluid on oil recovery is assessed for three different average permeabilities (very low: 0.005 md, low: 0.01 mD, and relatively high: 0.1 mD) and different injection rates (~ low: 7 rb/d, high: 32 rb/d, and very high: 48 rb/d) in reservoir barrels (rb/d) for both homogeneous and heterogeneous scenarios. A similar set of results is obtained to see the impact of injecting fluids constrained by a constant bottom-hole pressure (BHP ~ 8100 psi) instead of a constant injection rate.

3.1. Oil Rates

Figure 4 shows oil rates for four types of recovery processes, i.e. by primary depletion (no injection), by water injection, by surfactant injection, and by immiscible CO_2 injection. Each plot on left column of Figure 4 is for different reservoir matrix permeability in a homogeneous medium, whereas a corresponding plot on the right side of this figure is for heterogeneous medium with an average permeability equal to its corresponding homogeneous value. The results presented in Figure 4 are for an injection rate of ~7 rb/d for all three injection fluids. Figure 5 and Figure 6 show similar results as presented in Figure 4, but for higher injection rates of ~32 rb/d and ~48 rb/d, respectively. Figure 7 shows similar set of results as presented in Figure 4, but these results are obtained when the fluid injection is constrained by a constant BHP (~8100 psi) instead of a constant injection rate.

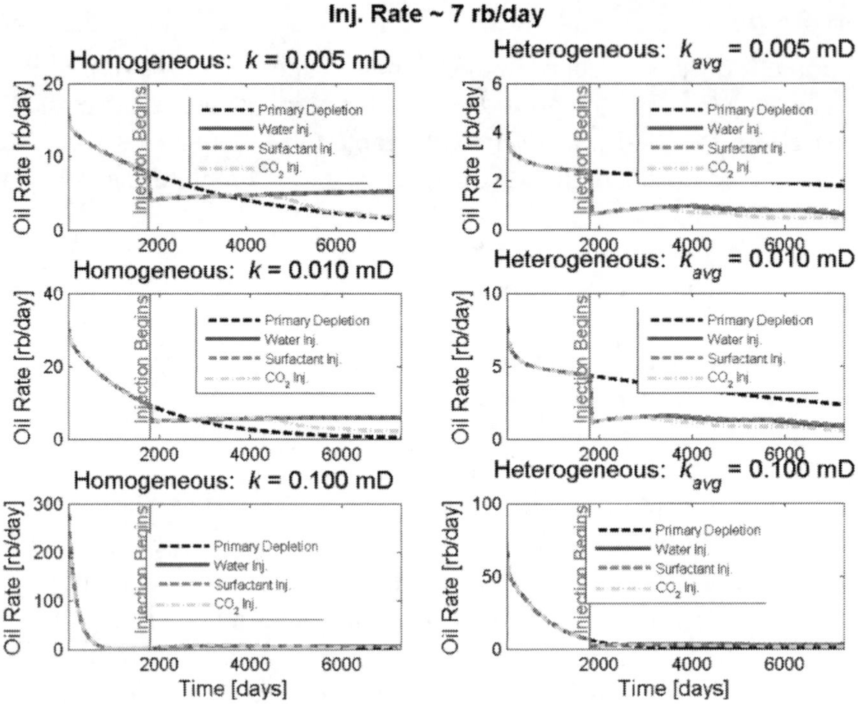

Figure 4: Oil rates from homogeneous and heterogeneous media with three improved recovery methods (injection rate ~ 7 rb/day). Primary depletion represents the base case where oil is produced without any fluid injection.

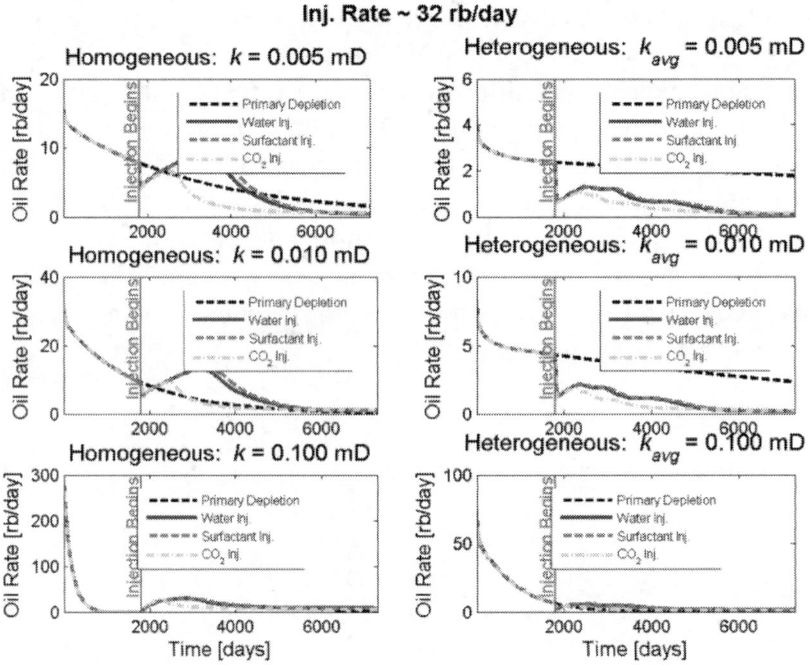

Figure 5: Oil rates from homogeneous and heterogeneous media with three improved recovery methods (injection rate ~ 32 rb/day). Primary depletion represents the base case where oil is produced without any fluid injection.

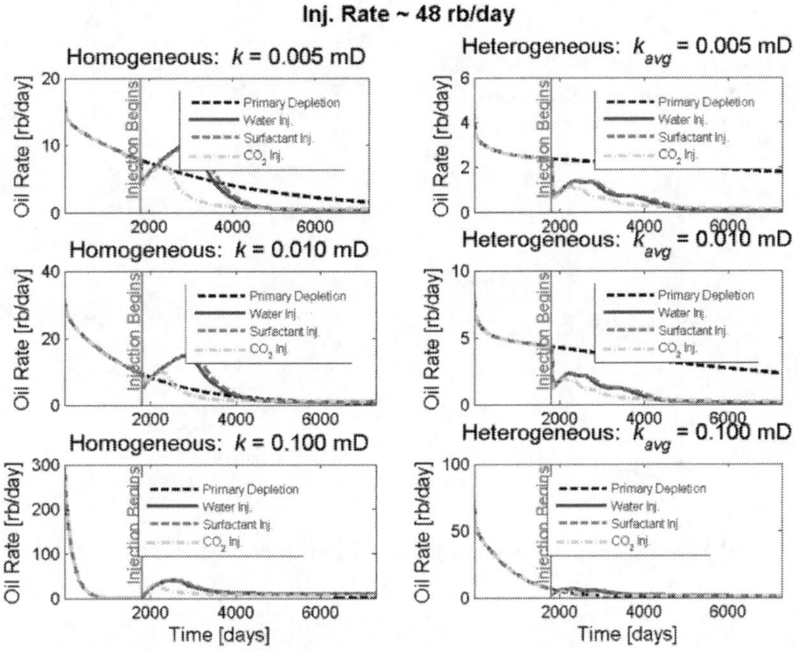

Figure 6: Oil rates from homogeneous and heterogeneous media with three improved recovery methods (injection rate ~ 48 rb/day). Primary depletion represents the base case where oil is produced without any fluid injection.

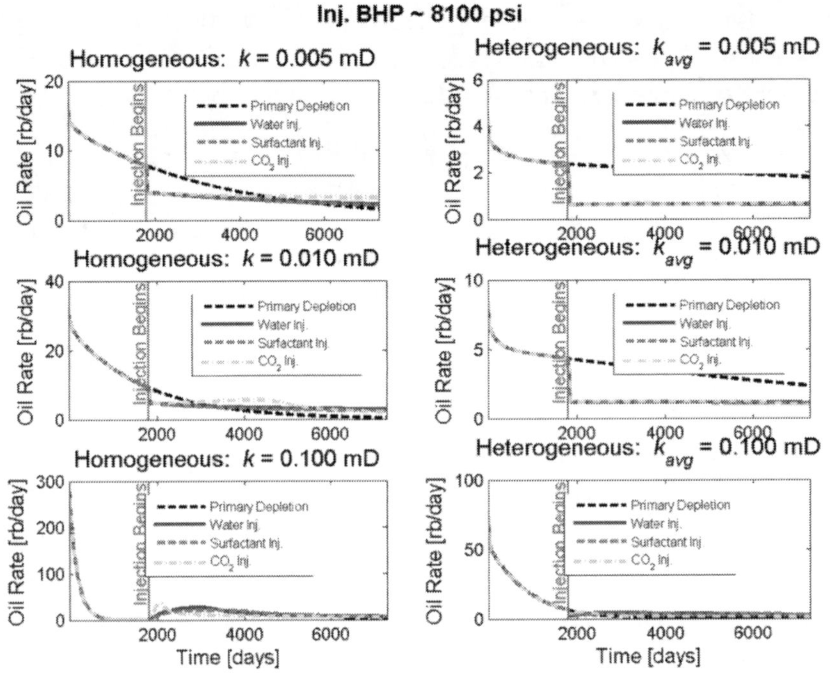

Figure 7: Oil rates from homogeneous and heterogeneous geological models with three improved recovery methods (injection BHP ~ 8100 psi). Primary depletion represents the base case where oil is produced without any fluid injection.

3.2. Oil Recovery

The oil rates plotted in Figure 4 through Figure 7 are used to find oil recovered as percent of the original oil in place (OOIP).

Figure 8, Figure 9, Figure 10, and Figure 11 show oil recovered as percent of the original oil in place (OOIP) for the corresponding oil rates plotted in Figure 4 through Figure 7, respectively. Among the recovery values from all scenarios, maximum oil is recovered (~1.2 % of OOIP; bottom left plot in Figure 10) by injecting surfactant (or water) at the largest rate of ~48 rb/d in a homogeneous medium with 0.1 mD permeability. The maximum oil recovered is ~65% more than the primary depletion (bottom left plot in Figure 14) by injecting surfactant (or water, with no noticeable difference compared to surfactant), whereas the oil recovered by immiscible CO_2 is about 25% less than that by water or surfactant, but it is 40% more than the primary depletion. This is because, as shown in Figure 6, the oil rates by water (or surfactant) get a larger bump in rates than by immiscible CO_2 initially, and the rate bump provided by water (or surfactant) is able to be sustained for a longer period than by immiscible CO_2. The reason for the higher bump provided by water (or surfactant) as opposed to immiscible CO_2 is that earlier CO_2 breakthrough (~2,200 days) is predicted in comparison to water (or surfactant) breakthrough (~2800 days). Minimum oil is recovered (~0.04 % of OOIP; top right plot in Figure 10) with the largest injection rate of ~48 rb/d in a heterogeneous medium with 0.005 mD average permeability. In this case, there is no substantial difference in oil recovered between the three injected fluids (top right plot in Figure 14), whereas the oil recovered by primary depletion (no injection) is almost 50% greater than the oil recovered by injection. The corresponding oil rates for the minimum oil recovered are shown by Figure 6, where it can be observed that the oil rates by water (or surfactant) get a larger bump than by immiscible CO_2 initially, however, the magnitude of oil rates ($< 2\ rb/d$) with all three fluids is not substantial enough to make any difference in the percent of oil recovered.

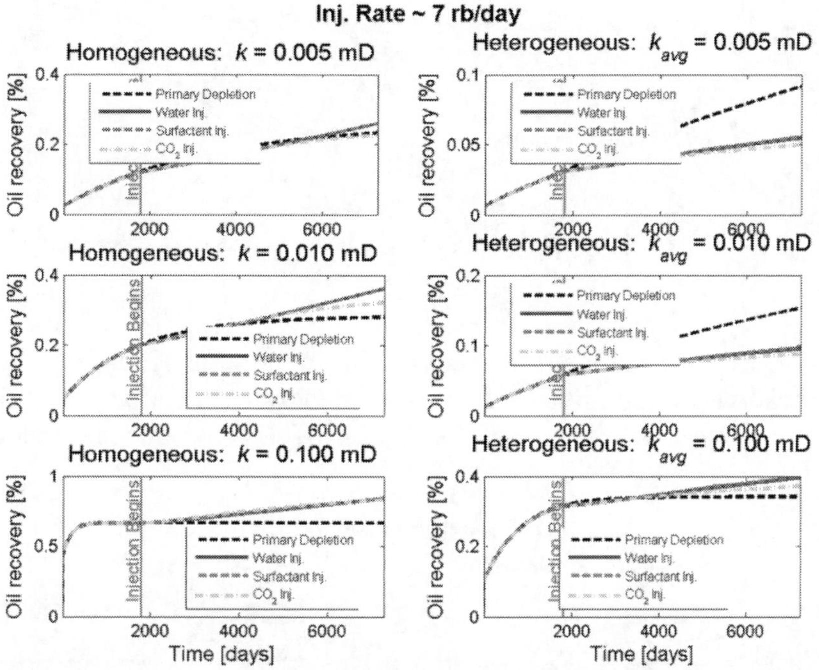

Figure 8: Percent cumulative recovery from homogeneous and heterogeneous geological models with three improved recovery methods (injection rate ~ $7\ rb/day$). Primary depletion represents the base case where oil is produced without any fluid injection.

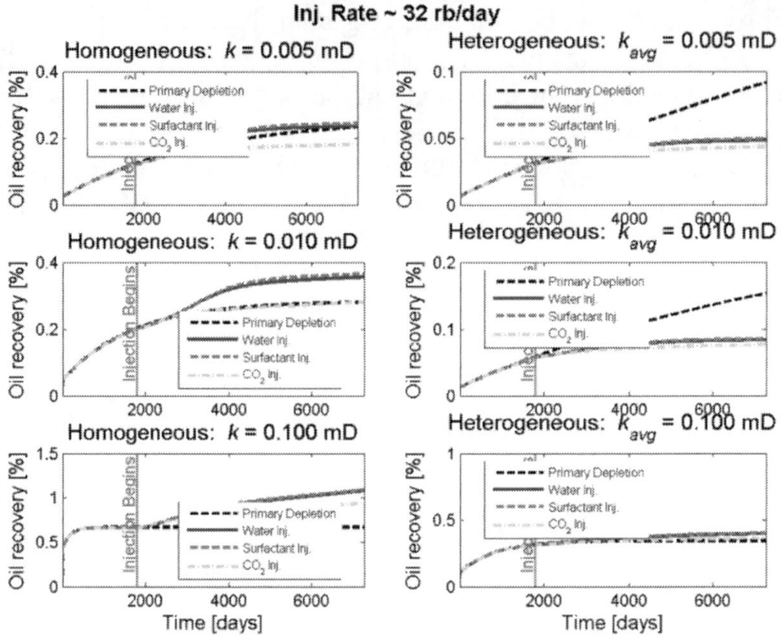

Figure 9: Percent cumulative recovery from homogeneous and heterogeneous geological models with three improved recovery methods (injection rate ~ 32 rb/day). Primary depletion represents the base case where oil is produced without any fluid injection.

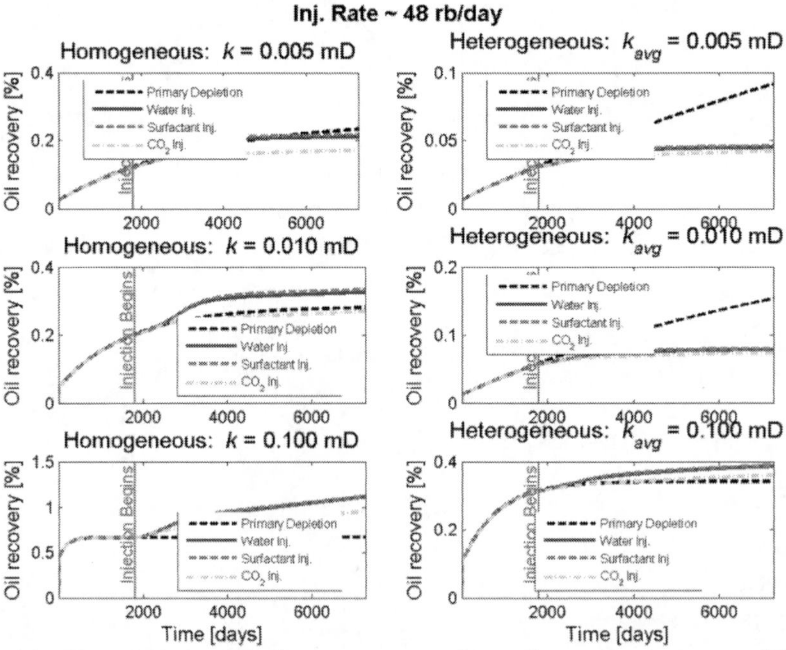

Figure 10: Percent cumulative recovery from homogeneous and heterogeneous geological models with three improved recovery methods (injection rate ~ 48 rb/day).

Primary depletion represents the base case where oil is produced without any fluid injection.

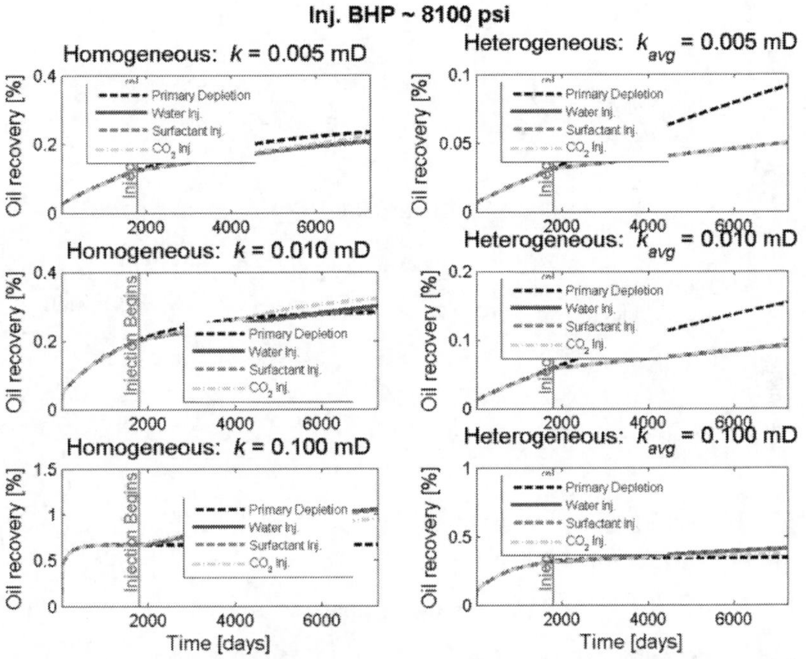

Figure 11: Percent cumulative recovery from homogeneous and heterogeneous geological models with three improved recovery methods (injection BHP ~ 8100 *psi*). Primary depletion represents the base case where oil is produced without any fluid injection.

3.3. Incremental Oil Recovery

To show the impact of injecting different fluids on recovery, incremental recovery (IR) with respect to primary depletion is calculated as shown:

$$Incremental\ recovery\ [\%] = \left(\frac{recovery_{inj} - recovery_{PD}}{recovery_{PD}}\right) \times 100 \qquad (7)$$

Where subscript inj represents the method of recovery by fluid injection (water, surfactant, CO_2), and PD is the method of recovery by primary depletion with no injection (base case). Figure 12, Figure 13, Figure 14, and Figure 15 show relative change in oil recovered as percent with respect to primary depletion (PD) for the corresponding oil recoveries plotted in Figure 8, Figure 9, Figure 10, and Figure 11, respectively.

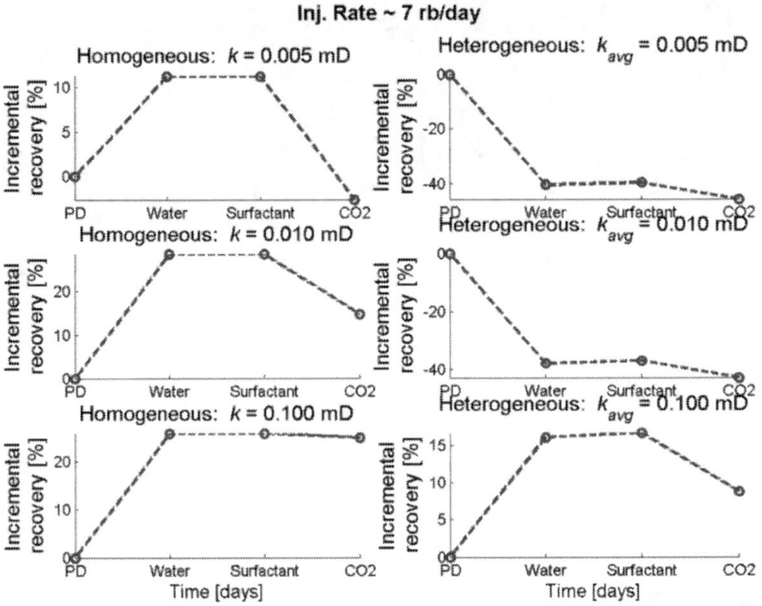

Figure 12: Incremental recovery (*IR* %) with respect to primary depletion (base case) for the recovery values plotted in Figure 8.

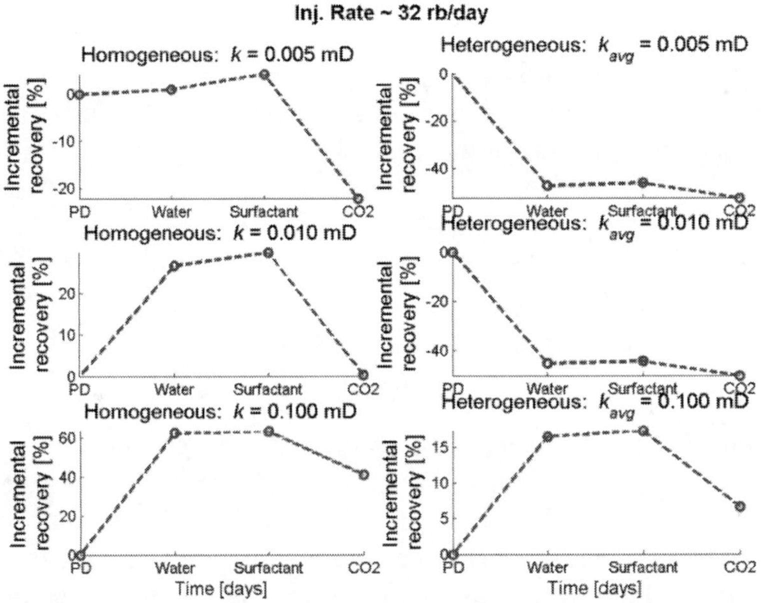

Figure 13: Incremental recovery (*IR* %) with respect to primary depletion (base case) for the recovery values plotted in Figure 9.

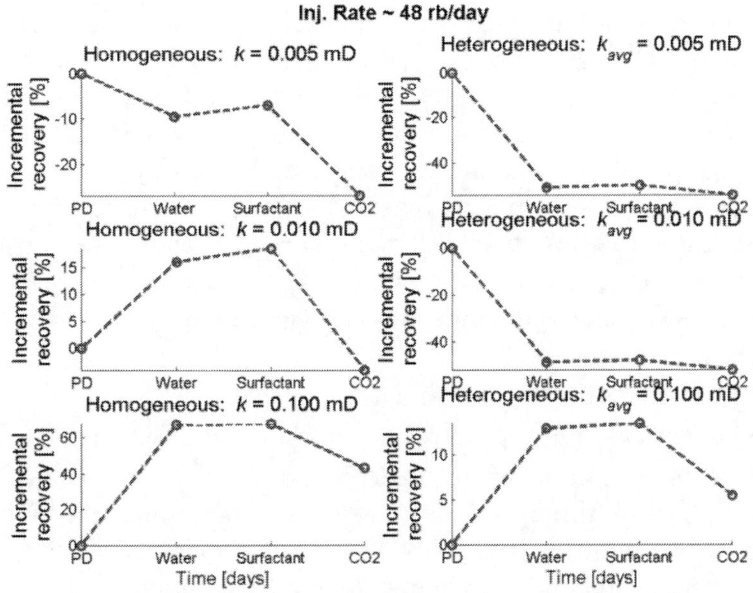

Figure 14: Incremental recovery (IR %) with respect to primary depletion (base case) for the recovery values plotted in Figure 10.

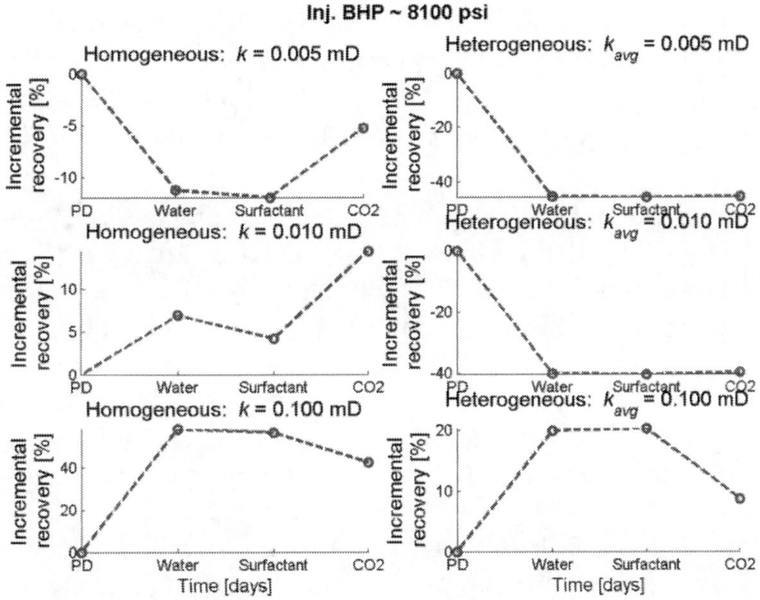

Figure 15: Incremental recovery (IR %) with respect to primary depletion (base case) for the recovery values plotted in Figure 11.

Impact of an Additional Lateral

To study the impact of number of laterals and hydraulic fractures on the incremental recovery, we replace a single lateral in the middle with two similar laterals (each intersecting six hydraulic fractures) at equal distance from the boundaries. For detailed results with two laterals, readers may refer Singh and Cai (2018). The results presented in Singh and Cai (2018) show that the percent of incremental recovery due to fluid injection is not affected by having an additional lateral in a homogeneous medium. Although the percent of IR is affected in heterogeneous medium due to relative change in spatial heterogeneity with two laterals as compared to a single lateral in the middle (Figure 2), the behavior of incremental recovery due to injection of each fluid (water, surfactant, or immiscible CO_2) is still the same; that is, the relative performance exhibited by each fluid in increasing fluid recovery with respect to the base case still shows the same behavior despite the relative change in heterogeneity.

3.4. Effect of Surfactant Concentration

The surfactant concentration used for this study is 50 kg/m^3 (or 5% by mass of water), which provides minimum surface tension and maximum miscibility factor of 1 as shown by Table IV. The minimum surface tension and maximum miscibility is obtained with surfactant concentration of 30 kg/m^3 (or 3% by mass of water). Any surfactant concentration above this value did not change the recovery values.

3.5. Effect of Capillary Pressure

Capillary pressure curve for water-oil and gas-liquid was obtained by using the laboratory-scale capillary pressure data from middle Bakken cores (Karimi and Kazemi, 2015) and extrapolating them by fitting a curve over the data. The expression obtained after extrapolating the data beyond the water saturation range of ~62-81% is shown as following:

$$p_c(psi) = \begin{cases} (-9.34 \times 10^{-42}).e^{143.3.S_w}, & \textit{water displacing oil} \\ (7.78 \times 10^3).e^{-4.91.S_o}, & \textit{gas displacing oil} \end{cases} \quad (8)$$

The capillary pressure at in-situ oil saturation of 80% as given by this curve is negligible in comparison to the initial reservoir pressure of ~8000 psi. Even after producing for 20 years, the pressure did not drop enough to be comparable to the highest possible value of capillary pressure by the above expression. Therefore, using the above curve as capillary pressure for the reservoir fluids in a model (mimicking properties of Bakken) did not affect the recovery estimates. The reason why capillary pressure did not play a role in affecting recovery is because the pore space that is inaccessible is due to its small permeability, and not because of low pressure support. The average pressure support in the reservoir even by the end of 20 years of production is much higher that can overcome capillary forces.

It should be noted that, although, there are some studies (e.g. Zhang et al., 2017) that show numerically that capillary pressure can have significant impact on oil recovery (upto 20 % higher), their methodology uses Young-Laplace equation for the estimates of capillary pressure, which usually gives impractical results for shales as discussed in a review by Singh (2016).

4. Discussions

Results suggest that fluid injection can improve recovery over the base case (with no injection), but, the recovery is not improved universally across all the studied scenarios. Fluid injection did not improve recovery (negative IR %) in 10 out of 24 cases studied, where 8 out of those 10 unsuccessful cases were from heterogeneous media with very low (0.005 mD) to low (0.01 mD) average matrix permeability. The remaining 2 unsuccessful cases are for very low (0.005 mD) matrix permeability in homogeneous media with very high injection rate (~48 rb/d), or when the production is constrained by a constant BHP, respectively. In the 14 successful or improved recovery (positive IR %) cases where injecting fluids improved the recovery, surfactant injection performed the best in 3 cases while immiscible CO_2 injection performed the best in 1 case. In the remaining 10 out of the 14 improved recovery cases, there was no clear winner among the three fluids; in most cases, surfactant and water performed almost equally, but better than immiscible CO_2.

4.1. Effect of Matrix Permeability

An increase in matrix permeability results in increasing oil recovery for all the recovery methods, including primary depletion, and for the three methods of recovery by fluid injection. This impact is consistent irrespective of injection rate or well constraint, and the relative change in oil recovery between different recovery methods increases with the increase in matrix permeability as shown in Figure 12, Figure 13, Figure 14, and Figure 15.

4.2. Effect of Injection Rate

An increase in injection rate leads to decrease in oil recovery matrix with very low (0.005 mD) to low (0.01 mD) matrix permeability, but it leads to increase in oil recovery for matrix with relatively higher permeability (0.1 mD). This can be seen in in Figure 12, Figure 13, Figure 14, and Figure 15, which show that at small injection rate (~ 7 rb/d) the recovery by injection for reservoir with very low (0.005 mD) to low (0.01 mD) matrix permeability is greater than recovery by primary depletion; but, with increase in injection rates (~ 32 rb/d and 48 rb/d) there is a relative drop in recovery for reservoir with very low (0.005 mD) to low (0.01 mD) matrix permeability. However, for reservoir with relatively high (0.1 mD) matrix permeability, the recovery increases relatively with increase in injection rate from 7 rb/d to 48 rb/d. This is possibly due to the fact that high injection rate (equivalent to relatively high convective flow and low diffusion) is useful only when the rock is sufficiently permeable that it allows the fluid flow; whereas low injection rate (equivalent to relatively low convective flow and high diffusion) is useful when the rock is extremely tight that the primary mode of flow is through diffusion.

4.3. EFFECT OF INJECTION CONSTRAINED BY CONSTANT BHP

An injector constrained by a constant BHP is relatively more effective in improving recovery in a reservoir that has a relatively high matrix permeability (0.1 mD) than in a reservoir having a very low (0.005 mD) to low (0.01 mD) matrix permeability. Injector constrained by a constant BHP is most effective for immiscible CO_2 injection, which tends to perform better with constant BHP than with constant injection rate. An injector constrained by a constant BHP performs better with the injection of immiscible CO_2 than with water or surfactant when the reservoir matrix permeability is very low (0.005 mD) to low (0.01 mD); whereas, effectiveness of immiscible CO_2 in case of reservoir with relatively high (0.1 mD) matrix permeability is not as good as water or surfactant, but it's still relatively better than its performance in the cases where injection is constrained by a constant rate.

4.4. EFFECT OF HETEROGENEITY

Injecting fluids in reservoirs with very low (0.005 mD) to low (0.01 mD) average matrix permeability does not lead to a better recovery. In fact, switching three fractures from producers to injectors for the purpose of injecting fluids in these very low (0.005 mD) to low (0.01 mD) permeability reservoirs lead to a net decrease in recovery relative to primary depletion; that is, producing through all six fractures with no injection leads to a higher recovery than by injecting fluids through half of the fractures. However, injecting fluids in reservoirs with relatively larger average matrix permeability of 0.1 mD does lead to a better recovery compared to primary depletion; lower injection rates are relatively more effective than a higher injection rate in increasing the recovery, whereas production constrained by a constant BHP is even better than low injection rates to increase recovery rate.

4.5. EFFECT OF TYPE OF FLUID

Results suggest that fluid injection improved recovery (positive IR %) in 14 out of 24 cases studied, where surfactant injection performed the best in 3 cases while immiscible CO_2 injection performed the best in 1 case. In the remaining 10 IR cases, there was no clear winner among the three fluids; in 9 cases out of 10, surfactant and water performed almost equally, but better than immiscible CO_2, while only in one case all the three injection fluids (water, surfactant, and immiscible CO_2) performed equally.

The 10 cases where injecting fluids did not improve recovery relative to primary depletion are:

1) Heterogeneous reservoirs with very low (0.005 mD) to low (0.01 mD) average matrix permeability irrespective of well constraint or injection rate [8 cases]
2) Very high injection rates (~48 rb/d) in reservoir with very low (0.005 mD) matrix permeability [1 cases], and
3) Production constrained by a constant BHP in reservoir with very low (0.005 mD) matrix permeability [1 case].

The 14 cases where injecting fluids improved the recovery relative to primary depletion are:

1) Surfactant injection performed the best in reservoir with very low (0.005 mD) matrix permeability with high injection rate (~32 rb/d) [1 case], and in low (0.01 mD) matrix permeability with high to very high injection rates (~32 rb/d and 48 rb/d) [2 cases]. The improvement in recovery by surfactant ranged between 1-3% higher than water, relative to primary depletion.
2) Immiscible CO_2 injection performed the best in reservoir with low (0.01 mD) matrix permeability when the production is constrained by a constant BHP [1 case].
3) In the remaining 10 IR cases, there was no clear winner among the three fluids; in 9 cases out of 10, surfactant and water performed almost equally, but better than immiscible CO_2, while only in one case all the three injection fluids (water, surfactant, and immiscible CO_2) performed equally.

In general, the oil rates by water (or surfactant) get a larger bump in rates initially than by immiscible CO_2 (as shown by Figure 4 through Figure 7) because of earlier breakthrough of immiscible CO_2 in comparison to water (or surfactant) by few years in most cases. This is because gas being a lower viscosity fluid than the water (or surfactant) will have a higher velocity (by Darcy's law) and that results in its earlier breakthrough.

A brief summary of above results is presented in form of matrix (Table VI and Table VII) to compare the performance of methods used for oil recovery in a homogeneous and heterogeneous medium, respectively. Scenarios marked as X is shown in red colored box to indicate that injection of fluid did not help improve recovery for that particular scenario. Also, scenarios marked with 'Water or Surf' (or 'Water or Surf or Immiscible CO_2') indicate that there was no single injection fluid that performed the best in that particular scenario, and that any one of the injection fluids mentioned in that scenario will result in a similar recovery as obtained by the rest of the fluids mentioned for that scenario.

Table VI: Recovery technique for each scenario that recovered the most oil in a homogeneous medium. Scenarios marked as X is shown in red colored box to highlight that injection of fluid did not improve recovery.

HOMOGENEOUS	*Low Injection Rate*	*High Injection Rate*	*Very High Injection Rate*	*Constrained by BHP*
Very Low Permeability	Water or Surf	Surfactant	X	X
Low Permeability	Water or Surf	Surfactant	Surfactant	Immiscible CO_2
High Permeability	Water or Surf or Immiscible CO_2	Water or Surf	Water or Surf	Water or Surf

Table VII: Recovery technique for each scenario that recovered the most oil in a heterogeneous medium. Scenarios marked as X is shown in red colored box to highlight that injection of fluid did not improve recovery.

HETEROGENEOUS	*Low Injection Rate*	*High Injection Rate*	*Very High Injection Rate*	*Constrained by BHP*
Very Low Permeability	X	X	X	X
Low Permeability	X	X	X	X
High Permeability	Water or Surf	Water or Surf	Water or Surf	Water or Surf

4.6. Relating Recovery with Mechanism of Injected Fluid

Recovery with Water and Immiscible CO_2

In general, the oil rates by water (or surfactant) get a larger bump in rates initially than by immiscible CO_2 (as shown by Figure 4 through Figure 7) because of earlier breakthrough of immiscible CO_2 in comparison to water (or surfactant) by few years in most cases. This is because gas being a lower viscosity fluid than the water (or surfactant) will have a higher velocity (by Darcy's law) and that results in its earlier breakthrough. Both water flooding and immiscible CO_2 flooding work by increasing the pressure that drives the oil out by immiscible displacement. Although, the relative permeability of gas with oil is different than the relative permeability of water with oil, it does not unilaterally dictate the oil recovery; this is because, relative permeability of oil with gas favors higher mobility of both the gas and the oil, which is good for driving the oil out faster, however, it also simultaneously results in faster breakthrough of gas due to its higher mobility. This is especially important in a tight-oil reservoir where the tight pore space dominates the reservoir, and even a slightly higher permeable pathways can lead to preferential pathway for the gas that may result in its breakthrough. In the current injection-production scheme, the gas is injected through the fractures, and it tends to breakthrough from the other fractures that are working as production zones. However, water being a heavier fluid with about ten times higher viscosity than the CO_2, is not likely to breakthrough as fast as gas because of its much lower velocity compared to CO_2 ($\propto \mu_{gas}/\mu_{water}$~0.1). Therefore, without early breakthrough, water is able to sustain higher pressure than CO_2 for a longer period of time.

Recovery with Surfactant

Unlike water and immiscible CO_2 flooding, surfactant flooding works by mixing with the oil to form a miscible solution that reduces the surface tension of the oil and helps bring down the capillary forces; therefore, reduced capillary forces push the oil out from the small pores that hold the oil due to high capillary pressure. Therefore, the success of a surfactant flooding in extracting oil depends not just on its chemical properties that can reduce the surface tension of oil (also referred as wettability alteration), but also on the amount of oil trapped by capillary forces. If the average reservoir pressure is sufficiently high, such as in the studied scenario (and also the scenario with most tight-oil formations, which are over pressured) the amount of oil that can be displaced by surfactant depends on the capability of the surfactant to first access the tight rock to mobilize additional oil for recovery. In conventional reservoirs, surfactant flooding generally performs better over water flooding because the surfactant is able to access the pores that are otherwise inaccessible due to high capillary pressure with water. Although, surfactant results in favorable fluid properties to mobilize oil, the reason surfactant did not perform well in tight-oil formation studied here is because the unswept oil is inaccessible due to small permeability, and not because of high capillary pressure. The average pressure support in the reservoir even by the end of 20 years of production is much higher that can overcome capillary forces. Even though the surfactant performs better in 2 cases out of 24 cases, the increase in recovery (1-3% higher than water) is not sufficiently higher to warrant its supremacy over water in tight-oil formations, primarily because it cannot access the tight pore space.

5. Summary and Conclusions

Production rates in tight-oil reservoirs decline rapidly within a year after the well is put to production. To arrest the declining production rates and recover additional oil from tight-oil resources, traditional methods of IOR (water flooding, immiscible CO_2 flooding, surfactant flooding) were investigated. To overcome the challenge of negligible fluid movement and sweep outside the fractures, the fluid was injected through a set of fractures and produced through another set of fractures intersecting the same lateral. We compare the estimates of oil recovery obtained by injecting water, surfactant, and immiscible CO_2, with the oil recovered by primary depletion (with no injection).

Results suggest that fluid injection did not improve recovery in 10 out of 24 cases studied, where 8 out of those 10 unsuccessful cases were from heterogeneous medium with very low (0.005 mD) to low (0.01 mD) average matrix permeability. The remaining 2 unsuccessful cases are for very low (0.005 mD) matrix permeability in homogeneous media with very high injection rate (~48 rb/d) and when the injection is constrained by a constant BHP, respectively. In the 14 successful or improved recovery cases, surfactant injection performed the best in 3 cases while immiscible CO_2 injection performed the best in 1 case. In the remaining 10 improved recovery cases, there was no clear winner among the three fluids; in 9 cases out of 10, surfactant and water performed almost equally, but better than immiscible CO_2, while only in one case all the three injection fluids (water, surfactant, and immiscible CO_2) performed similarly.

In conclusion, the results summarized in Table VI and Table VII suggest that an enhanced recovery method (e.g. water, immiscible CO_2, or surfactant) may not always improve recovery from tight-oil reservoirs, especially in a heterogeneous reservoir whose average matrix permeability is less than a threshold of 0.01 mD. Although, surfactant results in favorable fluid properties to mobilize oil, the reason surfactant did not perform well in tight-oil formation studied here is because the unswept oil is inaccessible due to small permeability, and not because of high capillary pressure. Therefore, it can be said that reservoir heterogeneity coupled with the average matrix permeability of a tight-oil reservoir seems to play the most significant role in exploring the potential of enhanced recovery.

There have been some lab-scale EOR studies (Li and Sheng, 2016, 2017; Li et al., 2017) on tight cores, but we should be cautious in directly adapting the results from lab-scale studies; that is because, unlike EOR from conventional reservoirs where reservoir heterogeneity is the primary variable that decides the performance of the recovery, there are multiple factors that may decide the performance of an EOR in tight-oil formations. One of those factors, besides other factors (e.g. fractures), is reservoir heterogeneity coupled with the average matrix permeability, which seems to play the most important role in the feasibility of an EOR method as discussed above. Most of the lab-scale studies look at performance of a certain EOR method on a core that is fairly homogeneous (with no permeability variation) and do not account for other important parameters (e.g. fractures) that may impact the performance of some EOR methods, specifically the EOR by gas injection. Therefore, we must be careful in adapting the results of those studies to support field scale response.

6. Acknowledgment

This research was supported in part by an appointment to the National Energy Technology Laboratory Research Participation Program, sponsored by the U.S. Department of Energy and administered by the Oak Ridge Institute for Science and Education.

7. References

Aloulou, F. (2016). World tight oil production to more than double from 2015 to 2040.

Andersen, O. (2017). Simplified models for numerical simulation of geological CO2 storage. The University of Bergen.

Delaihdem, D.K. (2013). Decline curve analysis and enhanced shale oil recovery based on Eagle Ford Shale data. Thesis. University of Alaska Fairbanks.

Deutsch, C.V., and Journel, A.G. (1997). GSLIB: Geostatistical Software Library and User's Guide (Oxford University Press, USA).

Dombrowski, Fonseca, Karanikas, Moya, and Lamantia (2015). Fluid injection in light tight oil reservoirs. US Patent US20130228337.

Dong, T., Harris, N.B., Ayranci, K., Twemlow, C.E., and Nassichuk, B.R. (2017). The impact of composition on pore throat size and permeability in high maturity shales: Middle and Upper Devonian Horn River Group, northeastern British Columbia, Canada. Marine and Petroleum Geology *81*, 220–236.

Flannery, J. (2006). Integrated Analysis of the Bakken Petroleum System, U.S. Williston Basin. pp. 138–145.

Gaswirth, S.B., and Marra, K.R. (2015). U.S. Geological Survey 2013 assessment of undiscovered resources in the Bakken and Three Forks Formations of the U.S. Williston Basin Province. AAPG Bulletin *99*, 639–660.

Gaswirth, S.B., Marra, K.R., Cook, T.A., Charpentier, R.R., Gautier, D.L., Higley, D.K., Klett, T.R., Lewan, M.D., Lillis, P.G., Schenk, C.J., et al. (2013). Assessment of undiscovered oil resources in the Bakken and Three Forks Formations, Williston Basin Province, Montana, North Dakota, and South Dakota, 2013 (Reston, VA: U.S. Geological Survey).

Ghaderi, S.M., Clarkson, C.R., Ghanizadeh, A., Barry, K., and Fiorentino, R. (2017). Improved Oil Recovery in Tight Oil Formations: Results of Water Injection Operations and Gas Injection Sensitivities in the Bakken Formation of Southeast Saskatchewan. (Society of Petroleum Engineers), p. SPE-185030-MS.

Gong, X. (2013). Assessment of Eagle Ford Shale Oil and Gas Resources. Thesis. Texas A&M University.

Hawthorne, S.B., Miller, D.J., Grabanski, C.B., Sorensen, J.A., Pekot, L.J., Kurz, B.A., Gorecki, C.D., Steadman, E.N., Harju, J.A., and Melzer, S. (2017a). Measured Crude Oil MMPs with Pure and Mixed CO2, Methane, and Ethane, and Their Relevance to Enhanced Oil Recovery from Middle Bakken and Bakken Shales. (Society of Petroleum Engineers), p.

Hawthorne, S.B., Jin, L., Kurz, B.A., Miller, D.J., Grabanski, C.B., Sorensen, J.A., Pekot, L.J., Bosshart, N.W., Smith, S.A., Burton-Kelly, M.E., et al. (2017b). Integrating Petrographic and Petrophysical Analyses with CO2 Permeation and Oil Extraction and Recovery in the Bakken Tight Oil Formation. (Society of Petroleum Engineers), p.

He, K., Xu, L., Gao, Y., Neeves, K.B., Yin, X., Bai, B., Ma, Y., and Smith, J. (2014). Validating Surfactant Performance in the Eagle Ford Shale: A Correlation between the Reservoir-on-a-Chip Approach and Enhanced Well Productivity. (Society of Petroleum Engineers), p.

Honarpour, M.M., Nagarajan, N.R., Orangi, A., Arasteh, F., and Yao, Z. (2012). Characterization of Critical Fluid PVT, Rock, and Rock-Fluid Properties - Impact on Reservoir Performance of Liquid Rich Shales. (Society of Petroleum Engineers), p.

Hustak, C., Dieva, R., Baker, R., MacIsaac, B., Frankiw, K., and Clark, B. (2017). Waterflooding a Multi-layered Tight Oil Reservoir Developed with Hydraulically Fractured Horizontal Wells. (Society of Petroleum Engineers), p. SPE-185021-MS.

Jørgensen, K. (2013). Implementation of a Surfactant Model in MRST With Basis in Schlumberger's Eclipse. Norwegian University of Science and Technology.

Karimi, S., and Kazemi, H. (2015). Capillary Pressure Measurement using Reservoir Fluids in a Middle Bakken Core. (Society of Petroleum Engineers), p.

Kenomore, M., Hassan, M., Dhakal, H., and Shah, A. (2017). Total organic carbon evaluation of the Bowland Shale Formation in the Upper Bowland of the Widmerpool Gulf. Journal of Petroleum Science and Engineering *150*, 137–145.

Li, L., and Sheng, J.J. (2016). Experimental study of core size effect on CH4 huff-n-puff enhanced oil recovery in liquid-rich shale reservoirs. Journal of Natural Gas Science and Engineering *34*, 1392–1402.

Li, L., and Sheng, J.J. (2017). Numerical analysis of cyclic CH4 injection in liquid-rich shale reservoirs based on the experiments using different-diameter shale cores and crude oil. Journal of Natural Gas Science and Engineering *39*, 1–14.

Li, S., and Luo, P. (2017). Experimental and simulation determination of minimum miscibility pressure for a Bakken tight oil and different injection gases. Petroleum *3*, 79–86.

Li, L., Sheng, J.J., Xu, J., and others (2017). Gas Selection for Huff-n-Puff EOR in Shale Oil Reservoirs Based upon Experimental and Numerical Study. In SPE Unconventional Resources Conference, (Society of Petroleum Engineers), p. SPE-185066-MS.

Lie, K.-A. (2016). An introduction to reservoir simulation using MATLAB: User guide for the Matlab Reservoir Simulation Toolbox (MRST). SINTEF ICT. http://www.sintef.no/Projectweb/MRST/Publications.

Liu, R., Liu, H., Li, X., Wang, J., and Pang, C. (2010). Calculation of Oil and Water Relative Permeability for Extra Low Permeability Reservoir. (Society of Petroleum Engineers), p.

Lund, L. (2014). Decline curve analysis of shale oil production: The case of Eagle Ford. Uppsala University.

Mavor, M. (2013). Reservoir Fluid Properties Required for Low Permeability Oil Reservoir Analysis, by Matt Mavor; #90186 (2013). In Hydrocarbon Charge Considerations in Liquid-Rich Unconventional Petroleum Systems, (Canada: AAPG), p.

Mirchi, V., Saraji, S., Goual, L., and Piri, M. (2015). Dynamic interfacial tension and wettability of shale in the presence of surfactants at reservoir conditions. Fuel *148*, 127–138.

Moghadasi, L., Guadagnini, A., Inzoli, F., and Bartosek, M. (2015). Interpretation of two-phase relative permeability curves through multiple formulations and Model Quality criteria. Journal of Petroleum Science and Engineering *135*, 738–749.

Rystad Energy (2014). SHALE COMPANIES HAVE PROVEN MORE EFFICIENT BUT NOT MORE PRODUCTIVE (Rystad Energy).

Sanchez-Rivera, D., Mohanty, K., and Balhoff, M. (2015). Reservoir simulation and optimization of Huff-and-Puff operations in the Bakken Shale. Fuel *147*, 82–94.

Sheng, J.J. (2015). Status of surfactant EOR technology. Petroleum *1*, 97–105.

Shuler, P.J., Lu, Z., Ma, Q., and Tang, Y. (2016). Surfactant Huff-n-Puff Application Potentials for Unconventional Reservoirs. (Society of Petroleum Engineers), p.

Singh, H. (2016). A Critical Review of Water Uptake by Shales. Journal of Natural Gas Science and Engineering *34*, 751–766.

Singh, H., and Cai, J. (2018). Screening improved recovery methods in tight-oil formations by injecting and producing through fractures. International Journal of Heat and Mass Transfer *116*, 977–993.

Steiner, S., Ahsan, S.A., Raina, I., Dasgupta, S., and Lis, G.P. (2016). Interpreting Total Organic Carbon TOC in Source Rock Oil Plays. (Society of Petroleum Engineers), p.

Tuero, F., Crotti M, M., and Labayen, I. (2017). Water Imbibition EOR Proposal for Shale Oil Scenarios. (Society of Petroleum Engineers), p.

US EIA (2016). Annual Energy Outlook 2016 (Government Printing Office).

Wachtmeister, H., Lund, L., Aleklett, K., and Höök, M. (2017). Production Decline Curves of Tight Oil Wells in Eagle Ford Shale. Nat Resour Res *26*, 365–377.

Wang, D., Butler, R., Liu, H., and Ahmed, S. (2011). Surfactant Formulation Study For Bakken Shale Imbibition. (Society of Petroleum Engineers), p.

Wang, D., Butler, R., Zhang, J., and Seright, R. (2012). Wettability Survey in Bakken Shale With Surfactant-Formulation Imbibition. SPE Reservoir Evaluation & Engineering *15*, 695–705.

Wang, D., Zhang, J., Butler, R., and Olatunji, K. (2016). Scaling Laboratory-Data Surfactant-Imbibition Rates to the Field in Fractured-Shale Formations. SPE Reservoir Evaluation & Engineering *19*, 440–449.

Witkowsky, J.M., Galford, J.E., Quirein, J.A., and Truax, J.A. (2012). Predicting Pyrite and Total Organic Carbon from Well Logs for Enhancing Shale Reservoir Interpretation. (Society of Petroleum Engineers), p.

Xu, F., Guo, X., Wang, W., Zhang, N., Jia, S., and Wang, X. (2011). Case Study: Numerical Simulation Of Surfactant Flooding In Low Permeability Oil Filed. (Society of Petroleum Engineers), p.

Yu, W., Lashgari, H., and Sepehrnoori, K. (2014). Simulation Study of CO2 Huff-n-Puff Process in Bakken Tight Oil Reservoirs. (Society of Petroleum Engineers), p.

Zhang, K., Seetahal, S., Alexander, D., Lv, J., Hu, Y., Lu, X., Zhang, D., Wu, K., and Chen, Z. (2016). Effect of Confinement on Gas and Oil Relative Permeability During CO2 Flooding in Tight Oil Reservoirs. (Society of Petroleum Engineers), p.

Zhang, Y., Yu, W., Sepehrnoori, K., and Di, Y. (2017). A Comprehensive Numerical Model for Simulating Fluid Transport in Nanopores. Sci Rep *7*.

Zhu, P., Balhoff, M.T., and Mohanty, K.K. (2017). Compositional modeling of fracture-to-fracture miscible gas injection in an oil-rich shale. Journal of Petroleum Science and Engineering *152*, 628–638.

Zuloaga-Molero, P., Yu, W., Xu, Y., Sepehrnoori, K., and Li, B. (2016). Simulation Study of CO2-EOR in Tight Oil Reservoirs with Complex Fracture Geometries. Sci Rep *6*.

Appendix

Surfactant Model (from Jørgensen, 2013)

$$f(N_c) = \begin{cases} N_c^{misc};\ k_\alpha^{misc}(s_\alpha);\ s_{\alpha,res}^{misc}, & max(N_c) \\ N_c^{immisc};\ k_\alpha^{immisc}(s_\alpha);\ s_{\alpha,res}^{immisc}, & min(N_c) \end{cases} \quad \text{(A-1)}$$

Where, $\alpha \in \{w, o\}$

The residual saturation values for the relative permeability are obtained by weighted average of residual saturation values at immiscible (without surfactant) and fully miscible conditions.

$$s_{\alpha,res}^{N_c} = m.\,s_{\alpha,res}^{misc} + (1-m).\,s_{\alpha,res}^{immisc} \qquad \text{(A-2)}$$

In addition to reducing the IFT and residual oil saturation, surfactant also changes the relative permeability due to miscible flow such that the effective relative permeability after adding surfactant is the weighted average of the relative permeabilities with immiscible (without surfactant) and with fully miscible conditions.

$$k_{r\alpha}(s_\alpha) = m.\,k_{r\alpha}^{misc}(\bar{s}_\alpha^{misc}, N_c) + (1-m).\,k_{r\alpha}^{immisc}(\bar{s}_\alpha^{immisc}, N_c) \qquad \text{(A-3)}$$

The discrete relative permeability values cannot be directly scaled (a continuous function) to match the new end-points, therefore, it require modified saturation values that are calculated as following.

$$\bar{s}_\alpha^{misc}(N_c) = \left(\frac{s_\alpha - s_{\alpha,res}^{N_c}}{1 - s_{\alpha,res}^{N_c} - s_{\alpha',res}^{N_c}}\right).\left(1 - s_{\alpha,res}^{misc} - s_{\alpha',res}^{misc}\right) + s_{\alpha,res}^{misc} \qquad \text{(A-4)}$$

$$\bar{s}_\alpha^{immisc}(N_c) = \left(\frac{s_\alpha - s_{\alpha,res}^{N_c}}{1 - s_{\alpha,res}^{N_c} - s_{\alpha',res}^{N_c}}\right).\left(1 - s_{\alpha,res}^{immisc} - s_{\alpha',res}^{immisc}\right) + s_{\alpha,res}^{immisc} \qquad \text{(A-5)}$$

Also, addition of surfactant to the water increases its viscosity in a way that the effective viscosity of water after adding surfactant is given as:

$$\mu_w^{surf}(c_s, p) = \frac{\mu_w(p)}{\mu_w(p_{ref})}.\,\mu_s(c_s) \qquad \text{(A-6)}$$

487521 Maximizing Uniformity of Hydraulic Fracture Stimulation of Horizontal Wells through Stress Shadow Balancing and Limited Entry Methods

Abstract:

The combined practice of horizontal drilling and hydraulic fracturing has made production of petroleum resources from ulta-low permeability shale formations both technically feasible and economically viable. However, a major concern about ongoing development of shale reservoirs is a trend towards increasing development intensity in order to try to increase gas/oil production. That is, there can be a tendency towards solutions that entail more well pads, more water, more chemicals, more sand, and therefore more potential for negative environmental and community impacts. Reversing a trend towards increasing development intensity requires a focus on developing approaches that make the most of every well that is drilled, striving towards an industry that is more efficient in terms of gas/oil recovery from a given level of development activity. And, in fact, there ought to be substantial gains that can be made through efforts to make the most of every foot of every well because one thing is clear about the current technology of horizontal drilling and hydraulic fracturing: It is still achieving results that are far from optimal. Of particular concern is the fact that data from producing wells show a tendency for 1/3 or more of the drilled length of horizontal wells to be completely unproductive. A likely cause is that fluid that is injected with the intention of generating hydraulic fractures from all perforation clusters â€" the clusters of holes through the well casing that allow fluid exchange between the well and the reservoir - is instead diverted to only a portion of the clusters leaving many sections of the well unstimulated.

Thus motivated, here we present a numerical parametric study showing two methods for optimizing uniformity of stimulation through design of the clusters of perforation holes that are used as the entry points for fluid from the wellbore into the formation. We show that non-uniform perforation cluster locations combined with designing the entry friction associated with the perforation clusters (â€œlimited entryâ€) can be optimized to produce the maximum surface area for an array of hydraulic fractures. Optimization is carried out using a reduced order, approximate simulator capable of carrying out an entire coupled simulation, which would normally take hours or days, in seconds to minutes on a desktop computer. Candidate designs are then tested using a fully-coupled, Planar 3D, multi-hydraulic fracture simulator that uses an Implicit Level Set Algorithm (ILSA). We find that for stages completed with 5 perforation clusters, optimizing the cluster spacing leads to over 70% increase in fracture surface area generated by the same injected volume. For uniform cluster spacing, optimal limited entry design alone is able to improve the fracture surface area by approximately 10%. When combined and optimized, improvements can approach a doubling of the fracture surface area leading to a potential for a similar order of increase in hydrocarbon recovery.

498730 Enhancing Unconventional Shale Porosity and Permeability By Subcritical Water Treatment

Abstract:

Utica and Marcellus shales are two large unconventional shales in the U.S. With the advance horizontal drilling technology along with state-of-the-art fracking fluid, only 28% of the reserve is recoverable from unconventional shales according to the U.S. Energy Information Administration (EIA). Among other challenges, low porosity (<5 %) and low permeability (2-50 nD) of Utica and Marcellus shales are the primary reasons for the low recovery. Increasing porosity and permeability without compromising the mechanical strength are necessary for enhanced gas recovery. Therefore, the fracking fluid needs to react with the shale minerals and kerogen in order to connect nano-pores and increase the porosity, permeability, and recovery of the shale.

Hydraulic fracturing uses water (~90%) as fracking fluid at temperatures in the range of 50-95 °C. Physical and chemical properties of hot compressed water (T=100-374 °C and corresponding saturation pressure) and supercritical water (T>374 °C and P>22.4 MPa) are very different than water below 100 °C. Thermodynamic properties of water (e.g., density, ionic product, and dielectric constant) changes significantly with the increase of water temperature. As a result, sub-and supercritical water behaves like a strong nonpolar solvent. Supercritical water also becomes highly reactive and corrosive depending on the density and temperature. Therefore, sub-and supercritical water are promising to react with kerogen as well as shale minerals.

In this study, hydrothermal treatment of both Utica and Marcellus shales were performed in a batch reactor. The reaction temperatures were set between 200-300 °C and the reaction times were tested for 1-6 h. The treated shales were analyzed with BET surface area analyzer, mercury induced capillary pressure (MICP) porosimeter, FE-SEM, XRD, and TEM. The process water was analyzed with ICP-OES, pH, and IC. Mass recovery was reduced with the increase of treatment temperature and time. Porosity, calculated from BET, showed a positive correlation with hydrothermal treatment temperature. For instance, porosity increases from 4.0 % to 10 %, when Marcellus shale was treated at 250 °C for 3 h. FE-SEM and TEM revealed the increase of pore sizes up to 100 nm, while XRD shows the decrease of several minerals from the solid matrix of the treated shales. ICP-OES confirmed the mass loss in the form of inorganics due to the hydrothermal treatment.

EERC

Energy & Environmental Research Center

RESEARCH TO INVESTIGATE THE POTENTIAL FOR ENHANCED OIL RECOVERY IN THE BAKKEN PETROLEUM SYSTEM

American Institute of Chemical Engineers
2017 Annual Meeting
November 2, 2017
Minneapolis, Minnesota

James Sorensen, Steven Hawthorne, Beth Kurz, Lawrence Pekot, Steve Smith, Lu Jin, Charles Gorecki, Ed Steadman, and John Harju
Energy & Environmental Research Center

Critical Challenges. **Practical Solutions.**

PARTNERS IN EERC BAKKEN PROJECTS

NETL
North Dakota oil & gas research program
Marathon Oil
Continental Resources
XTO Energy
HESS
Liberty Resources
Baker Hughes
CMG Computer Modelling Group Ltd
Schlumberger
Kinder Morgan Inc.
INGRAIN Digital Rock Physics Lab
DMR
North Dakota Petroleum Council

EERC

Critical Challenges. **Practical Solutions.**

BAKKEN PETROLEUM SYSTEM LITHOLOGY

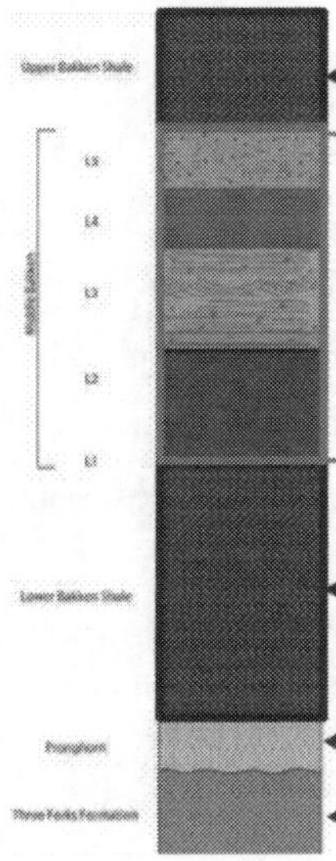

Drilling Targets

Upper Bakken Shale: Brown to black, organic-rich.
- **Bakken source rock**

Middle Bakken: Variable lithology (up to nine lithofacies), ranging from silty sands to siltstones and tight carbonates
- **Bakken tight reservoir rock (horizontal drilling target)**

Lower Bakken Shale: Brown to black, organic-rich
- **Bakken source rock**

Drilling Targets

Pronghorn Member: Mixed sandstone, siltstone, dolomite, and shale.

Three Forks Formation: Interbedded dolostone/limestone, siltstone/mudstone, shale, evaporites.

THE ROCKS WITHIN THE SYSTEM ARE COMPLEX

Upper Shale

Middle Bakken Lithofacies

Lower Shale

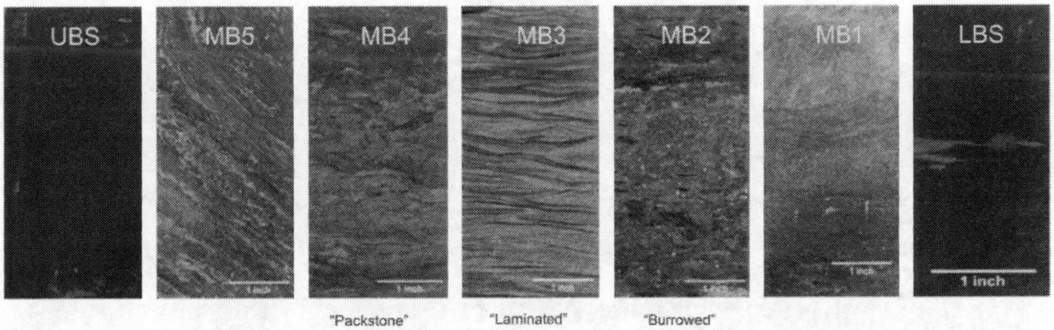

"Packstone" "Laminated" "Burrowed"

MICP Pore Throat Size Ranges for Bakken Rocks	
Lower Bakken Shale	3–25 nm
Middle Bakken	3–250 nm
Upper Bakken Shale	3–25 nm

EERC

SIZE OF THE PRIZE – BAKKEN AND THREE FORKS

OOIP Estimates

300 Bbbl
(Flannery and Kraus, 2006)

900 Bbbl
(Continental Resources, 2011)

EERC

5

Technically Recoverable Reserve Estimates

7.4 Bbbl
(USGS, 2013)

24 Bbbl
(Continental Resources, 2011)

- Currently, only a 3%–10% recovery factor.
- Small improvements in recovery could yield over a billion barrels of oil.
- Can CO_2 and/or rich gas work in the Bakken?

CHALLENGES OF CO_2 AND RICH GAS EOR IN THE BAKKEN – FOCAL POINTS OF EERC RESEARCH

- Fractures acting as fast flow pathways, limiting the ability of the working fluid to interact with the matrix.
- High vertical heterogeneity of the lithofacies complicates our understanding of flow regimes (fractures and matrix).
- Multiphase fluid flow behavior varies substantially depending on the size of the pore throats.
  - Fluid viscosity and density are much different in nanoscale pores than in macroscale pores.
- How does the sorptive capacity of the organic carbon materials affect working fluid mobility and EOR?

EERC EFFORTS TO ADDRESS BAKKEN EOR CHALLENGES

CO_2, ethane, methane…

EERC

Critical Challenges. **Practical Solutions.**

MMP BY CAPILLARY RISE-VANISHING INTERFACIAL TENSION

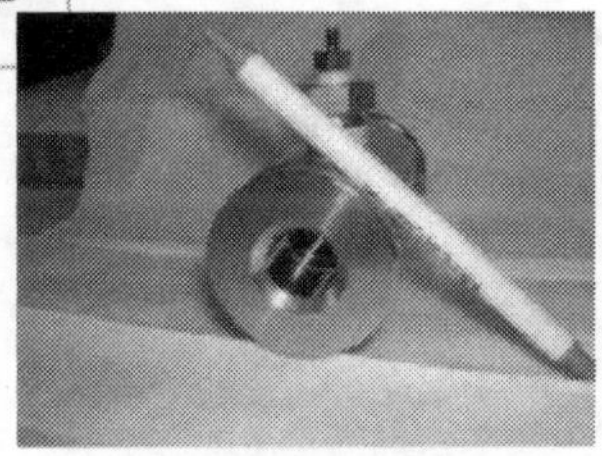

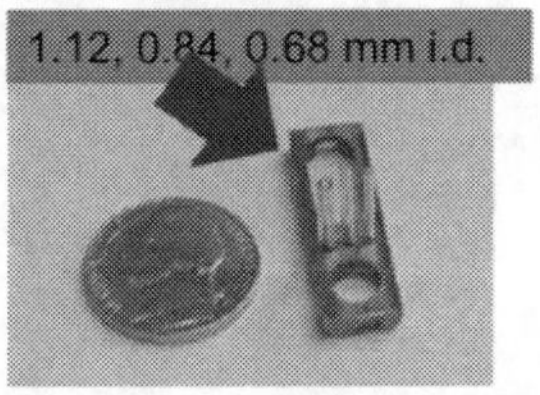

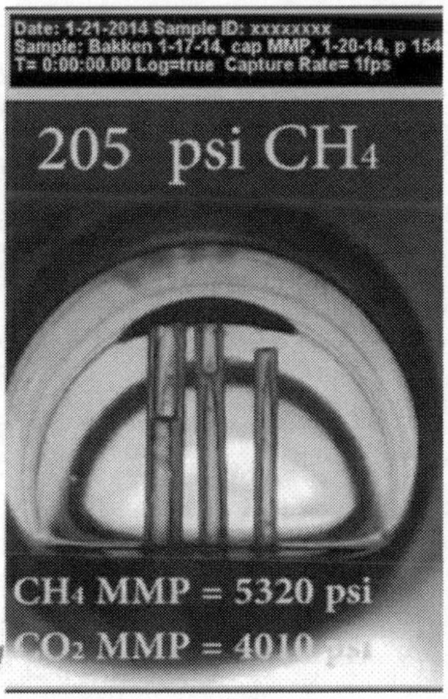

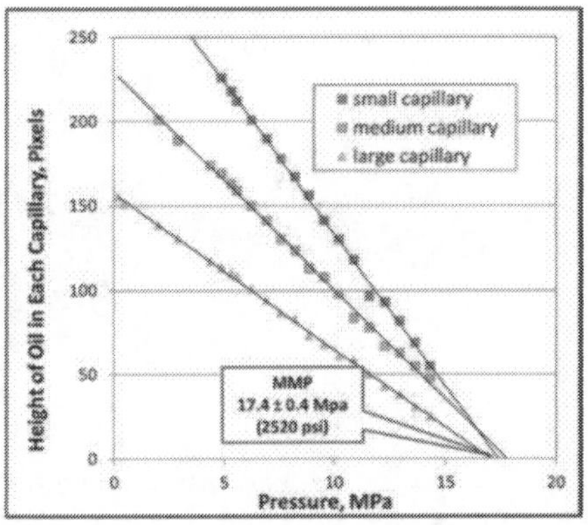

Patent pending

EERC

Critical Challenges. **Practical Solutions.**

MMP STUDIES OF POTENTIAL EOR WORKING FLUIDS

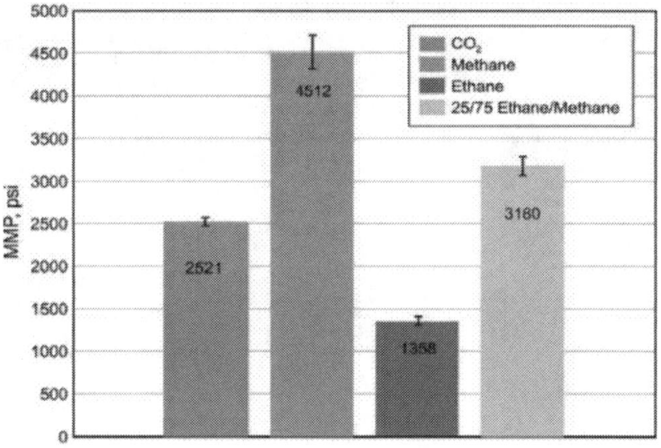

Ethane is very effective at achieving lower MMPs, methane is not! API 41.5 Crude (Bakken), 110 ° C

EERC

Critical Challenges. Practical Solutions.

PORE SIZE AFFECTS FLUID PHASE BEHAVIOR

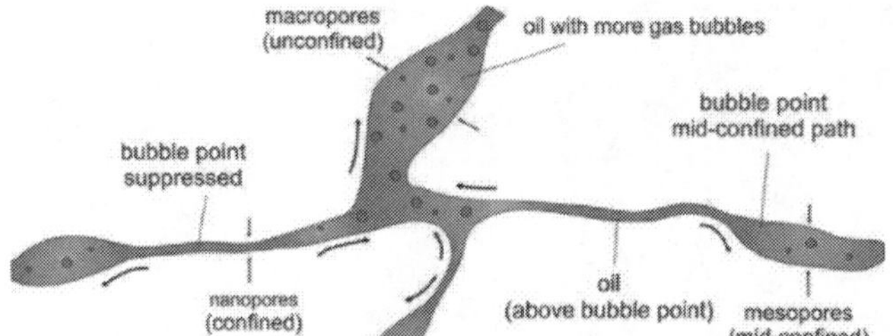

Conceptual pore network model showing different phase behavior in different pore sizes for a bubblepoint system with phase behavior shift.

Source: Alharthy, N.S., Nguyen, T.N., Teklu, T.W., Kazemi, H., and Graves, R.M., 2013, SPE 166306, Colorado School of Mines, and Computer Modelling Group Ltd.

To predict hydrocarbon phase behavior in the Bakken, we need to understand:

- Size and geometry of the pore throat networks.
- Distribution of those pore throat networks.
- Working fluid–oil interactions at Bakken conditions.
  - Miscibility
  - Hydrocarbon species selectivity

EERC

10

Critical Challenges. Practical Solutions.

ADVANCED CHARACTERIZATION

- Cores and well logs come from four well locations (yellow triangles).
- Samples represent:
  - Middle Bakken reservoir lithofacies.
  - Upper and Lower Bakken shale source rocks.
- Samples provided by Marathon and North Dakota Geological Survey.

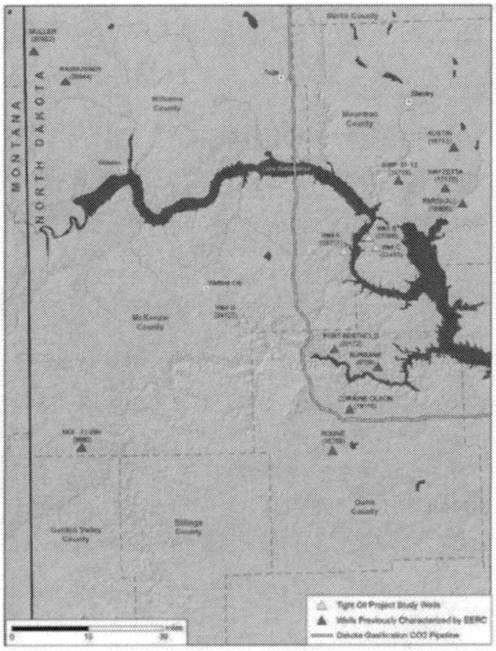

EERC

FRACTURE AND PORE ANALYSIS TECHNIQUES

Initial Macro- and Microfracture Characterization

- **Whole core and thin section analysis.**
- **Conventional SEM** methods were used for macro- and microscale fracture analysis.

Micro- and Nanoscale Fracture and Pore Analysis

- **Field emission (FE)–SEM**, **micro-CT scanning**, and **focused ion beam (FIB)–SEM** conducted by Ingrain Inc. were used to characterize micro- and nanoscale fractures and pores.

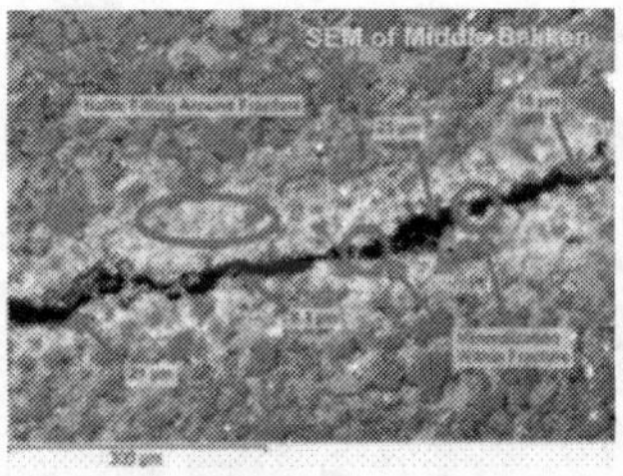

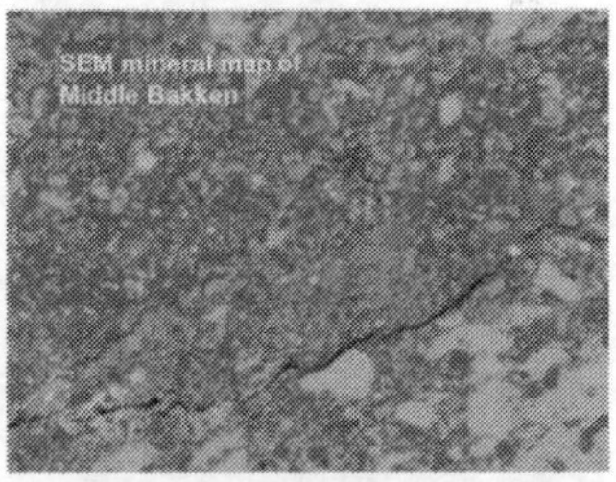

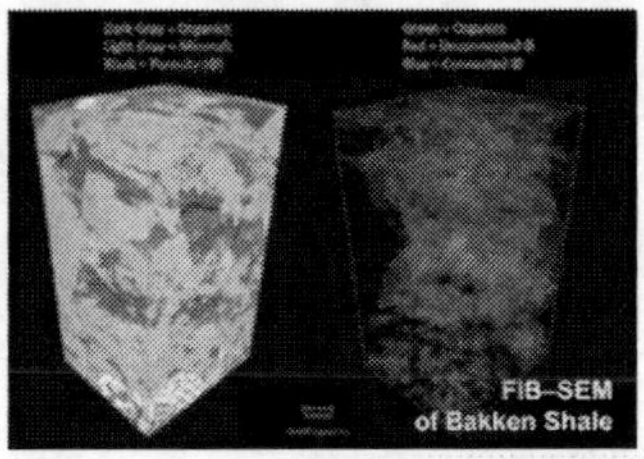

EERC

Critical Challenges. **Practical Solutions.**

UPPER BAKKEN SHALE

MICRO- AND NANOSCALE FESEM AND FIB–SEM

Green = organics
Red = unconnected Φ
Blue = connected Φ

The amount of connected and unconnected pore space is roughly equal.

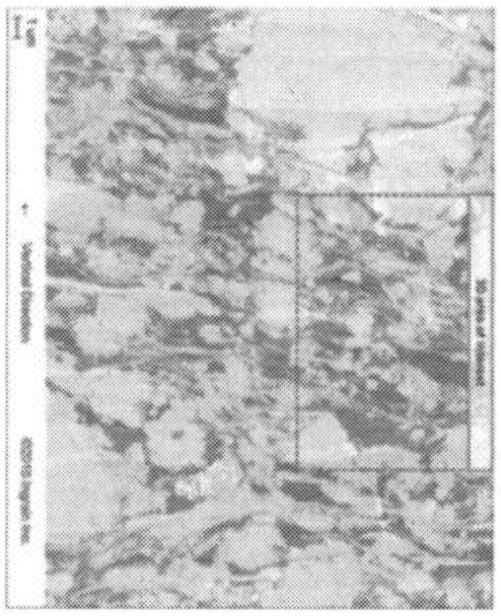

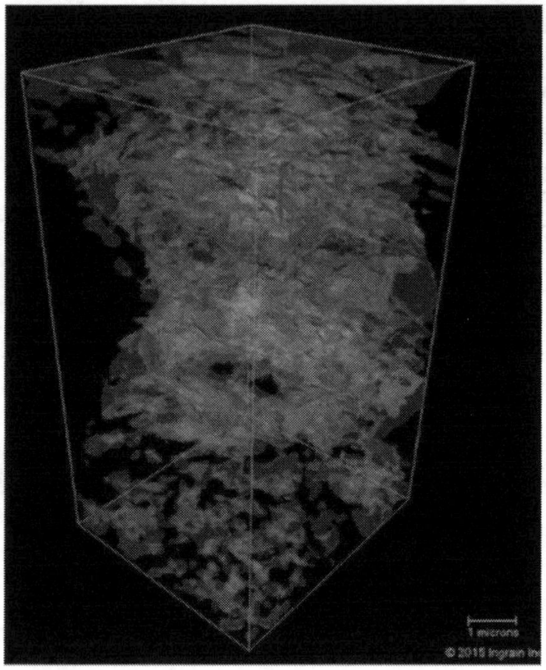

EERC

13

FESEM AND FIB–SEM OF MIDDLE BAKKEN

14

OIL EXTRACTION FROM BAKKEN ROCKS IN THE LAB

- Thirteen core plugs covering all three units were selected to investigate the EOR effects of four potential EOR gases: N_2, CO_2, CH_4, and C_2H_6.
- 24-hour oil extraction was conducted for the rock samples using different gases.
- All experiments were carried out under reservoir conditions (5000 psi, 230°F).

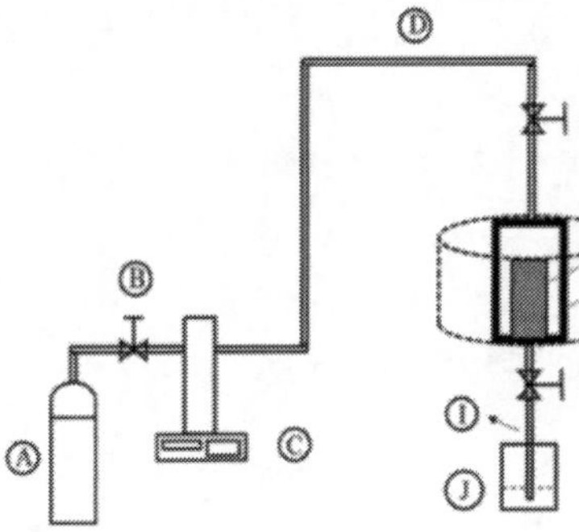

Legend
A: CO_2 tank;
B: Shut off valve;
C: ISCO pump;
D: Connecting tubing;
E: Rock core;
F: Extraction vessel;
G: ISCO SFX-210 heated extractor;
H: Outlet control valve;
I: Flow control restrictor;
J: Collection solvent vial;

- Rock is "bathed" in gases, not swept with gases as would be the case in confined flow-through tests.

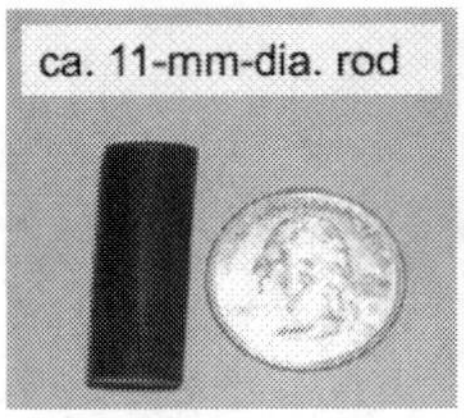

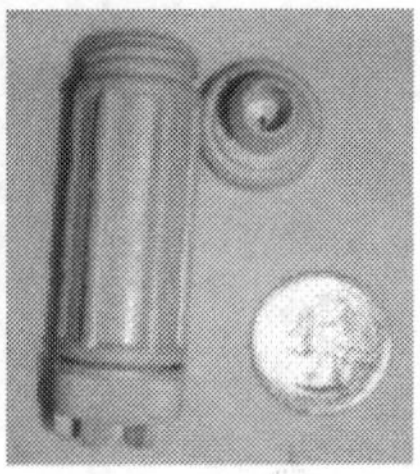

Laboratory **CO_2 Oil Recovery from Six Bakken Lithofacies from a McKenzie County Well (24 hours).**

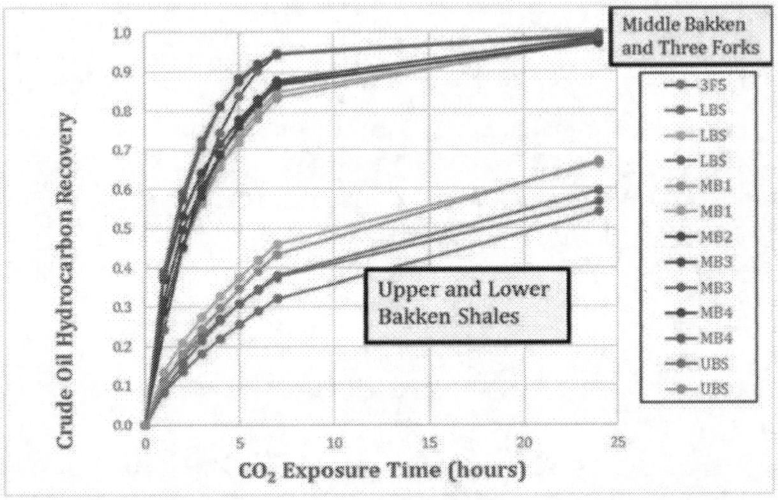

Results also show that CO_2 selectively mobilized lower-molecular-weight hydrocarbons.

EERC

Critical Challenges. **Practical Solutions.**

OIL EXTRACTION TESTS ON MIDDLE BAKKEN USING DIFFERENT WORKING FLUIDS

24-hour oil recovery using different gases for the Middle Bakken cores.

Ethane (C_2H_6) has the best performance while N_2 has the lowest oil recovery.

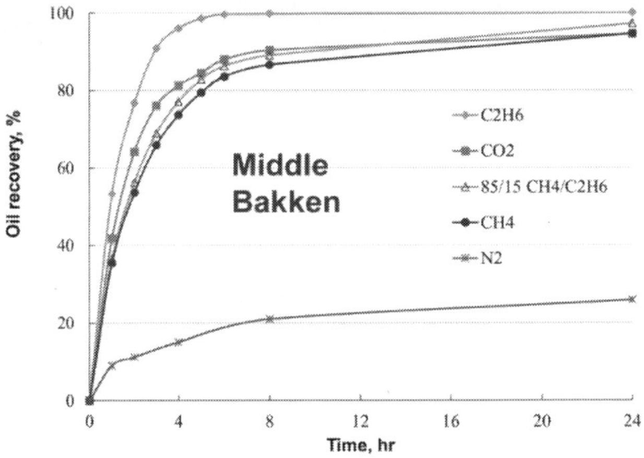

EERC

17

EERC BAKKEN RESERVOIR-SCALE SIMULATION RESULTS

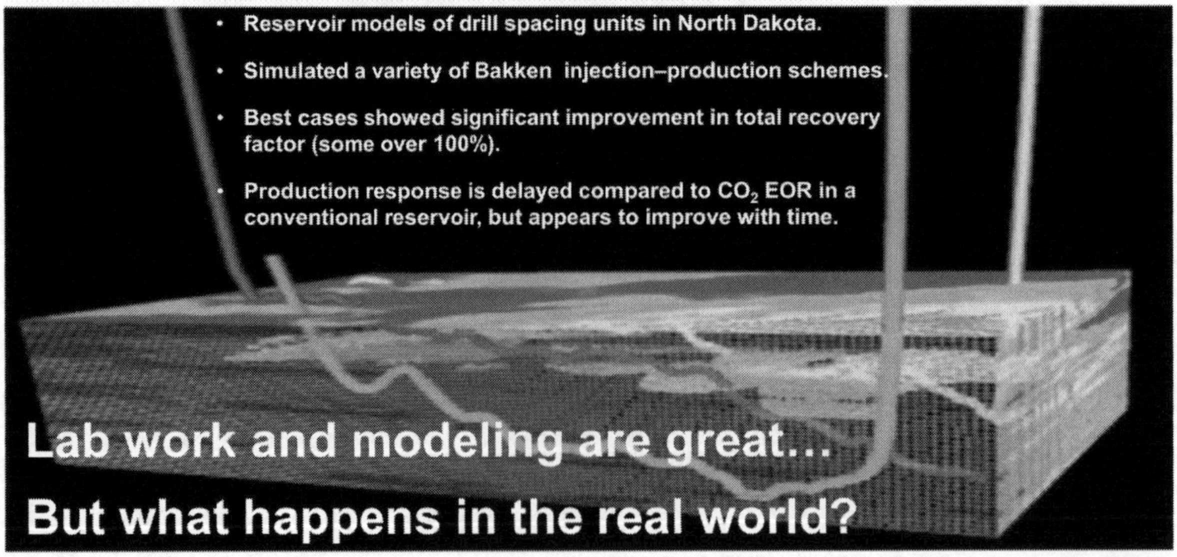

BAKKEN CO_2 STORAGE AND ENHANCED RECOVERY PROGRAM – FIELD INJECTION TEST

EERC

Critical Challenges. Practical Solutions.

TEST CONCEPT AND HYPOTHESIS

Past pilot-scale CO_2 injection tests into horizontal, hydraulically fractured Bakken wells have shown little to no effect on oil mobilization.

- CO_2 likely moved so quickly through fractures that it did not have enough contact time, or became too dispersed, to interact with stranded oil in the matrix.

Hypotheses to be tested in a vertical well:

1. CO_2 can be injected into an unstimulated Bakken reservoir.
2. The injected CO_2 can interact with the in-place fluids, resulting in subsequent mobilization of hydrocarbons and storage of CO_2.

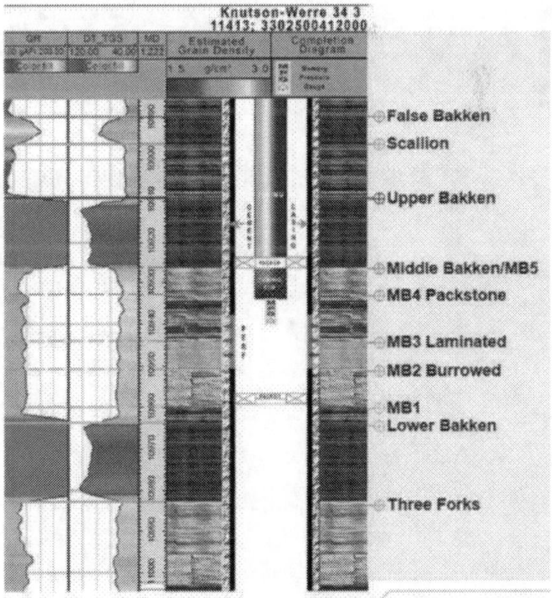

EERC

20

Critical Challenges. Practical Solutions.

CO_2 INJECTION TEST KEY RESULTS

- No oil production prior to injection.
- Injected 100 tons of CO_2 over 4 days into unstimulated Middle Bakken reservoir.
- After two weeks of soaking, oil flowed to surface.
- Composition of the oil shifted toward lower-molecular-weight hydrocarbons after the injection.
- This is evidence that CO_2 penetrated the matrix and mobilized oil.

EERC

Critical Challenges. **Practical Solutions.**

ACCOMPLISHMENTS TO DATE

- New knowledge of effects of CO_2 and rich gas on MMP.
- Gained insight regarding the nature and role of microfractures in fluid mobility.
- Lab work demonstrated the ability of CO_2 to permeate shales and tight nonshale rocks and mobilize oil.
- Generated insight on the key characteristics and mechanisms controlling CO_2 permeation and oil mobility.
  - Diffusion
  - Sorption
  - Total organic content
  - Thermal maturity
- Applied those insights and data on developing improved models and designed and conducted a field test.

EERC

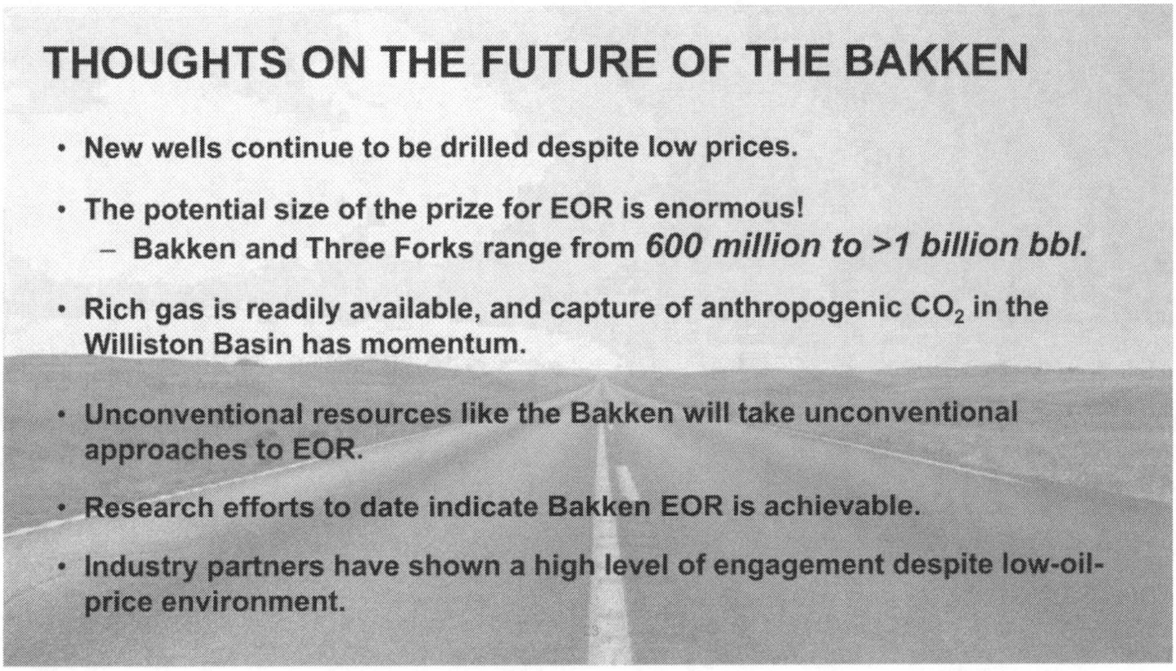
THOUGHTS ON THE FUTURE OF THE BAKKEN
• New wells continue to be drilled despite low prices.
• The potential size of the prize for EOR is enormous!
– Bakken and Three Forks range from *600 million to >1 billion bbl.*
• Rich gas is readily available, and capture of anthropogenic CO_2 in the Williston Basin has momentum.
• Unconventional resources like the Bakken will take unconventional approaches to EOR.
• Research efforts to date indicate Bakken EOR is achievable.
• Industry partners have shown a high level of engagement despite low-oil-price environment.

Thanks!
Jim Sorensen
Principal Geologist
Energy & Environmental Research Center
(701) 777-5287
jsorensen@undeerc.org

ACKNOWLEDGMENT

This material is based upon work supported by the U.S. Department of Energy National Energy Technology Laboratory under Award No. DE-FE0024454.

Disclaimer

This presentation was prepared as an account of work sponsored by an agency of the United States Government. Neither the United States Government, nor any agency thereof, nor any of their employees, makes any warranty, express or implied, or assumes any legal liability or responsibility for the accuracy, completeness, or usefulness of any information, apparatus, product, or process disclosed or represents that its use would not infringe privately owned rights. Reference herein to any specific commercial product, process, or service by trade name, trademark, manufacturer, or otherwise does not necessarily constitute or imply its endorsement, recommendation, or favoring by the United States Government or any agency thereof. The views and opinions of authors expressed herein do not necessarily state or reflect those of the United States Government or any agency thereof.

EERC

Critical Challenges. **Practical Solutions.**

496798 Evaluation of Reinjection As a Means to Increase Petroleum Production from the Utica/Point Pleasant Unconventional Play: Impact of Reinjection Fluid Composition, Pressure, and Pore Size on Hydrocarbon Bubble Point Behavior

Abstract:

The development of U.S. unconventional resources has had a dramatic impact on global energy supply, with the U.S. becoming the worldâ€™s leading producer of petroleum and natural gas. This change has been beneficial to the public lowering energy prices, while increasing energy security. Unconventional (or shale) resources differ from conventional reservoirs, as these plays consist of interbedded carbonate and hydrocarbon containing shale layers consisting of micropore and mainly nanopore structures.

In such small pores, strong pore-fluid and fluid-fluid interactions exist which lead to complex fluid phase behavior. These pores experience increased capillary forces, electrostatic interactions, and surface chemistry (known as confinement) which cause hydrocarbon physical properties to deviate from their bulk. These forces result in less hydrocarbon productivity in comparison to larger pore structures of conventional reservoirs, resulting in a significant decline in unconventional well productivity. Methods which improve hydrocarbon recovery from unconventional reservoirs is of interest to industry, government, and the public as such technologies will reduce overall shale development efficiency.

Secondary treatments using chemical complexes (after hydraulic fracturing) are a potentially attractive methodology to increase hydrocarbon production from unconventional reservoirs. The Utica/Point Pleasant play of Eastern Ohio is particularly well positioned for such secondary chemical treatments as the play has several coal-fired power plants (which could supply CO_2) and produces significant quantities of natural gas (NG) containing natural gas liquids (NGLs). Both CO_2 and NG/NGLs reinjection could be used as a means to increase petroleum production from this important U.S. energy play.

Ohio University (OHIO) is investigating CO_2 and NG/NGLs reinjection as a means to increase petroleum production from unconventional resources, with a particular focus on the Utica/Point Pleasant play. This study is investigating hydrocarbon bubble point behavior in pores with and without confinement by simulating downhole conditions using a high pressure differential calorimeter (HP-DSC). This presentation will present results from this study including the influence of the pore size, pressure and reinjection fluid composition on the phase behavior of the hydrocarbons inside micro and nanopores. Further, HP-DSC results evaluating reinjection composition and pressure on bubble point behavior from Point Pleasant well core will also be presented.

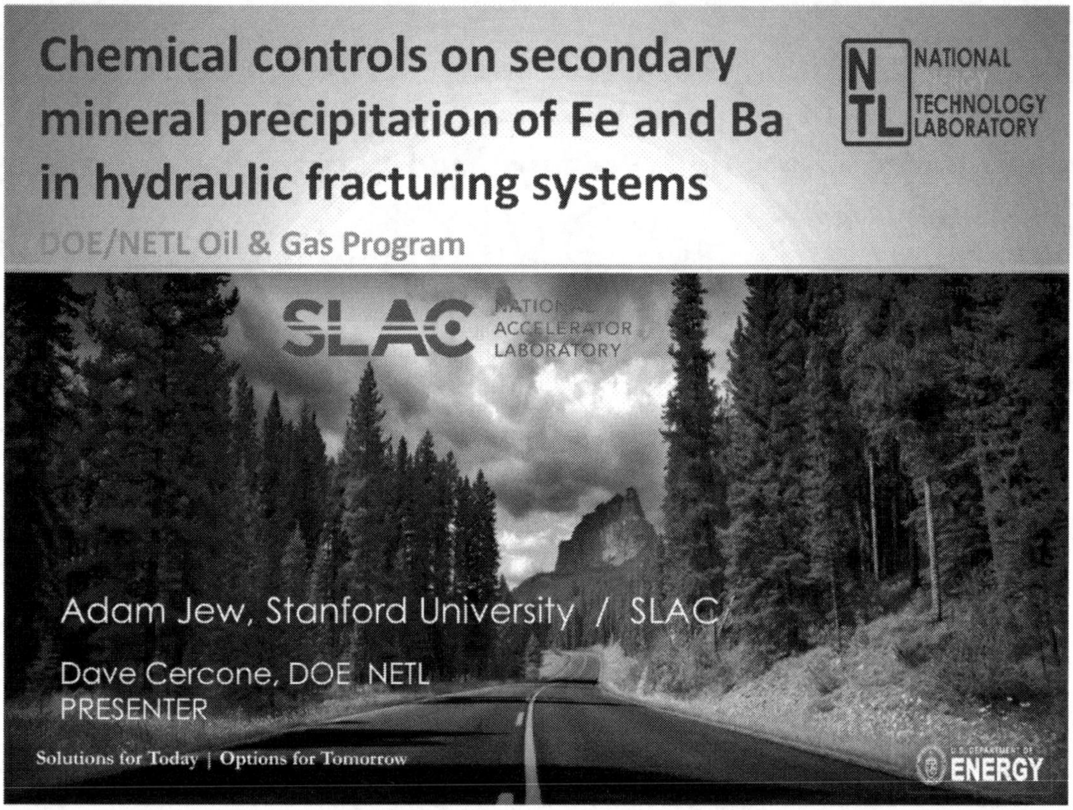
Chemical controls on secondary mineral precipitation of Fe and Ba in hydraulic fracturing systems
DOE/NETL Oil & Gas Program
NATIONAL TECHNOLOGY LABORATORY
SLAC NATIONAL ACCELERATOR LABORATORY
Adam Jew, Stanford University / SLAC
Dave Cercone, DOE NETL
PRESENTER
Solutions for Today | Options for Tomorrow
ENERGY

NETL – MARCELLUS EE LAB
SLAC NATIONAL ACCELERATOR LABORATORY

Motivation

- Iron and Barium are two very important scale materials
- Fracture fluid-shale interactions can damage recovery
- Unconventional extraction: Low efficiency
- Organics, Ionic Strength (I.S.), and pH can strongly impact secondary mineral production

Marcellus Shale Energy and Environment Laboratory during drilling operations

3

UNCONVENTIONAL RECOVERY FACTORS

SHALE GAS: 2 TO 19%; AVERAGE 13%

SHALE OIL: LESS THAN 2 %

NOT GOOD ENOUGH !!!
WE NEED TO STUDY

4

Major conclusions on scale precipitation

- pH is the most important chemical parameter for Fe and Ba precipitation followed by I.S.
- pH strongly affects the type of Fe(III) phase precipitated but does not effect Ba phase
- Bitumen released into solution can override the inhibition effect of low pH on Fe precipitation
- The majority of organics (including the scale inhibitor Ethylene Glycol) have no effect or slightly enhance barite scale production
- In high carbonate shales barite can infill microcracks when present but precipitates on the surface without microcracks

5

Iron Precipitation Experiments

- 3 week incubation
- 80°C
- Shale: High Carbonate or High Clay

Fracture fluid (Based on NETL's Marcellus Gas Well E, Greene Co., PA)

1 g ground shale

6

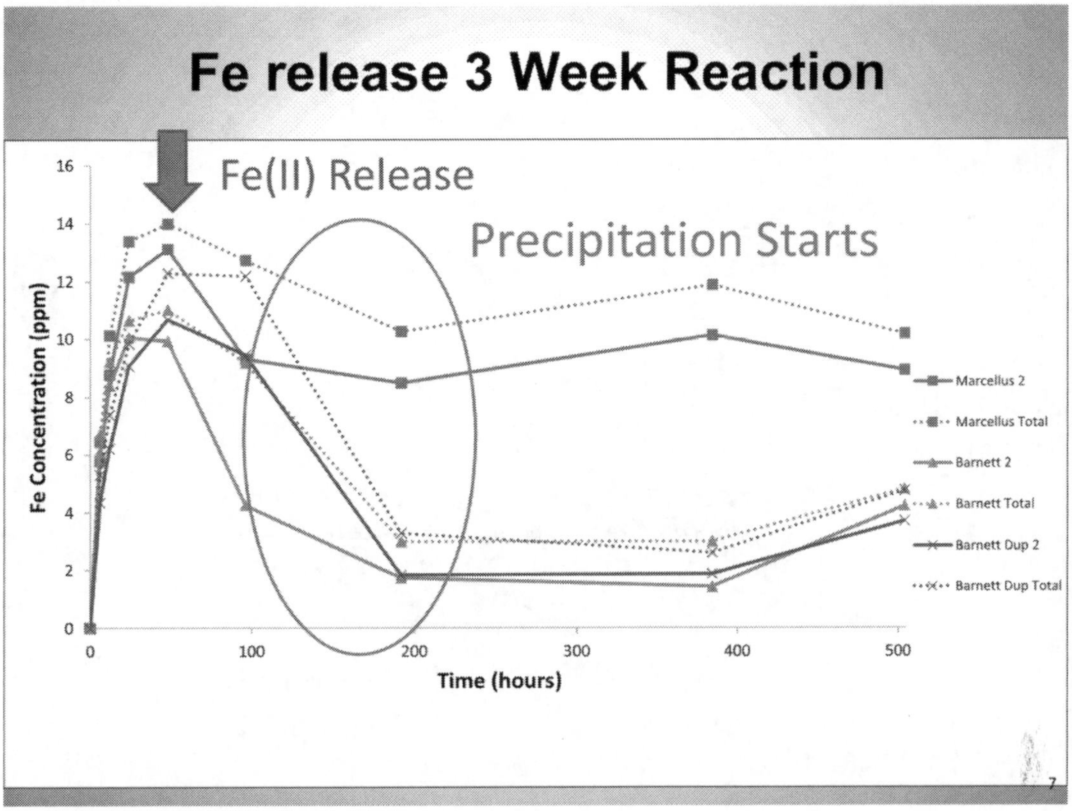
Fe release 3 Week Reaction
Fe(II) Release
Precipitation Starts
Fe Concentration (ppm)
Time (hours)
Marcellus 2
Marcellus Total
Barnett 2
Barnett Total
Barnett Dup 2
Barnett Dup Total
7

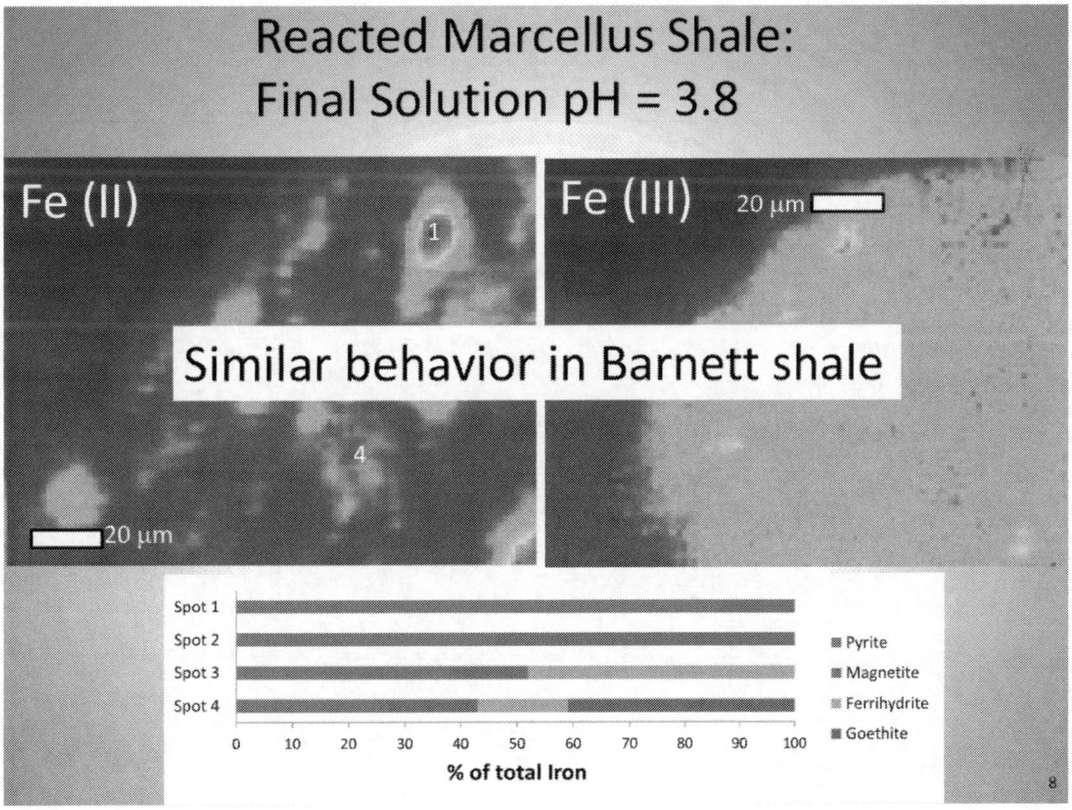
Reacted Marcellus Shale:
Final Solution pH = 3.8
Fe (II)
Fe (III)
20 μm
20 μm
Similar behavior in Barnett shale
Spot 1
Spot 2
Spot 3
Spot 4
Pyrite
Magnetite
Ferrihydrite
Goethite
% of total Iron
8

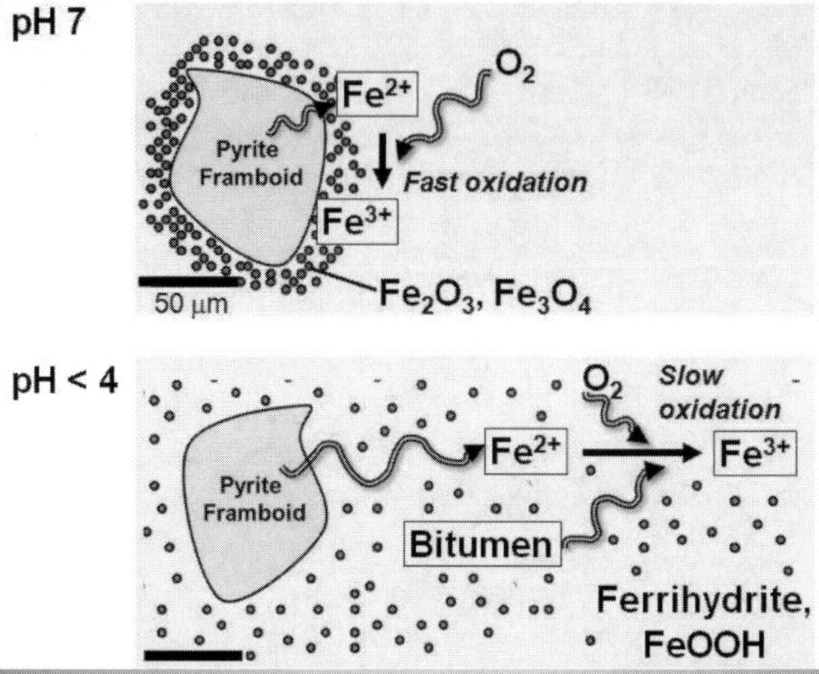

9

Effect of Bitumen on Fe precipitation

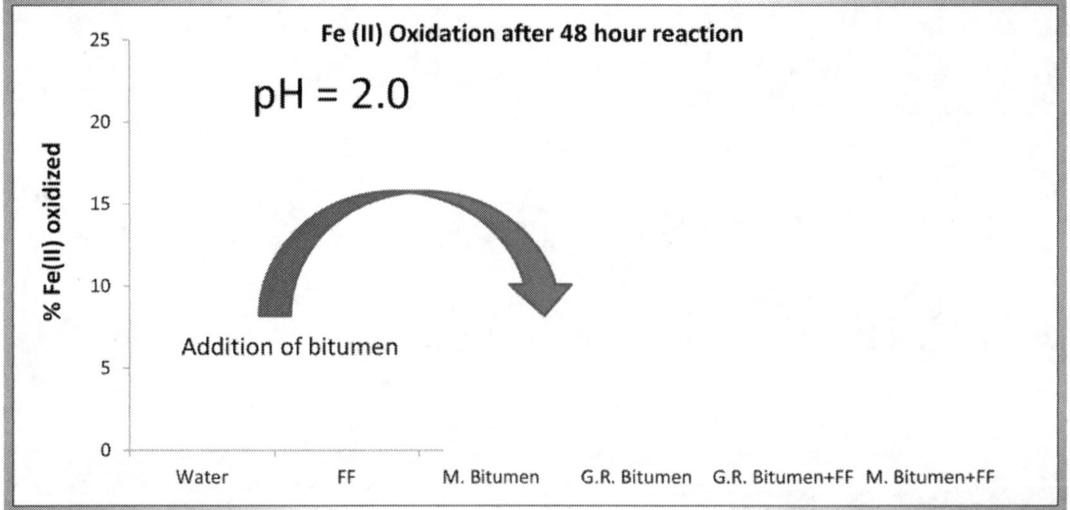

10

Reacted Eagle Ford Shale: Final Solution pH = 6.8

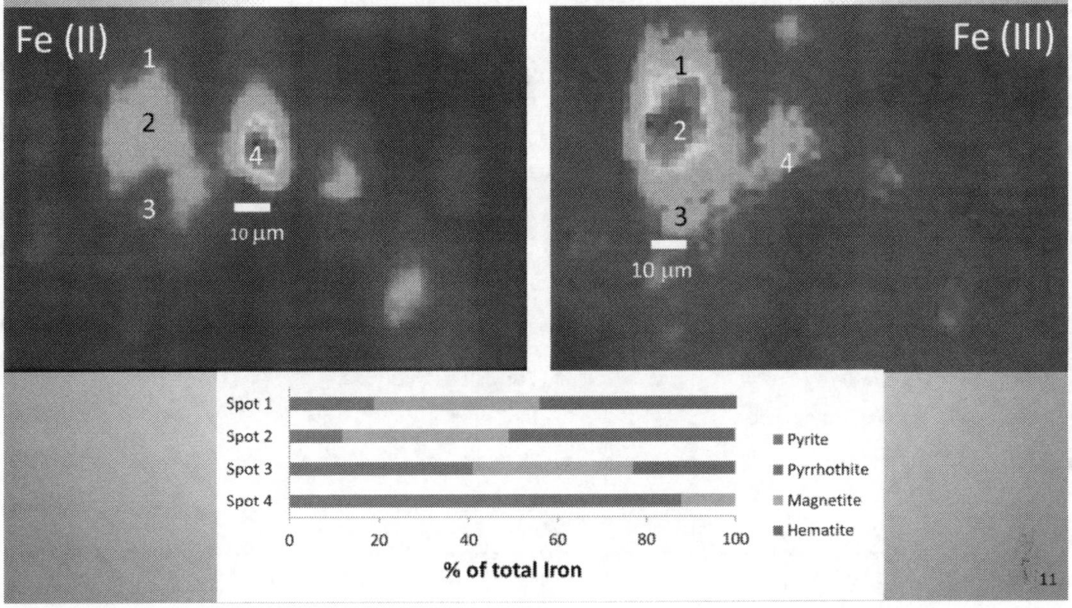

11

Barium Precipitation Experimental Approach

Systematic investigation:

- Individual organics in solution (no shale)
- Various pH conditions
- Various I.S. conditions

- 40 mL batch reactors
- Ba/SO_4 0.01 mM (1:1 $Ba:SO_4$)
- **+ Organics**

- 80°C
- pH: -0.25 (15% HCl) to 7
- 0.06 mM to 2.3 M NaCl
- 7 days, 24 hr sampling
- Filter size: 20 nm

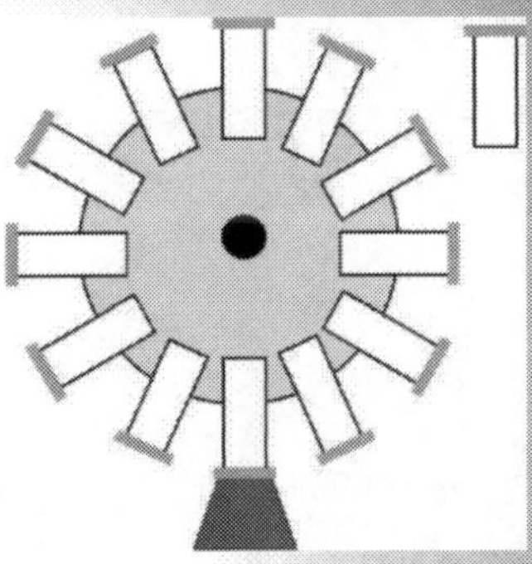

70 rpm End/End Tumbler

12

Organics investigated

	Organic	Purpose
Natural	Acetate	Natural
	Citric Acid	Natural, Iron Control
	Bitumen	Natural
Additives	Kerosene	Friction Reducer
	Ethylene Glycol	Winterizing agent, anti-scaling
	Polyethylene glycol	Biocide
	2-ethyl hexanol	Corrosion inhibitor
	Guar Gum	Gellant
	Glycol ether	Corrosion inhibitor
	Ammonium persulfate	Breaker
	Methanol	Corrosion inhibitor
	Glutaraldehyde	Biocide
Flowback / Produced	Malonate	Produced water
	N-alkane	Produced water
	Cyclohexane	Produced water

13

Effect of pH on barite precipitation rates

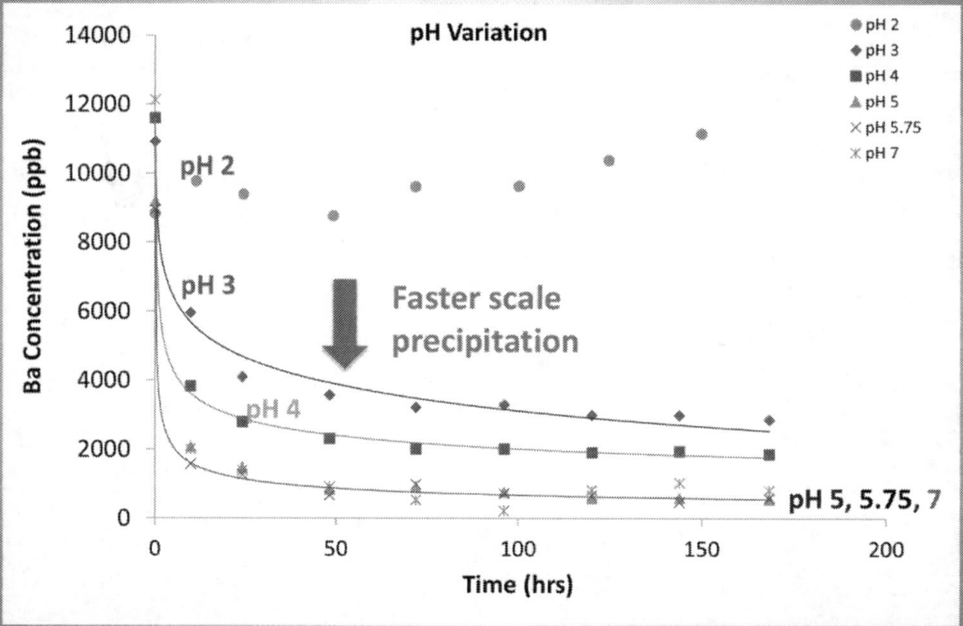

- Precipitation is fastest at neutral pH

14

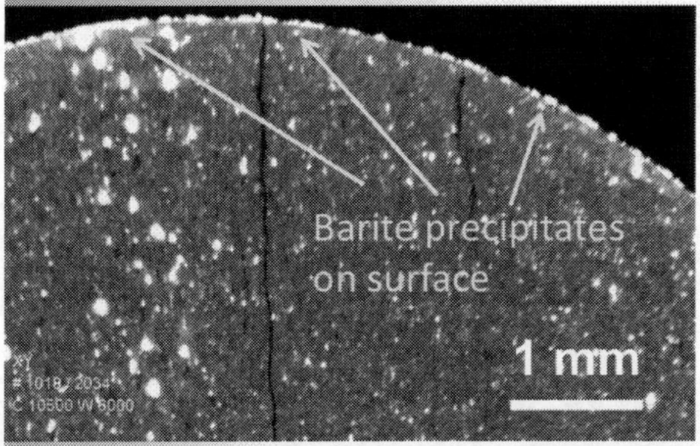

Marcellus shale core plug with Ba addition (S.I. = 1.3) 3 wt.% carbonate

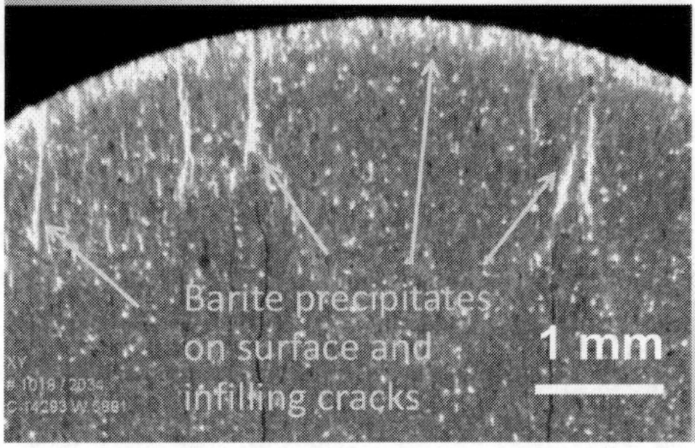

Eagle Ford shale core plug with Ba addition (S.I. = 1.3) 54 wt.% carbonate

$$S.I. = \frac{Q_{exp}}{K_{Barite}}$$

15

Effect of ionic strength

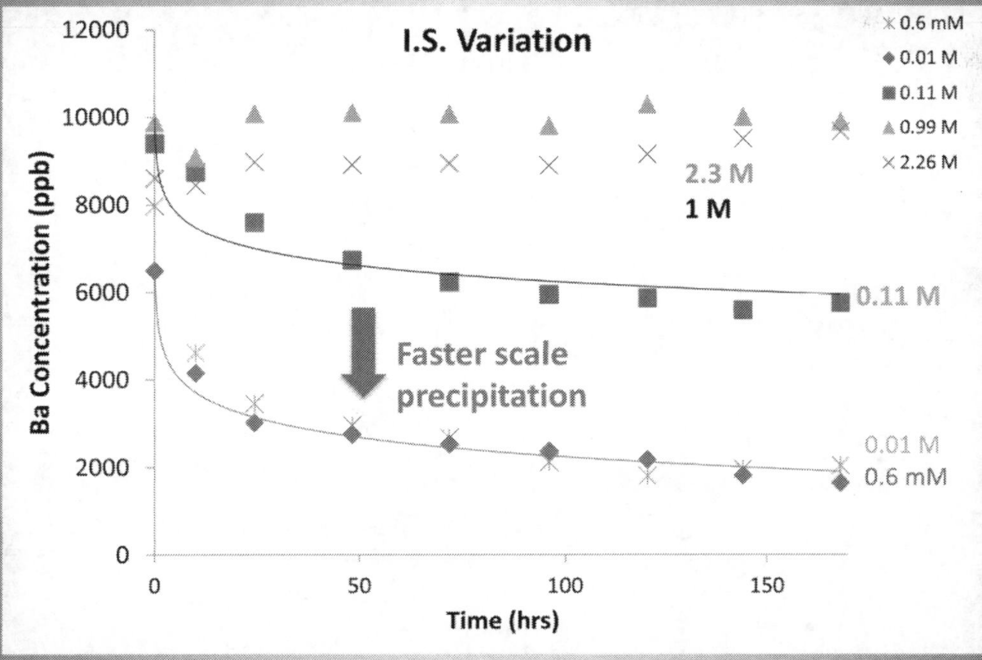

- Precipitation faster at low ionic strength

16

Effect of select organics

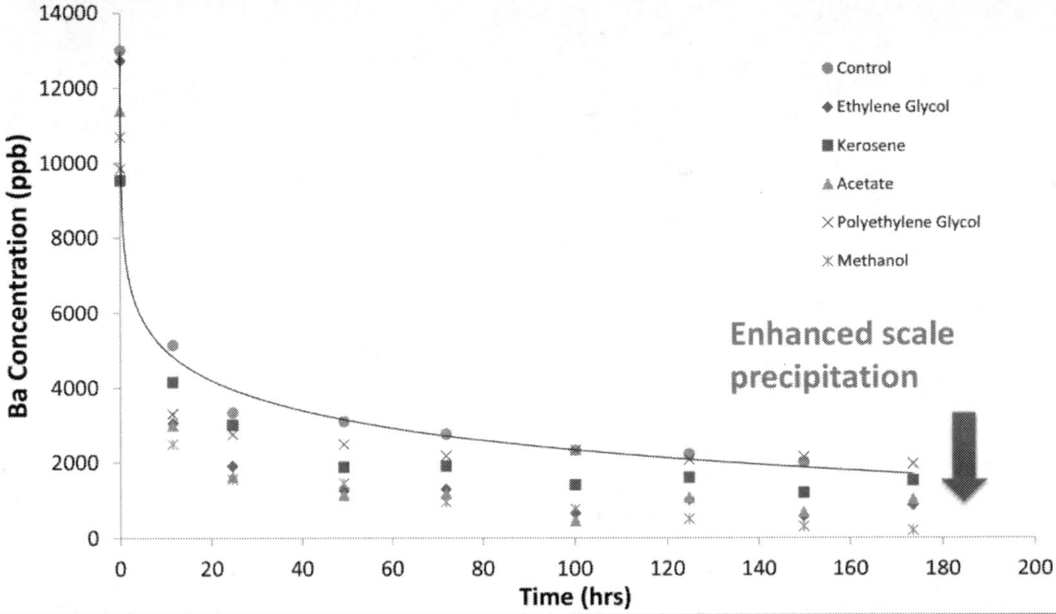

17

Effect of select organics

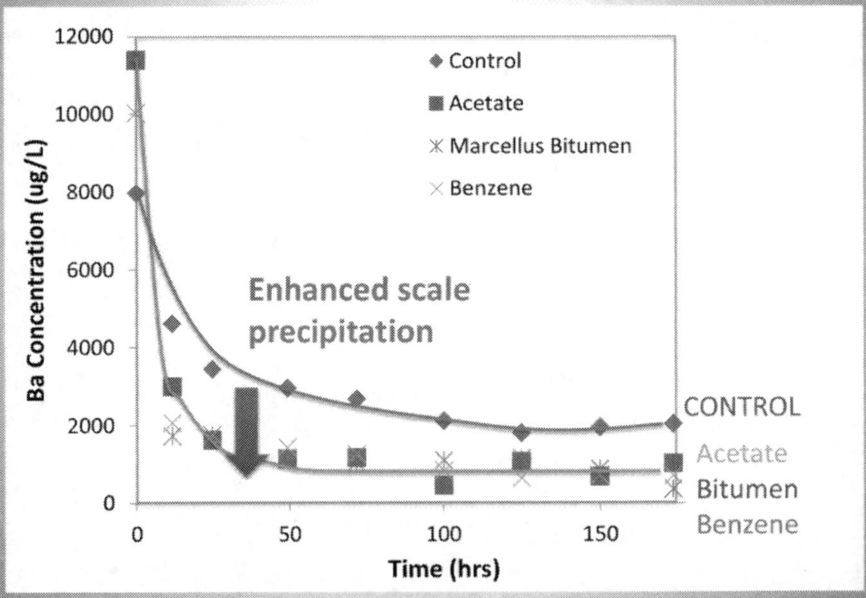

- Abundant natural organics enhance scale formation

18

Effect of organics

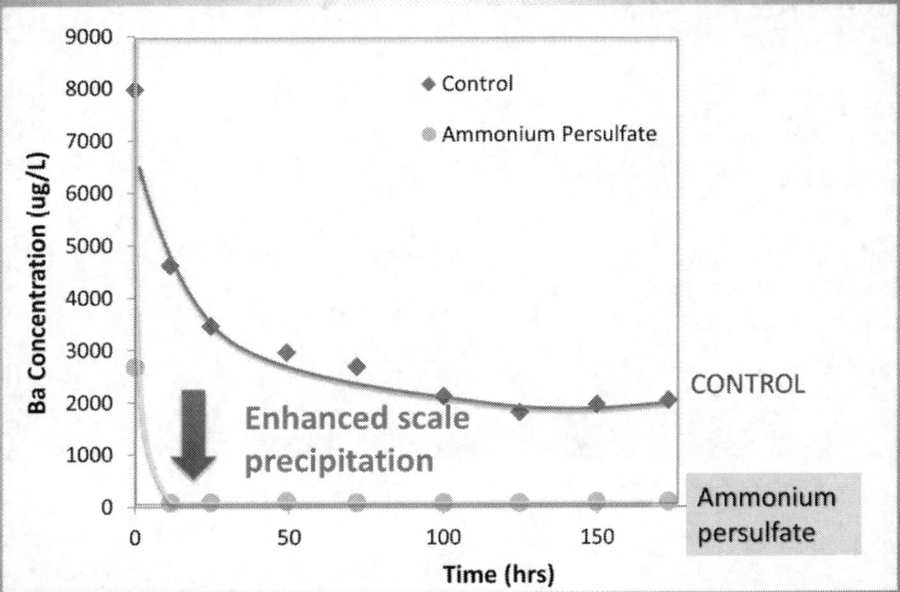

- Ammonium persulfate en[illegible] precipitation in 24 hours

19

Effect of chelating organics

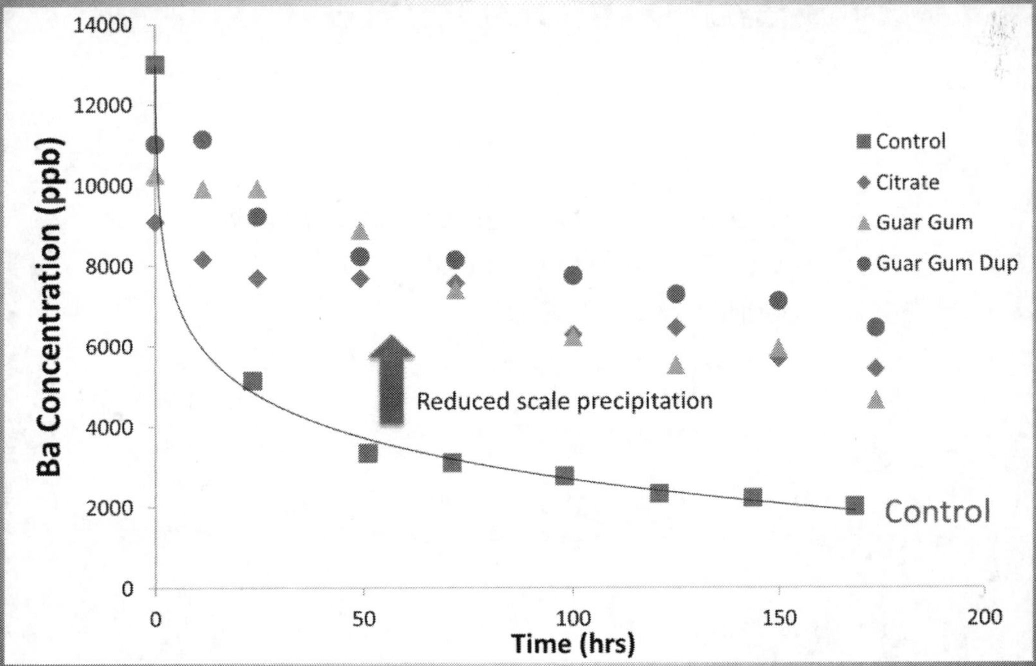

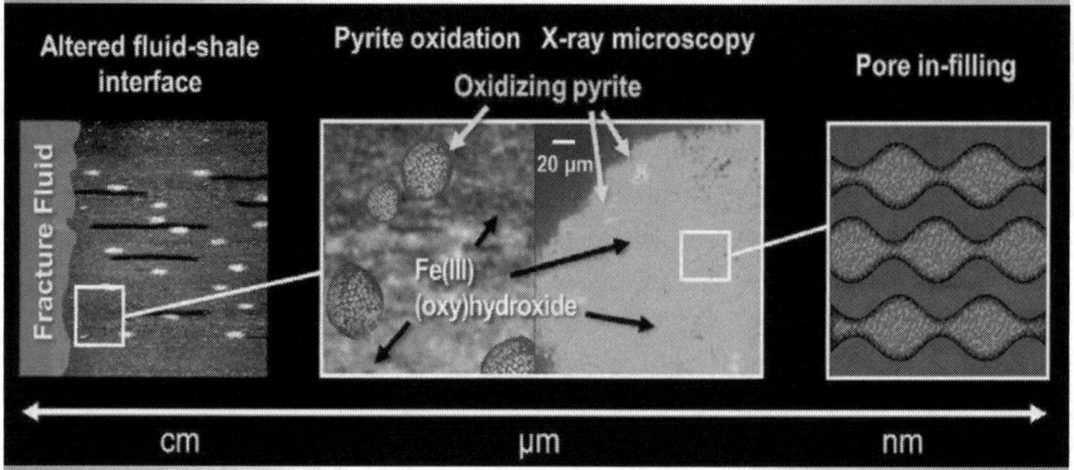

- Though [illegible] differ, this behavior is [illegible] for Barite
- Where [illegible] occurs is [illegible] for affecting [illegible]

21

Acknowledgements

- Stanford Environmental Measurements Laboratory-1:
    - Doug Turner and Guangchao Li
- SSRL is supported by DOE-BES
- A portion of this work was funded by SLAC LDRD

- Funding:
- National Energy Technology Laboratory Grant # FWP100211

22

Fracture Fluid Recipe

Ingredient	Purpose	Concentration (by mass)	Percentage of Ingredient
Water	Base Fluid	1995.6674 g	99.783 %
Ethylene Glycol	Scale Inhibitor, Iron Control, Breaker	0.4113 g	0.021%
Kerosene	Friction Reducer	0.4815 g	0.024%
Guar Gum	Dry Gellant	0.5807 g	0.029%
2-Ethyl hexanol	Corrosion Inhibitor for Acid	0.0093 g	0.0005%
Glycol ether	Corrosion Inhibitor for Acid	0.0031 g	0.0002%
Polyethylene glycol	Biocide	0.3974 g	0.020%
Hydrochloric acid	Acid	2.4493 g	0.122%

Final solution pH = 2

3 week shale (sand-sized particles) reaction

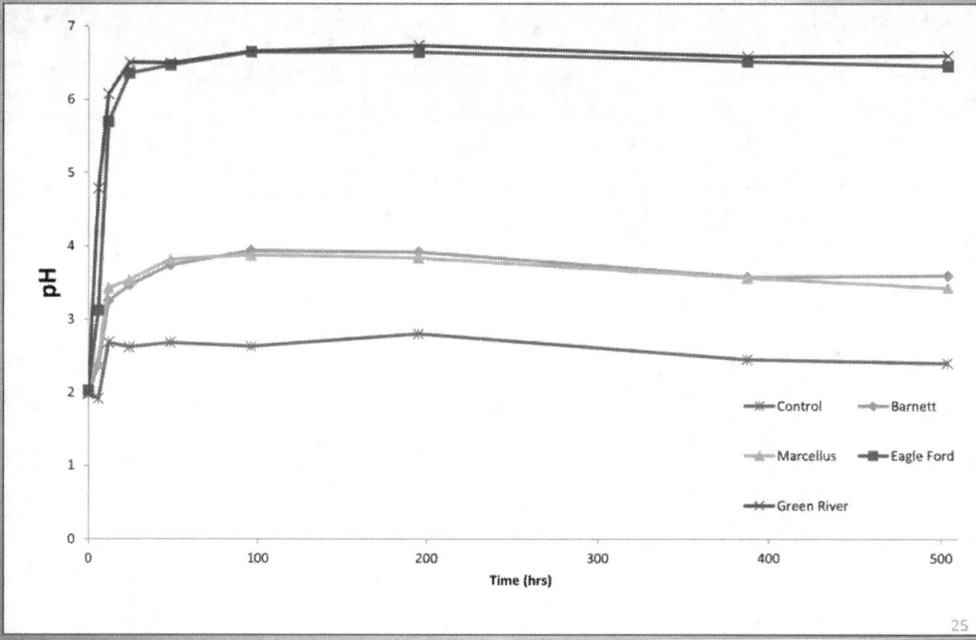

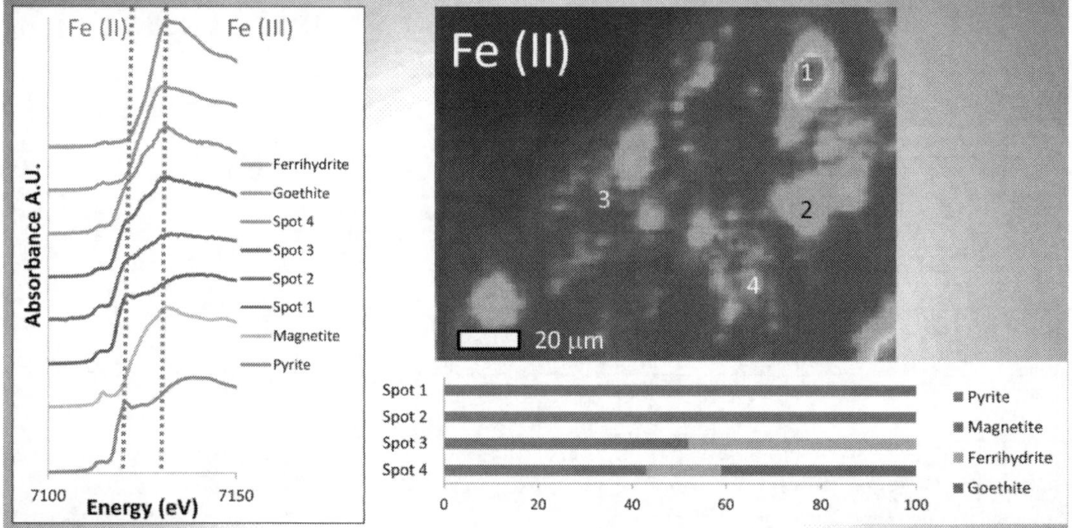

496769 Advanced Supercritical Water-Based Process Concepts for Treatment and Beneficial Reuse of Produced Water Generated By Oil/Gas Production

Abstract:

Both U.S. energy and economic security rely upon continued development of U.S. unconventional shale plays, which require access to suitably clean water resources. Produced water (brine) is the nation's largest industrial waste stream, with approximately 22 billion barrels generated annually by the U.S. oil and gas sector. The recent surge and continued development of unconventional U.S. shale plays will result in greater volumes of produced water. In addition, an additional brine waste stream will be created by future sequestration of CO_2 emissions generated by fossil-based power plants into saline aquifers. Utilizing this brine in development of unconventional shale resources represents a beneficial reuse of this waste stream. This approach can not only lower stress on local watersheds but can also address public/local government concerns regarding long-term ground water contamination potential and seismic activity associated with Class II salt water disposal wells (SWDs).

However, direct reuse of brines in shale development activities currently is limited due to constituents found in this waste stream. Specifically, the high levels of dissolved solids found in the brine may cause scaling within the shale or within production casings, reducing well productivity. Current brine treatment technologies including membrane- and thermal-based processes are ineffective in treating brine containing concentrations of dissolved solids greater than 80,000 ppm due to fouling or cost/sizing, respectively.

To address these limitations, Ohio University (OHIO) with funding from the U.S. Department of Energy's (DOE) National Energy Technology Laboratory (DE-FE0026315), has been developing an advanced supercritical water (SCW)-based process for treatment and beneficial reuse of brine waste streams. This SCW process offers an advantageous media for brine treatment, as its lower fluid density and decreased hydrogen bond strength provides a means to simultaneously remove dissolved solids and hydrocarbons.

Previous SCW-based brine treatment systems have been plagued by internal scaling, resulting in inefficient heat transfer, plugging, and process downtime. To address this issue OHIO has been developing both externally- and internally-heated SCW reactor design concepts, which utilize advantageous fluid dynamics and electrically driven Joule-heating mechanisms, respectively. OHIO's new SCW reactor designs offer the potential to provide a field deployable brine treatment process and a resultant product which may be reused in shale development or other beneficial reuse applications, thereby supporting goals of NETL's Strategic Center for Oil and Natural Gas.

To evaluate process potential, OHIO has been conducting both experimental investigations using prototype SCW reactors and process simulations/techno-economic assessments using Aspen Plus^{TM}. Removal of dissolved solids has been investigated at pressures ranging from 23-28 MPa, respectively, demonstrating clean water product with dissolved solids content ranging from 500-2,800 mg/L from feedstock containing

up to 180 g/L. Tests have also been conducted with produced water generated from an unconventional well located in the Point Pleasant Shale play in eastern Ohio. This presentation will review experimental and techno-economic study results from this U.S. DOE supported project.

493428 Liquid CO_2-in-Mineral Oil Emulsions Stabilized By Siloxane-Long Chain Alkyl Surfactants and Application As a Waterless Hydraulic Fracturing Fluid

Abstract:

CO_2 has long been considered an attractive fracturing fluid for water-sensitive formations, but the very low viscosity of CO_2 (e.g. 0.1 mPa s at 25°C and 2500 psig) inhibits the formation of large hydraulic fractures and makes it difficult to transport high concentrations of large proppant particles into fractures. The CO_2-rich waterless emulsions described in this study may be a practical way of dramatically increasing the apparent viscosity of CO_2 for fracturing applications while offering an easy way to add both the surfactant and the proppant to the mineral oil at ambient pressure (rather than blending surfactant and/or proppant with high pressure CO_2 at the surface).

A water-free, high pressure CO_2-in-oil emulsion (sometimes referred to as foam) of liquid CO_2 droplets within continuous films of mineral oil can be stabilized by a novel, completely hydrophobic, oil-soluble surfactant. The surfactant comprised of an oleophilic (CO_2-phobic) alkyl segment and a CO_2-philic oleophobic polydimethylsiloxane segment adsorbs at the CO_2-oil interface despite the very low interfacial tension without surfactant present. The surfactant increases the oil viscosity by 40% at a concentration of 2 wt % aiding emulsion stabilization. The emulsions were formed by either mixing with an impeller (CO_2 quality or volume fraction up to 60 vol%) or by simultaneously co-injecting the liquids through a packed bed of 22 Darcy unconsolidated sand (CO_2 quality up to 90 vol% with CO_2 droplet sizes in the 20 â€“ 150 micron range). These 50-90% quality CO_2-in-mineral oil emulsions exhibited an apparent viscosity of about 3 - 18 cP at 25°C and 2500 psia with a falling ball and capillary viscometer. Ultimately, these emulsions could be used to greatly reduce water requirements and minimize wastewater produced in hydraulic fracturing.

496595 Laser Induced Breakdown Spectroscopy (LIBS): AÂ Potential Technique for in-Situ Geochemical Characterization of Unconventional Shales

Abstract:

Shale gas is an important asset in the United States and therefore, understanding the geochemical composition of shale formations is of utmost interest and importance. Knowledge of mineralogy and overall chemical constituency can inform engineering and production decisions in the reservoirs and are important factors that influence decisions about drilling locations. The gas capacity can be correlated based on the organic and inorganic content of the shale. Currently, rock core must be retrieved from depth and analyzed using X-ray fluorescence (XRF) or other analytical techniques outside of the wellbore. While these techniques provide excellent data, most of them do not measure all the required elements simultaneously and the analyses are limited to the ambient conditions at the surface. In this study, laser induced breakdown spectroscopy (LIBS) was used as an alternative technique for the analysis of tight shale. LIBS is a fast analytical technique that can provide elemental information of every element on the Periodic Table (including C, H, N, O) both at surface and in-depth of a given sample. Additionally, LIBS can operate in extreme conditions which makes it an attractive tool for in-situ analysis of the core at various drilling depths. Shale samples taken from a Marcellus shale well at depths ranging from 7498-7551 feet were analyzed using a 266 nm Nd:YAG nanosecond laser (LIBS instrument) and an 81 x 81 grid pattern that covered an 8 x 8 mm area. The data collected from these experiments were used to construct 2D elemental maps for each sample. This study shows how the spatial elemental (including C and H) composition varies as a function of the sample depth and illustrates the benefits of LIBS over existing analytical techniques.

497658 Advancing the Production of Olefins and Aromatics from Natural Gas Via Methanol: Chemical Looping for Syngas Generation

Abstract:

Due to technological advances in the shale gas industry, the supply of natural gas in the United States has greatly increased (37% since 2008[1]) and its price has significantly dropped ($2.52/MMBtu in 2016[2]). The shale gas boom has not only revitalized the U.S. economy, but also rejuvenated interests in C1 catalysis toward the production of chemicals typically derived from petroleum refining[3]. Olefins (ethylene, propylene, butane isomers, butadiene) and aromatics (benzene, toluene, xylenes) are examples of valuable chemicals in high demand.

As diesel continues to command a larger portion of the crude oil usage, less naphtha will be available for petrochemical production. Given its availability, low cost, and hydrogen-rich composition, natural gas is an attractive feedstock for the on-purpose production of chemicals[4]. Conversion of natural gas to a methanol intermediate for further upgrading to chemicals is a potential avenue for chemicals production and is a promising commercial candidate for natural gas based refineries. Currently, methanol is almost entirely produced from syngas.

However, a judicious utilization of natural gas resources necessitates a systematic analysis of all available processing routes. To this aim, a gas to olefins and aromatics (GTOA) superstructure has been developed[5,6] and includes several olefins and aromatics production technologies. Previous works investigated reforming as the primary method to convert natural gas to syngas and demonstrated the profitability of GTOA processes. Recently, the benefits of chemical looping[7,8] for syngas generation have been observed in lowering the breakeven oil prices for a gas to liquids (GTL) process[9]. In this work, we incorporate chemical looping into the GTOA superstructure as an alternative to reforming and study its impact on chemicals production.

Process synthesis of the GTOA superstructure forms a large-scale nonconvex mixed-integer nonlinear model (MINLP). A tailored deterministic global optimization branch-and-bound algorithm[10] is used to solve the MINLP and determine optimal natural gas refineries with the highest profit. Several case studies are investigated to compare chemical looping against reforming for natural gas conversion. The effect of different plant capacities and product ratios are explored as well. Key topological selection of process technologies will be discussed.

1. U.S. Energy Information Administration. "Monthly Energy Review – March 2017". Available at: http://www.eia.gov/totalenergy/data/monthly/pdf/mer.pdf. Accessed April 2017.

2. U.S. Energy Information Administration. "Henry Hub Natural Gas Spot Price". Available at:

http://www.eia.gov/dnav/ng/hist/rngwhhdm.htm. Accessed April 2017.

3. Mokrani T., Scurrel M. Gas conversion to liquid fuels and chemicals: the methanol route-catalysis and processes development. Catalysis Reviews 2009, 51, 1-145

4. Floudas, C. A.; Niziolek, A. M.; Onel, O.; Matthews, L. R. Multi-scale systems engineering for energy and the environment: Challenges and opportunities. AIChE Journal 2016, 62 (3), 602-623.

5. Onel, O.; Niziolek, A. M; Floudas, C. A Optimal Production of Light Olefins from Natural Gas via the Methanol Intermediate. Industrial & Engineering Chemistry Research 2016, 55 (11), 3043-3063.

6. Niziolek, A.M.; Onel, O.; Floudas, C.A. Production of Benzene, Toulene, and Xylenes from Natural Gas via Methanol: Process Synthesis and Global Optimization. AIChE Journal 2016, 62 (5), 1531-1556.

7. Luo, S.; Zeng, L.; Xu, D.; Kathe, M.; Chung, E.; Deshpande, N.; Qin, L.; Majumder, A.; Hsieh, T.-L.; Tong, A.; Sun, Z.; Fan, L.-S. Shale gas-to-syngas chemical looping process for stable shale gas conversion to high purity syngas with a H2:CO ratio of 2:1. Energy & Environmental Science 2014, 7 (12), 4104–4117.

8. de Diego, L. F.; Ortiz, M.; García-Labiano, F.; Adánez, J.; Abad, A.; Gayán, P. Hydrogen production by chemical-looping reforming in a circulating fluidized bed reactor using Ni-based oxygen carriers. Journal of Power Sources 2009, 192 (1), 27–34.

9. Tso, W.W.; Niziolek, A.M.; Onel, O.; Demirhan, C.D.; Floudas, C.A.; Pistikopoulos, E.N.; Enhancing Natural Gas-to-Liquids (GTL) Processes Through Chemical Looping for Syngas Production: Process Synthesis and Global Optimization. *In preparation*.

10. Baliban, R. C; Elia, J. A; Misener, R.; Floudas, C. A Global optimization of a MINLP process synthesis model for thermochemical based conversion of hybrid coal, biomass, and natural gas to liquid fuels. Computers & Chemical Engineering 2012, 42, 64-86.

497121 Minimizing the Economic Impact of Amine Scrubbing Using High Fidelity Modeling and Optimization

Abstract:

Minimizing the economic impact of amine scrubbing using high fidelity modeling and optimization

R.C. Pattison, C. Tsay, Y. Zhang, P.T. Frailie, G.T. Rochelle, and M. Baldea

McKetta Department of Chemical Engineering, The University of Texas at Austin, Austin, TX 78712

In an effort to minimize carbon emissions from existing power generation units, amine scrubbing technologies have garnered significant attention owing to the energetically favorable set of reactions between carbon dioxide and amine solvents. A large body of research has been dedicated to this field, with a primary focus on characterizing the reaction kinetics, mass transfer properties, and thermodynamic properties for various solvents [1-2]. Process modeling and simulation studies have also been carried out, with a focus on predicting the performance and cost of the absorption and stripping units [3]. However, relatively little research has been focused on the holistic economic analysis and optimization of the entire process using detailed unit models.

Owing to the complexity of these systems and the numerous design decisions that impact process performance, identifying the most economically compelling designs is extremely challenging, and calls for the use of modern optimization tools. These must be used to consider all the design decisions and constraints simultaneously, such that capital cost and energy use are minimized while simultaneously guaranteeing that the process meets required removal specifications and operational constraints.. While detailed rate-based mass transfer separation models are available in various software programs, the corresponding equations are typically highly nonlinear, coupled and stiff; they are difficult to solve (particularly when embedded within process flowsheets), and thus, not suitable for process-wide optimization.

In this presentation, we introduce a novel framework for plant-wide design optimization of amine-based carbon capture processes. We describe a set of experimentally-validated first-principles and empirical rate-based models of the absorption and stripping columns, along with empirical but detailed capital cost estimates for the process units. The numerical underpinnings of our approach are derived from the pseudo-transient continuation methods and the relaxation-based flowsheet optimization algorithms introduced in our previous work [4-5].

We demonstrate our framework with two extensive case studies, considering amine-based carbon capture plants having different carbon dioxide removal requirements (90% and 99%) from flue gas. The design optimization objective function is designed to minimize the present value (based on both capital and operating expenses) of the plant in the two cases while ensuring that the removal requirements are met and operational constraints are satisfied. The results show significant cost reduction when compared to a heuristics-guided design.

References

[1] Frailie, P.T.; Plaza, J.; Van Wagener, D.; Rochelle, G.T. Modeling piperazine thermodynamics. *Energy Procedia*, **2011**, 4, 35-42.

[2] Sherman, B.; Frailie, P.T.; Li, L.; Salta, N.; Rochelle, G.T. Thermodynamic and Kinetic Modeling of Piperazine/ 2-Methylpiperazine. *Energy Procedia*, **2014**, 63, 1243-1255.

[3] Frailie, P.T. *Modeling of carbon dioxide absorption/stripping by aqueous methyldiethanolamine/ piperazine*. Doctoral Dissertation, 2014.

[4] Pattison, R.C.; Baldea, M. Equation-oriented flowsheet simulation and optimization using pseudo-transient models. *AIChE J.*, **2014**, 60, 4104-4123.

[5] Pattison, R.C.; Tsay, C.; Baldea, M. Pseudo-transient models for multiscale, multiresolution simulation and optimization of intensified reaction/separation/recycle processes: Framework and a dimethyl ether production case study. *Comp. & Chem. Eng.*, **2017.**

IDAES
Institute for the Design of Advanced Energy Systems

Optimal model synthesis for solvent-based CO_2 capture systems by simultaneously using multi-scale data

Paul T. Akula[a], John Eslick[b], Debangsu Bhattacharyya[a], David Miller[b]

[a] *Department of Chemical and Biomedical Engineering, West Virginia University, Morgantown, WV*
[b] *National Energy Technology Laboratory, Pittsburgh, PA*

AIChE Annual Meeting, Minneapolis, MN
November 2nd , 2017

Outline

- **Background**
- **Model Development**
  - **Packed Column**
  - **Wetted- Wall Column (WWC)**
- **Parameter Estimation**
  - **Sequential**
  - **Simultaneous**
- **Challenges**
- **IDAES Platform**
- **Results**
- **Conclusion**

IDAES NATIONAL ENERGY TECHNOLOGY LABORATORY Sandia National Laboratories Carnegie Mellon West Virginia University U.S. DEPARTMENT OF ENERGY

2

Solvent-Based CO_2 Capture System

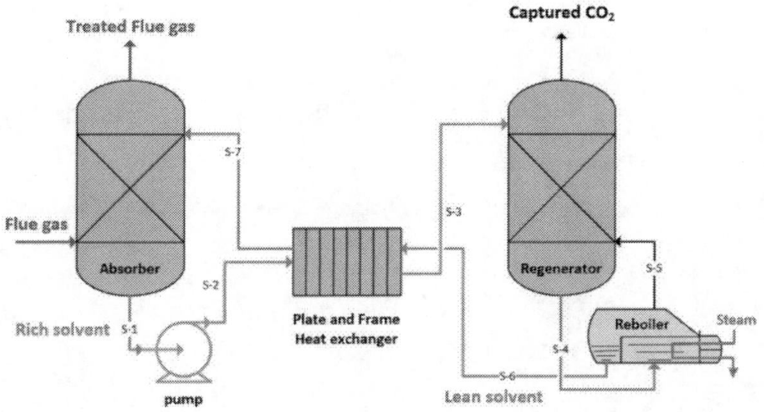

IDAES NATIONAL ENERGY TECHNOLOGY LABORATORY Sandia National Laboratories Carnegie Mellon WestVirginiaUniversity U.S. DEPARTMENT OF ENERGY

3

Packed Column

Mass Balance

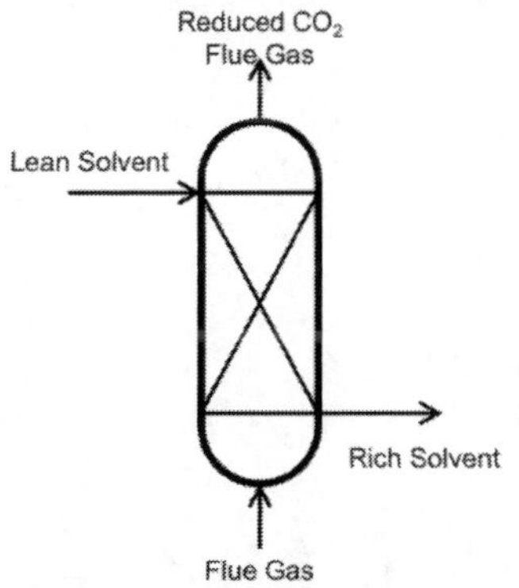

Vapor Phase

$$\varepsilon_V \frac{\partial C_{i,V}}{\partial t} = -u_V \frac{\partial C_{i,V}}{\partial z} - N_{i,V} \qquad i = 1....n_c$$

Liquid Phase

$$\varepsilon_L \frac{\partial C_{i,L}}{\partial t} = u_L \frac{\partial C_{i,L}}{\partial z} + N_{i,L} \qquad i = 1....n_c$$

Mass transfer rate

$$N_{i,V} = K_{V,i} a_e C_{tot,V} \left(y_i - y_i^{*,eq} \right) + N_{tot,V} \left(y_i \right) \qquad i = 1....n_c$$

$$N_{i,V} = N_{i,L}$$

- 1D PDE-based model with plug flow
- Rate-based
- Enhancement factor approach
- NRTL model for liquid phase nonideality

IDAES NATIONAL ENERGY TECHNOLOGY LABORATORY Sandia National Laboratories Carnegie Mellon WestVirginiaUniversity U.S. DEPARTMENT OF ENERGY

4

Packed Column

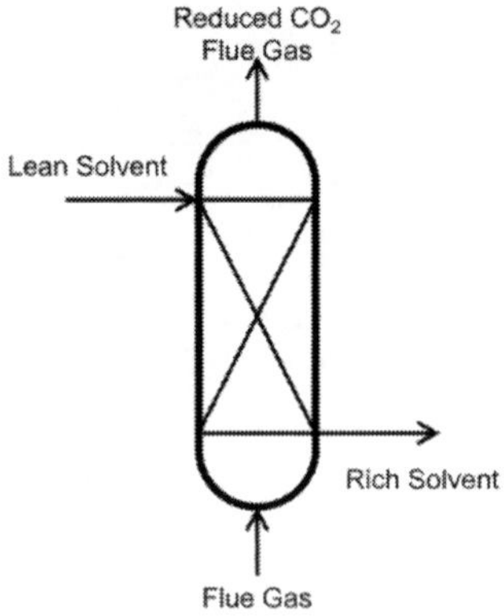

Energy Balance

Vapor Phase

$$\varepsilon_V C_{tot,V} C_{p,V} \frac{\partial T_V}{\partial t} = -u_V C_{tot,V} C_{p,V} \frac{\partial T_V}{\partial z} + Q_{h,V}$$

Liquid Phase

$$\varepsilon_L C_{tot,L} C_{p,L} \frac{\partial T_L}{\partial t} = u_L C_{tot,L} C_{p,L} \frac{\partial T_L}{\partial z} - Q_{h,L}$$

Heat transfer rate

$$Q_{h,V} = a_e h_{VL} (T_L - T_V)$$

$$Q_{h,L} = a_e h_{VL} (T_L - T_V) + N_{CO_2,V} \Delta H_{CO_2} - N_{H_2O,g} \Delta H_{H_2O}$$

Packed Column

Vapor Liquid Equilibrium

For MEA and H2O

$$y_i^* P = \gamma_i x_i^* p_i^s; \quad i = \text{MEA,H2O}$$

For CO2

$$y_i^* P = \gamma_i x_i^* \left(\frac{H_i}{\gamma_i^\infty} \right); \quad where\ \gamma_i^\infty = \lim_{x_i \to 0} \gamma_i\ for\ i = \text{CO2};$$

Henry's constant

$$\ln \left(\frac{H_i}{\gamma_i^\infty} \right) = \sum_s \ln \left(\frac{H_{is}}{\gamma_{is}^\infty} \right) \frac{x_s (V_{cs})^{2/3}}{\sum_{s'} x_{s'} (V_{cs'})^{2/3}} \qquad s = s' = \text{MEA,H2O}$$

Kinetics

For CO_2

$$\frac{1}{K_{V,CO_2}} = \frac{1}{k_{V,CO_2}} + \frac{m_{He_{CO_2}}}{k_{L,CO_2} \boxed{E}}$$

- Physical equilibrium
- Chemical equilibrium
- diffusivity

Enhancement factor

$$E_{abs} = \frac{\sqrt{k_{rx,CO_2} D_{L,CO_2}}}{k_{L,CO_2}}$$

$$E_{desp} = 1 + \frac{\left(D_{L,MEACOO^-} / D_{L,CO_2} \right) \sqrt{K} C_{MEA,L}^*}{\left[1 + 2 D_{L,MEACOO^-} \sqrt{\frac{K C_{CO_2}^*}{D_{L,MEA}}} \right] \left(\sqrt{C_{CO_2,L}^*} + \sqrt{C_{CO_2,L}^{int}} \right)}$$

Wetted-Wall Column (WWC)

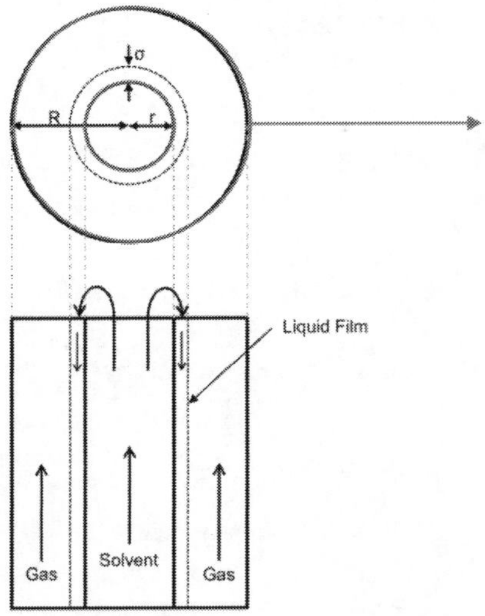

Equivalent Column diameter

$$D = 2\sqrt{(R+r)(R-r)}$$

Specific Interfacial Area

$$a_p = a_e = \frac{\text{interfacial area}}{\text{volume of annulus}} = \frac{2(r+\delta)}{(R^2-r^2)}$$

Void fraction and hydraulic diameter

$$\varepsilon_P = 1, \quad d_h = \frac{4A_c}{P} = 2(R-r)$$

Liquid hold up

$$\varepsilon_L = \frac{\text{volume of liquid film}}{\text{volume of annulus}} = \frac{(r+\delta)^2 - r^2}{R^2 - r^2}$$

Average liquid velocity

$$u_L = \frac{\rho g \delta^2}{3\mu}$$

IDAES NATIONAL ENERGY TECHNOLOGY LABORATORY Sandia National Laboratories Carnegie Mellon West Virginia University U.S. DEPARTMENT OF ENERGY

7

Sub-Models Containing Estimated Parameters

Equilibrium constant

$$K_1 = \frac{C_{RNHCOO^-} C_{RNH_3^+}}{C^*_{CO_2,L} C^2_{RNH_2}}$$

$$K_2 = \frac{C_{RNH_3^+} C_{HCO_3^-}}{C^*_{CO_2,L} C_{RNH_2}}$$

Reaction Equilibrium Model
(WWC & Packed Column)

$$K_{eq,i} = \exp\left(a_i + \frac{b_i}{T} + c_i \ln T\right)$$

Mass transfer Model
(WWC & Packed Column)

$$k_{v,CO_2} = \frac{C_v}{\sqrt{\varepsilon_p - \varepsilon_L}} \frac{\sqrt{a}}{d_h} D_{v,CO_2} \left(\frac{u_v}{a v_v}\right)^{3/4} \left(\frac{\mu_v}{\rho_v D_{v,CO_2}}\right)^{1/3}$$

$$k_{L,CO_2} = 12^{1/6} C_L \left(\frac{u_L}{\varepsilon_L}\right)^{1/2} \left(\frac{D_{v,CO_2}}{d_h}\right)^{1/2}$$

Interfacial area Model
(Packed Column Only)

$$\left(\frac{a_e}{\beta_1}\right)^{\beta_2} = \frac{\rho_L}{\sigma_L} u_L^{\beta_3}$$

IDAES NATIONAL ENERGY TECHNOLOGY LABORATORY Sandia National Laboratories Carnegie Mellon West Virginia University U.S. DEPARTMENT OF ENERGY

8

Typical Approach To Parameter Estimation: Sequential

1st Step Optimization
(smaller scale apparatus or eqp.)

$$\min_{\theta_1}(y-\eta)^T\Sigma_1^{-1}(y-\eta)$$

s.t.

$$\dot{\eta} = f(\eta, u, \theta_1, \theta_2)$$
$$g(\eta, u, \theta_1, \theta_2) \leq 0$$
$$u_{L1} \leq u \leq u_{U1}$$
$$y_{L1} \leq y \leq y_{U1}$$

Small scale (Wetted wall Column)

- Mass transfer
- Diffusivity
- Reaction kinetic Parameters

2nd Step Optimization
(larger scale apparatus or eqp.)

$$\min_{\theta_2}(y-\varphi)^T\Sigma_2^{-1}(y-\varphi)$$

s.t.

$$\dot{\varphi} = f(\varphi, u, \theta_1, \theta_2)$$
$$h(\varphi, u, \theta_1, \theta_2) \leq 0$$
$$u_{L2} \leq u \leq u_{U2}$$
$$y_{L2} \leq y \leq y_{U2}$$

Large scale (Packed Column)

- Interfacial area
- Holdup
- Mass transfer
- Diffusivity
- Reaction kinetic Parameters

IDAES NATIONAL ENERGY TECHNOLOGY LABORATORY Sandia National Laboratories Carnegie Mellon West Virginia University U.S. DEPARTMENT OF ENERGY

9

Limitations of the Sequential Approach

- Mass Transfer and chemical reactions occur simultaneously.
- Experiment must isolate chemical reactions & mass transfer.
- Small scale data may not be representative of large scale equipment.

IDAES NATIONAL ENERGY TECHNOLOGY LABORATORY Sandia National Laboratories Carnegie Mellon West Virginia University U.S. DEPARTMENT OF ENERGY

10

Our Approach: Simultaneous

$$\min_{\theta_1,\theta_2}(y-\eta)^T\Sigma_1^{-1}(y-\eta)+(y-\varphi)^T\Sigma_2^{-1}(y-\varphi)$$

s.t.

$$\dot{\eta}=f(\eta,u,\theta_1,\theta_2)$$
$$\dot{\varphi}=f(\varphi,u,\theta_1,\theta_2)$$
$$h(\varphi,u,\theta_1,\theta_2)\le 0$$
$$g(\eta,u,\theta_1,\theta_2)\le 0$$
$$min(u_{L1},u_{L2})\le u\le max(u_{U2},u_{U2})$$
$$min(y_{L1},y_{L2})\le y\le max(y_{U2},y_{U2})$$

Data:
Small scale (Wetted wall Column)
Large scale (Packed Column)

Simultaneous Parameter Estimation

$$\min_{\boldsymbol{\theta}\in\{\theta_1,\theta_2...\}}\sum_{i=1}^{NG}w_i\left(\sum_{j=1}^{ND}(\varepsilon_{1,i,j}^T\sigma_{1,i,j}^{-2}\varepsilon_{1,i,j}+\varepsilon_{2,i,j}^T\sigma_{2,i,j}^{-2}\varepsilon_{2,i,j})\right)$$

subject to:

WWC model $\forall i,\forall j$

packed Column $\forall i,\forall j$

$\boldsymbol{\theta}^L\le\boldsymbol{\theta}\le\boldsymbol{\theta}^U$

where:

$\boldsymbol{\theta}$ = vector of estimated parameters

σ = standard deviation of data points

NG = number of data set

ND = number of data points in data set

w_i = weight for data set i

$$\varepsilon_{1,i}^T\sigma_{1,i}^{-2}\varepsilon_{1,i}=\frac{(CO_2flux_{wwc,i}^{exp}-CO_2flux_{wwc,i}^{cal})^2}{\sigma_{CO_2flux,wwc,i}^2}$$

$$\varepsilon_{2,i}^T\sigma_{2,i}^{-2}\varepsilon_{2,i}=\frac{(Loading_{abs,i}^{exp}-Loading_{abs,i}^{cal})^2}{\sigma_{loading,abs,i}^2}$$

Data

Bench scale -WWC
Pilot scale - NCCC

} Wide range of operating conditions {

- Liquid flowrate
- Gas flowrate
- Solvent loading
- Temperature
- Flue gas concentration

Data Sources
- National Carbon Capture Center (NCCC) Pilot Plant
- Bench Scale WWC[1]

[1]Experimental data retrieved from Dugas, R.E . (2009). *Carbon Dioxide Absorption, Desorption, and Diffusion in Aqueous Piperazine and Monoethanolamine*. (UT Austin Phd Thesis)

Challenges

Size
- Number of equations (absorber ~15,000, WWC~6,000)
- Number of data points (absorber data from NCCC ~30 set, WWC data ~200 set)
- Total number of equation (simultaneously considered):~1,650,000

Nature
- Highly nonlinear equations
- Different scales (WWC height~9 cm and diameter~ 1.5 cm and
  NCCC absorber height ~20 m and diameter~ 0.64 m)

Capabilities Required
- Commercial simulators: primarily sequential
- Process and property sub-models
- Large # of equations
- Large amount of experimental data

IDAES NATIONAL ENERGY TECHNOLOGY LABORATORY Sandia National Laboratories Carnegie Mellon WestVirginiaUniversity U.S. DEPARTMENT OF ENERGY

Core Pyomo Capabilities Enabling IDAES

- **Hierarchical modeling**
  - Blocks
  - External functions

PYOMO ➡ **Property block & Unit models**

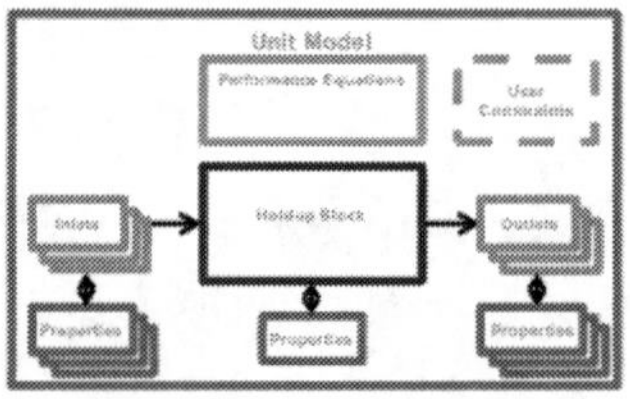

- **(Partial) differential equations** PYOMO DAE ➡ **Dynamic process models, optimal control**

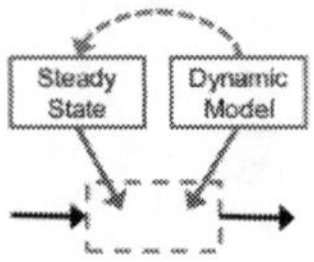

- **Stochastic programming** PySP ➡ **Uncertainty quantification**

- **Generalized Disjunctive Programming [Pyomo.GDP]** ➡ **Superstructure optimization**

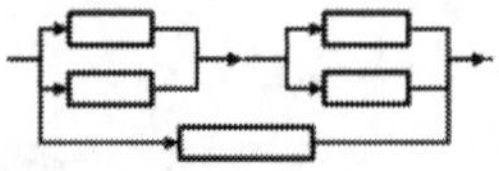

IDAES NATIONAL ENERGY TECHNOLOGY LABORATORY Sandia National Laboratories Carnegie Mellon WestVirginiaUniversity U.S. DEPARTMENT OF ENERGY

Pyomo: Extensible Object-oriented Algebraic Modeling in Python

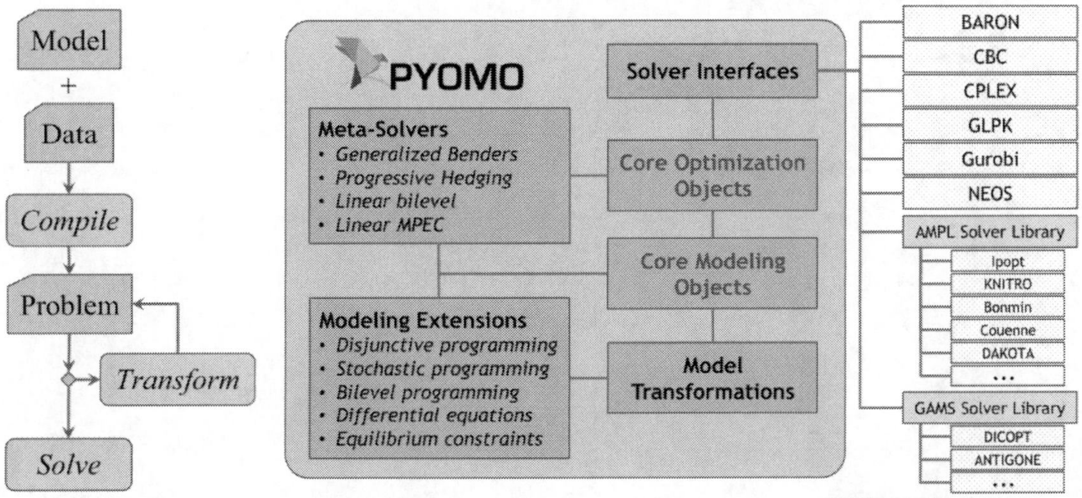

15

Results

WWC & Packed Column

	Before opt	After opt
OBJECTIVE	**14.35**	**0.2058**

Model	Parameters	Before opt	After opt
Equilibrium reaction	a_1	198.4	200.0
	b_1	-1986	-1803
	c_1	-32.71	-32.98
	a_2	166.6	164.0
	b_2	-2102	-2200
	c_2	-26.51	-25.95
Mass transfer	Cv	0.357	0.1495
	CL	0.5	0.4903
Interfacial area	β_1	157.0	39.87
	β_2	8.621	23.95
	β_3	1.333	0.9962

16

Results

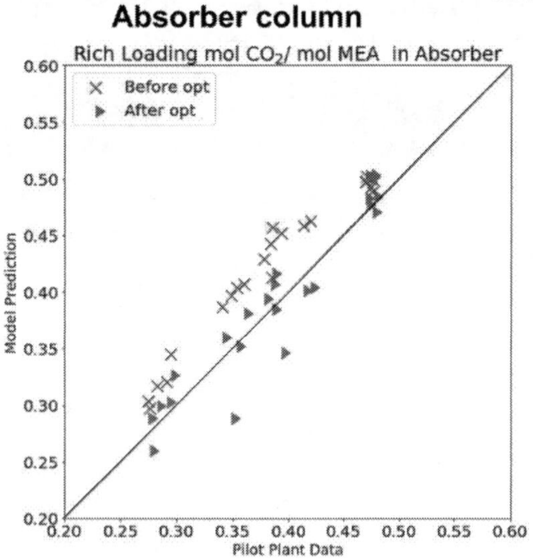

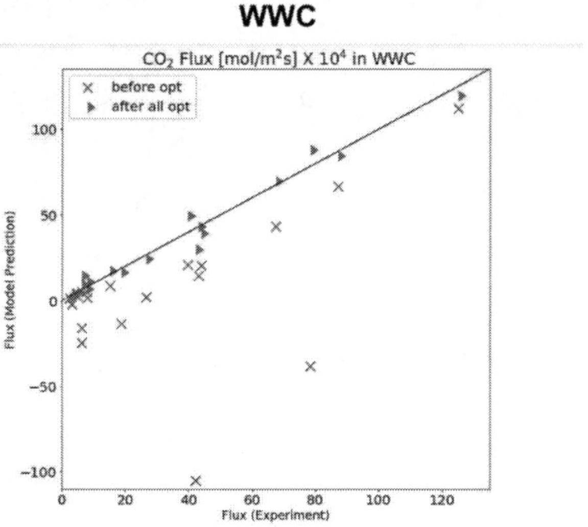

Conclusion

- Process models for packed bed and wetted wall columns
- IDAES framework demonstrated for parameter estimation
  - Complex process models
  - Multi-scale data
- Superior fitting

Continuing work:

- Larger subsets of model parameters with larger data sets
- Pyomo decomposition
  - Enable larger data sets and more complex models

ACKNOWLEDGEMENT & DISCLAIMER

19

Acknowledgement

This research was funded by the Lawrence Berkeley National Laboratory through Subcontract NO.7268587 as part of the U.S. DOE's Institute for the Design of Advanced Energy Systems (IDAES).

Disclaimer

This presentation was prepared as an account of work sponsored by an agency of the United States Government. Neither the United States Government nor any agency thereof, nor any of their employees, makes any warranty, express or implied, or assumes any legal liability or responsibility for the accuracy, completeness, or usefulness of any information, apparatus, product, or process disclosed, or represents that its use would not infringe privately owned rights. Reference herein to any specific commercial product, process, or service by trade name, trademark, manufacturer, or otherwise does not necessarily constitute or imply its endorsement, recommendation, or favoring by the United States Government or any agency thereof. The views and opinions of authors expressed herein do not necessarily state or reflect those of the United States Government or any agency thereof.

IDAES NATIONAL ENERGY TECHNOLOGY LABORATORY Sandia National Laboratories Carnegie Mellon West Virginia University U.S. DEPARTMENT OF ENERGY

19

THANK YOU FOR YOUR ATTENTION

20

IDAES NATIONAL ENERGY TECHNOLOGY LABORATORY Sandia National Laboratories Carnegie Mellon West Virginia University U.S. DEPARTMENT OF ENERGY

20

494599 The Value of Chemical Looping Combustion in Future Electricity Systems

Abstract:

Power generation via Chemical Looping Combustion (CLC) is inherently low in carbon dioxide emissions while maintaining the advantages of conventional thermal power plants such as dispatchability, flexibility, or firmness of capacity. The energy penalty incurred by the carbon dioxide removal process is performed at a fraction of other carbon capture techniques [1], such as post-combustion capture using amine solvents.

In this work, we quantify the role and value of three different CLC technologies and their integration into the future electricity system; CLC with natural gas coupled to a combined cycle (NGCC-CLC), integrated gasification combined cycle with CLC (IGCC-CLC) and CLC with in situ gasification and combustion of solid fuel (iG-CLC). We develop a national-scale capacity expansion and unit commitment model based on a mixed-integer linear program (MILP), minimising total system cost while limiting carbon emission, and maintaining security, and operability requirements [2]. Detailed operational constraints (e.g., start-up, ramp rates, minimum stable generation) on an hourly time discretisation enable the analysis of individual power plant performance characteristics and an assessment of their system-level impact.

We investigate the effects of the integration of CLC power plants into a 2030 and 2050 electricity system of the United Kingdom on the optimal capacity mix, the operational and system cost, and dispatch patterns. All three CLC power generation technologies prove valuable to the electricity system, as their deployment reduces system-wide investment and operational cost. Power generation from intermittent renewables becomes less economical as CLC technologies become available. Consequently, the requirements for carbon-intense reserve capacity, energy storage and overseas electric interconnection reduces. NGCC-CLC operating at virtually zero-carbon dioxide emission and moderate operation cost compares most favourably to IGCC-CLC and iG-CLC. The most cost-effective operational strategy for CLC power plants is in a steady manner, however, their ability for flexible operation can complement intermittent renewable power generation.

Additionally, we present the sensitivities to capital and operating expenditure and efficiency. A reduction in the efficiency of NGCC-CLC power plants from 46 % to 42 % reduces their annual average utilisation rate by only 1 %. For iG-CLC, the sensitivity to the cost of the oxygen carrier material is small; a variation of -50 % to +70 % causes a deviation in the average utilisation factor of -2 % to 3 %. For iG-CLC and IGCC-CLC, a 20-40 % increase in capital cost can reduce optimal capacity deployment levels considerably. NGCC-CLC deployment is more robust against investment cost increase. The results emphasises the high value of CLC power generation in a carbon-constrained system and can indicate and guide a prioritisation of future research efforts.

[1] C. Ekstrom, F. Schwendig, O. Biede, F. Franco, and G. Haupt. Technoeconomic evaluations and benchmarking of pre-combustion CO_2 capture and oxy-fuel processes developed in the European ENCAP project. Energy Procedia,

1(1):4233-4240, 2009.
[2] C. F. Heuberger, I. Staffell, N. Shah, and N. Mac Dowell. Levelised Value of Electricity - A Systemic Approach to Technology Valuation. In 26th European Symposium on Computer Aided Process Engineering, volume 38, pages 721-726, 2016.

498958 Superstructure-Based Optimization of Membrane-Based Carbon Capture Systems

Abstract:

This work aims to optimize the plant design, process configuration and operating conditions of a post-combustion membrane-based CO_2 capture plant. A new superstructure-based mathematical model is presented, modeling discrete and continuous decision variables to minimize the cost of electricity (COE). The proposed approach tackles a major issue in carbon capture and storage (CCS) for the power generation industry, involving the simultaneous optimization of process configurations and operating conditions (gas temperature, pressure ratio, sweep feed, etc.).

The development of efficient and cost-effective CCS systems is important to mitigate the CO_2 emission in fossil fuel-based power generation. Available technologies for post-combustion carbon capture include liquid-solvents, gas-permeation, and solid sorbent systems. Solvent and Sorbent â€“ based systems can considered mature technologies for CCS, while membrane systems are generally less well-studied.

This work focuses on membrane-based systems for CCS applied to gas permeation, where the membranes with high CO_2 selectivity separate the CO_2 from other components in the flue gas. The main driving force of the process is the partial pressure between the feed and the permeate sides. Membrane systems often consist of multiple stages of membranes and compressors with intercooling. Most of the applications consider membrane models discretizing the membrane area and solving the separation problem for each membrane-section. This often results in a large-scale model typically analyzed through process simulation-based frameworks (ASPEN, gPROMs, etc.)[1] developed primarily for accurately modeling performance as opposed to optimization. Therefore, the number of configurations that can be optimized by conventional simulation-based frameworks is limited.

Significant efforts have been made for developing multi-stage systems, where previous studies focused on optimizing the number of stages and their operation[2,3], instead of analyzing advanced process configurations[4]. Advanced process configurations can lead to important savings in operating and capital costs. This work analyzes advanced process configurations while considering rigorous process models in a mathematical optimization framework.

A set of streams, and multiple processes (membrane stages, mixers, splitters, pumps and compression units with intercooling) is considered. Discrete design decisions determine the plant layout by selecting the optimal number of stages (where each stage consists of a membrane, compressor, pump, and heat exchanger units) and the process configuration. At the operating level, continuous decision variables involve the optimal flows, temperatures, concentrations, pressure ratio, and sweep feed. The proposed model is formulated as a mixed integer nonlinear programming model that minimized the cost of electricity to derive an optimal design and operation of a post-combustion carbon capture plant. The optimization of several process configurations led to lower CCS costs.

[1] Eslick J. C., Tong C., Lee A., Dowling A. W., Mebane D. S. (2015). Optimization under uncertainty with rigorous process models. Proceedings AIChE Annual Meeting 2015.

[2] Hasan M. M., Baliban R. C., Elia J. A., Floudas C. A. (2012). Modeling, Simulation, and Optimization of Postcombustion CO_2 Capture for Variable Feed Concentration and Flow Rate. 1. Chemical Absorption and Membrane Processes. Industrial and Engineering Chemistry Research. 51, 15642-15664.

[3] Arias A. M., Mussati M. C., Mores P. L., Scenna N. J. (2016). Optimization of multi-stage membrane systems for CO_2 capture from flue gas. International Journal of Greenhouse Gas Control. 53, 371-390.

[4] Merkel T. C., Lin H., Wei X., Baker R. (2010). Power plant post-combustion carbon dioxide capture: An opportunity for membranes. Journal of Membrane Science. 359, 126-139.

Imperial College London

Bio-energy with carbon capture and storage (BECCS): Opportunities for efficiency improvement

Mai Bui[a,b], Mathilde Fajardy[a,b], Niall Mac Dowell[a,b]

[a]Centre for Environmental Policy, Imperial College London, London SW7 1NA UK
[b]Centre for Process Systems Engineering, Imperial College London, London SW7 2AZ UK

cep Centre for Process Systems Engineering

2nd November 2017

EPSRC Engineering and Physical Sciences Research Council

Imperial College London

Meeting the below 2°C and 1.5°C COP21 targets

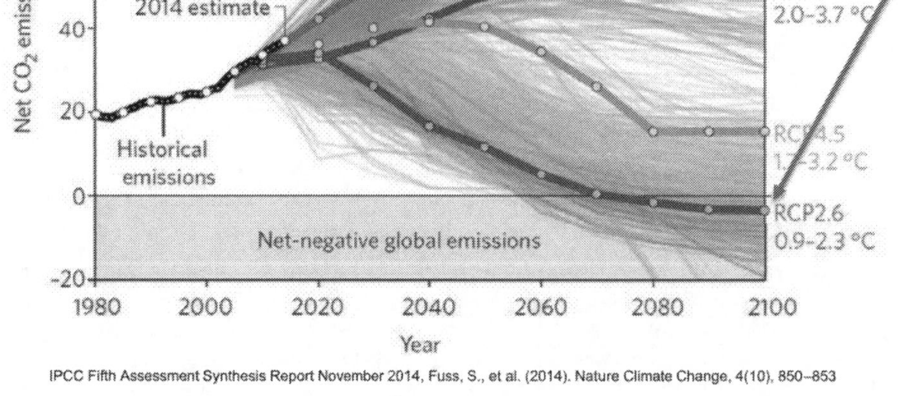

- Mitigation costs 138% more without CCS
- Without CCS, <50% of IAM models solve the problem
- Most feasible solutions use BioEnergy + CCS substantially

→ BECCS is a key negative emissions technology

IPCC Fifth Assessment Synthesis Report November 2014, Fuss, S., et al. (2014). Nature Climate Change, 4(10), 850–853

2

Imperial College London

Negative emissions technology: BECCS

Atmospheric CO_2 is captured and permanently stored → Negative CO_2 balance

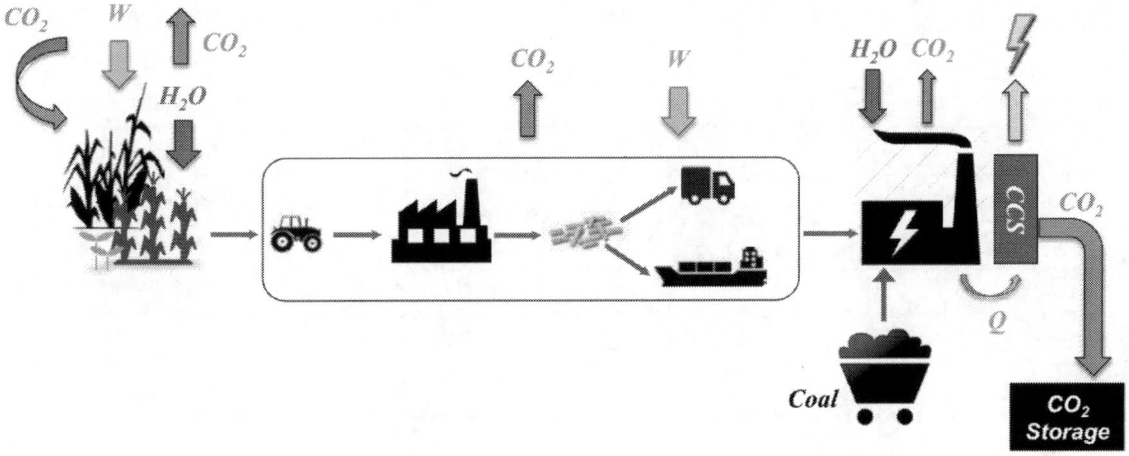

Reference: Fajardy, M. & Mac Dowell, N. (2017). Can BECCS deliver sustainable and resource efficient negative emissions? Energy and Environmental Science, 10 (6), 1389-1426.

3

Imperial College London

Fuel properties and composition

Property	Coal	Wood	Herbaceous
Ash content (wt% dry)	8.5 – 10.9	0.4 – 4.0	5 – 7.5
Moisture content (wt% wet)	6 – 10	5 – 60	15 – 25
Lower heating value, LHV (MJ/kg)	26.0 – 28.3	18.4 – 20.0	17.1 – 17.5
C (wt% dry)	76 – 87	47 – 52	45 – 47
H (wt% dry)	3.5 – 5.5	5.8 – 6.7	5.7 – 6.0
O (wt% dry)	2.8 – 11.3	38 – 46	40 – 46
Cl (wt% dry)	< 0.1	0.01 – 0.05	0.09 – 0.97
S (wt% dry)	0.5 – 3.1	0.02 – 0.10	0.05 – 0.2
N (wt% dry)	0.8 – 1.5	0.1 – 0.8	0.4 – 1.04

Efficiency penalty due to: Biomass (moisture & low heating value) + CO_2 capture process requires heat

4

Imperial College London

Fuel properties and composition

Property	Coal	Wood	Herbaceous
Ash content (wt% dry)	8.5 – 10.9	0.4 – 4.0	5 – 7.5
Moisture content (wt% wet)	6 – 10	5 – 60	15 – 25
Lower heating value, LHV (MJ/kg)	26.0 – 28.3	18.4 – 20.0	17.1 – 17.5
C (wt% dry)	76 – 87	47 – 52	45 – 47
H (wt% dry)	[illegible]	[illegible]	5.7 – 6.0
O (wt% dry)	[illegible]	[illegible]46	40 – 46
Cl (wt% dry)	< 0.1	0.01 – 0.05	0.09 – 0.97
S (wt% dry)	0.5 – 3.1	0.02 – 0.10	0.05 – 0.2
N (wt% dry)	0.8 – 1.5	0.1 – 0.8	0.4 – 1.04

SO_X and NO_X emissions

Reference: Bui, M., Fajardy, M. & Mac Dowell, N. (2017). Bio-Energy with CCS (BECCS) performance evaluation: Efficiency enhancement and emissions reduction. Applied Energy, 195, 289–302.

5

Imperial College London

Heating and condensation of gas phase

Solids and liquids formed at various stages

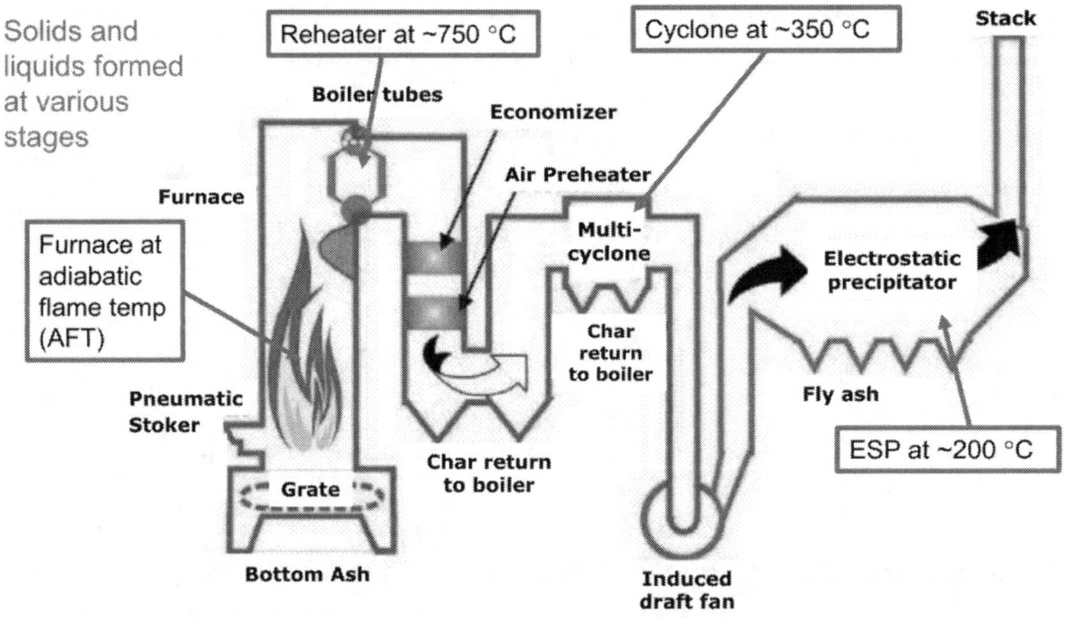

Image: Overend, Ralph P. Direct combustion of biomass, Encyclopedia of Life Support Systems

6

Imperial College London

Flue gas condensation in a power plant

Unit	Temperature (°C)	Solid/liquid removed (%)	Solid/liquid carried to next step (%)
Furnace[1]	AFT	80	20
Reheater[2]	750	15 (fouling)	85
Multi-cyclone[3]	350	80	20
ESP[4]	200	99	1

[1] Biomass combustion ash is typically 80% bottom ash, 20% fly ash (Jensen et al., 2013; Hansen et al., 2014)

[2] Biomass ash fouling at ~750 °C based on T_{15} (varies with Cl & K content) (Mansikkasalo et al., 2014)

[3] Multi-cyclones: separation efficiency of 80-95% (US EPA, 2016)

[4] ESP: separation efficiency of 99-99.9% (US EPA, 2016)

References: Jensen, P. A., *et al.* (2013). Characterization and quantification of deposits build up and removal in straw suspension fired boilers. Energinet.dk project no. 7217
Hansen, S. B., *et al.* (2014). Deposit probe measurements in large biomass-fired grate boilers and pulverized-fuel boilers. Energy & Fuels, 28 (6), 3539–3555.
Mansikkasalo, J., *et al.* (2014). High power generation from recovery boilers, what are the limits? [Online]. 2014 International Chemical Recovery Conference, Finland.
USEPA (2016). Air Pollution Control Technology Fact Sheets. United States Environmental Protection Agency (USEPA) Technology Transfer Network.

7

Imperial College London

Automation of FactSage: Macros

FactSage

FactSage 7.0 is a thermo-chemical modelling tool – accesses thermodynamic data from the FACT and SGTE databases

Macros can be used to automate a large number of simulations:

- Incremental temp increase
- Removal of a particular phases from the reaction

Designed macros used to simulate the following:

- Combustion Process: (200°C to AFT)
- Flue gas condensation divided into:
  i. Furnace to Reheater (AFT to 750°C)
  ii. Reheater to Multi-cyclone (750°C to 400°C)
  iii. Multi-cyclone to ESP (400°C to 200°C)
  iv. ESP to Stack (200°C to 120°C)

Can observe flue gas composition changes during heating and cooling of the flue gas

8

Imperial College
London

Flue gas combustion-condensation graph

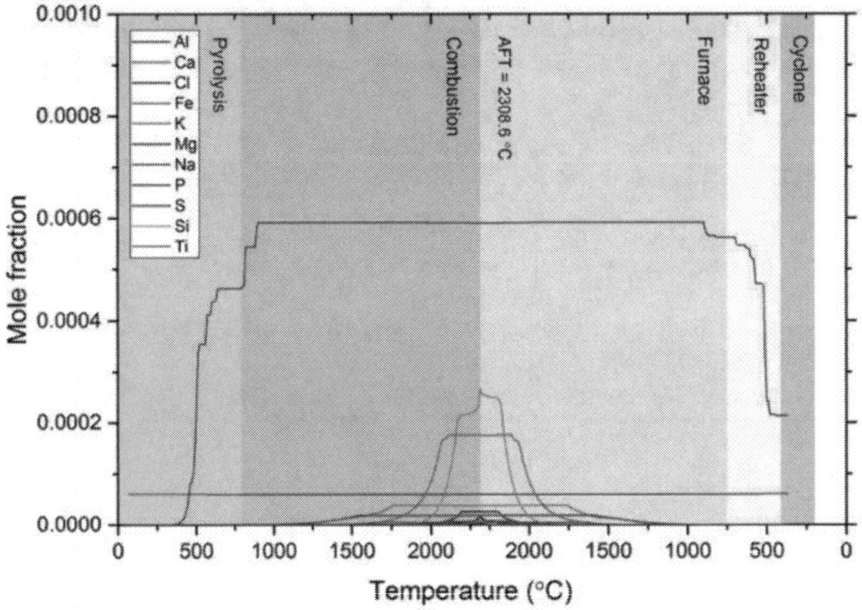

9

Imperial College
London

Energy flow in a thermal power plant

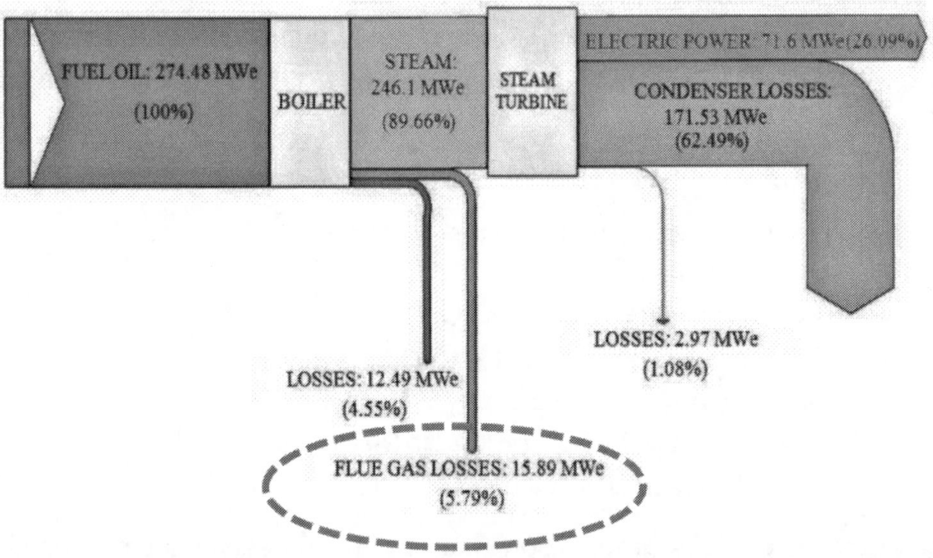

Irrazabal Bohorquez *et al.*: Repowering: An option for refurbishment of old thermal power plants in Latin-American countries. Proceedings of ASME Turbo Expo 2010: Power for Land, Sea and Air GT2010 (June 14-18, 2010, Glasgow, UK). DOI: 10.1115/GT2010-23058
http://www.sankey-diagrams.com/tag/power-plant/

10

Imperial College London

Flue gas heat recovery to improve power plant efficiency

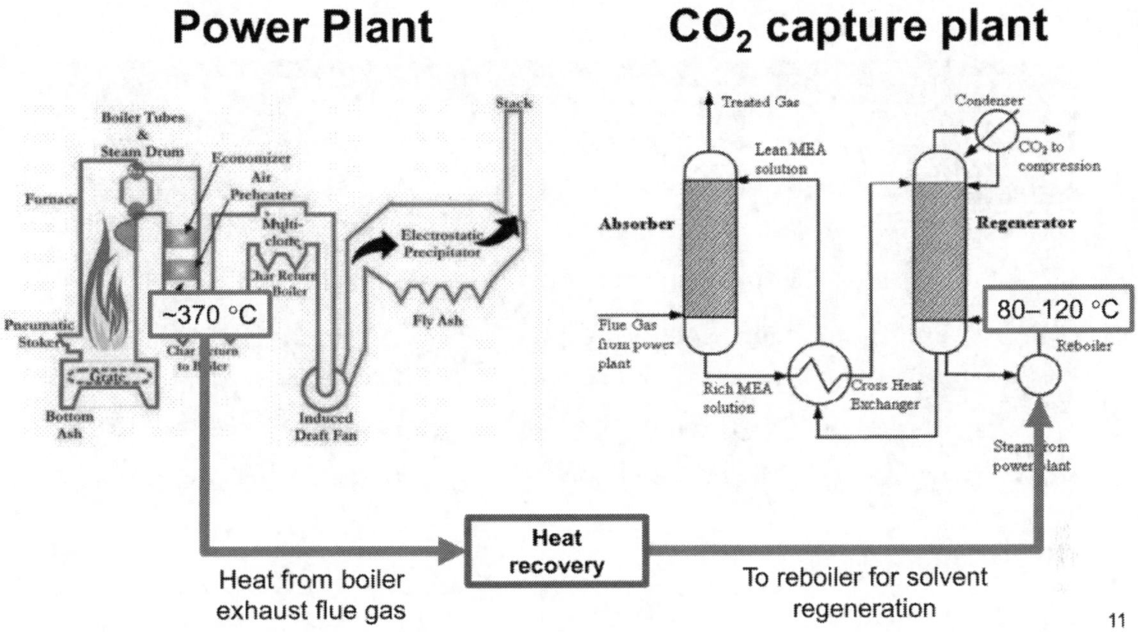

11

Imperial College London

BECCS performance evaluation study: Objectives

- Assess the performance of a BECCS system using a 500 MW ultra-supercritical pulverized fuel power plant with 90% CO_2 capture.
- The effects of the following factors were evaluated:
  i. Coal types: high sulphur 2.51 wt% and medium sulphur 0.86 wt%
  ii. Biomass type (wheat straw and clean wood chips)
  iii. Variable moisture content of the biomass (5%, 16%, 50%)
  iv. Biomass co-firing proportion %
  v. Solvent selection (MEA, Cansolv, biphasic solvents)
- Key performance indicators include:
  - Combustion adiabatic flame temperature (AFT)
  - Pollutant emissions e.g. SO_X and NO_X
  - Recoverable heat from the flue gas
  - Energy efficiency
  - Carbon intensity

12

Imperial College London

Co-firing scenarios modelled in FactSage

Scenario	Coal type	Biomass type
A	Medium sulphur coal	Wheat straw 5% moisture
B	Medium sulphur coal	Wheat straw 16% moisture
C	Medium sulphur coal	Wood chip 5% moisture
D	Medium sulphur coal	Wood chip 50% moisture
E	High sulphur coal	Wheat straw 5% moisture
F	High sulphur coal	Wheat straw 16% moisture
G	High sulphur coal	Wood chip 5% moisture
H	High sulphur coal	Wood chip 50% moisture

13

Imperial College London

Biomass Co-firing: Adiabatic Flame Temperature (AFT)

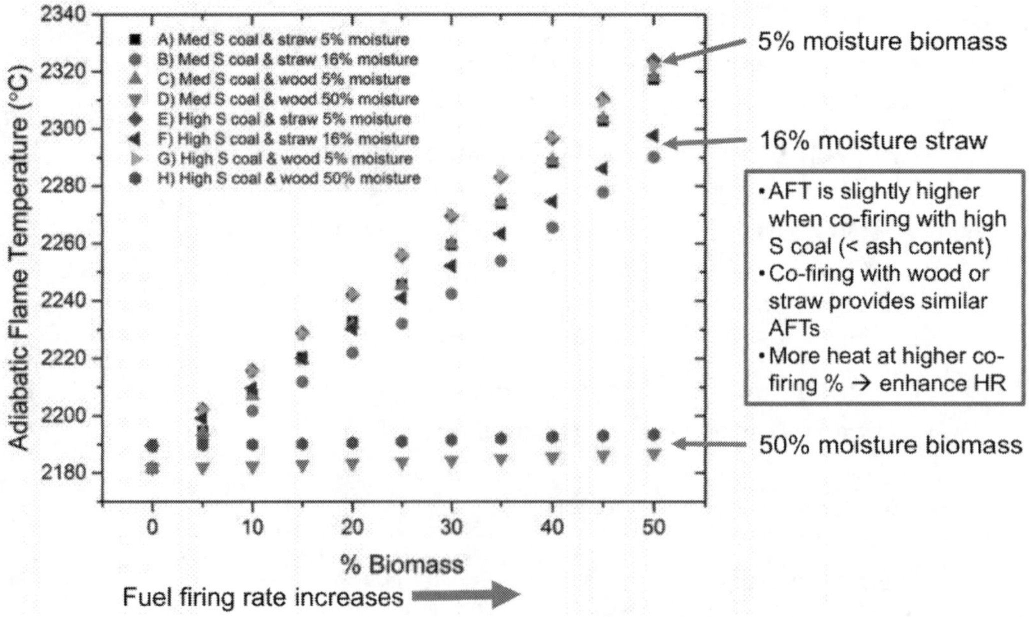

- AFT is slightly higher when co-firing with high S coal (< ash content)
- Co-firing with wood or straw provides similar AFTs
- More heat at higher co-firing % → enhance HR

Reference: Bui, M., Fajardy, M. & Mac Dowell, N. (2017). Applied Energy, 195, 289–302.

14

Imperial College
London

Coal co-firing with biomass : SO_x emissions

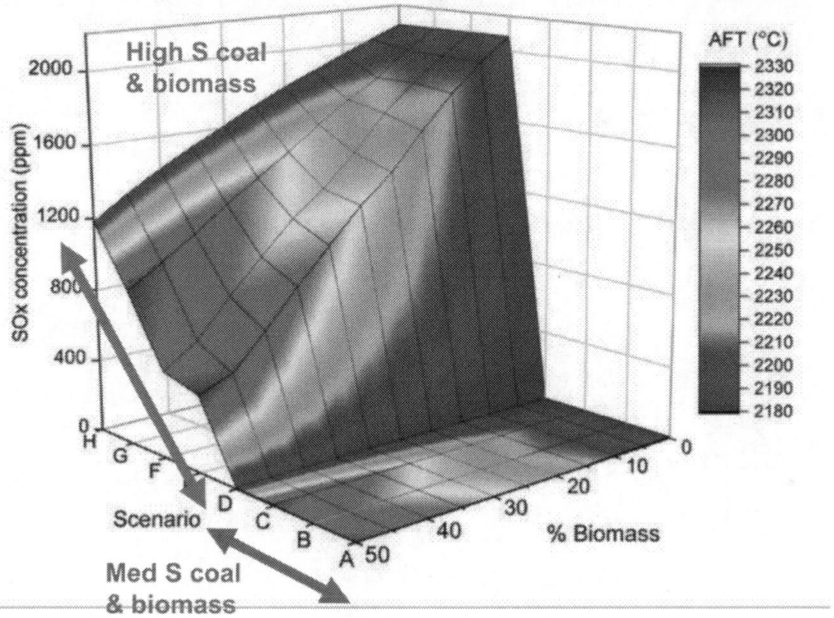

Reference: Bui, M., Fajardy, M. & Mac Dowell, N. (2017). Applied Energy, 195, 289–302.

Imperial College
London

Coal co-firing with biomass : SO_x emissions

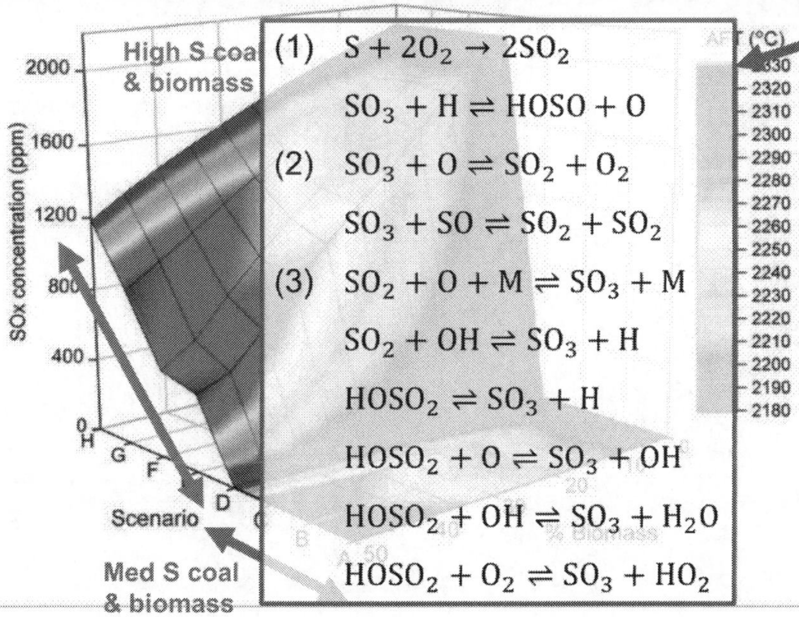

(1) $S + 2O_2 \rightarrow 2SO_2$

$SO_3 + H \rightleftharpoons HOSO + O$

(2) $SO_3 + O \rightleftharpoons SO_2 + O_2$

$SO_3 + SO \rightleftharpoons SO_2 + SO_2$

(3) $SO_2 + O + M \rightleftharpoons SO_3 + M$

$SO_2 + OH \rightleftharpoons SO_3 + H$

$HOSO_2 \rightleftharpoons SO_3 + H$

$HOSO_2 + O \rightleftharpoons SO_3 + OH$

$HOSO_2 + OH \rightleftharpoons SO_3 + H_2O$

$HOSO_2 + O_2 \rightleftharpoons SO_3 + HO_2$

- SO_X formation mechanism involves equilibrium reactions
- SO_2 formation thermodynamically favoured
- Temperature ↓ : Equilibrium shifts towards SO_3 formation (iron oxide can catalyse reaction 2)

Reference: Bui, M., Fajardy, M. & Mac Dowell, N. (2017). Applied Energy, 195, 289–302.

Imperial College London

Coal co-firing with biomass : SO_x emissions

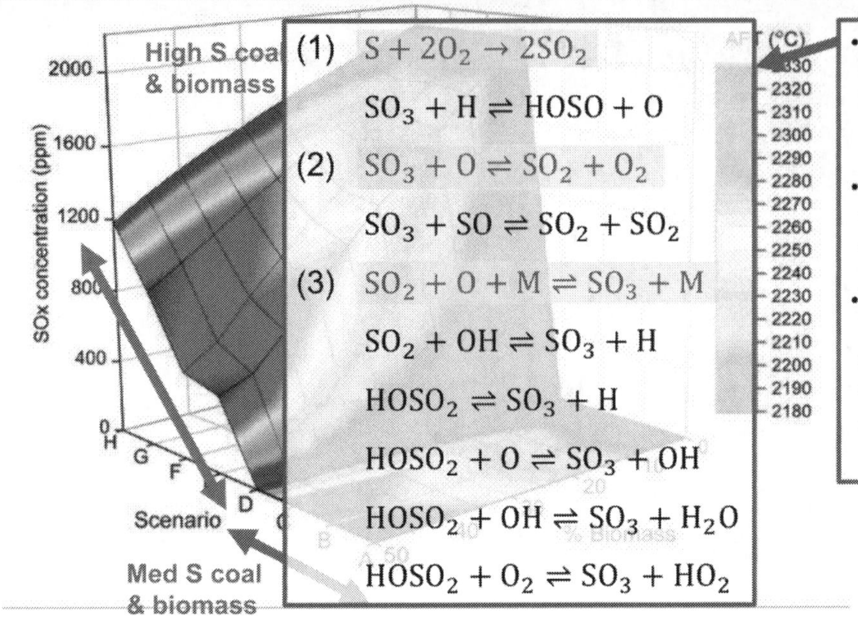

(1) $S + 2O_2 \rightarrow 2SO_2$

$SO_3 + H \rightleftharpoons HOSO + O$

(2) $SO_3 + O \rightleftharpoons SO_2 + O_2$

$SO_3 + SO \rightleftharpoons SO_2 + SO_2$

(3) $SO_2 + O + M \rightleftharpoons SO_3 + M$

$SO_2 + OH \rightleftharpoons SO_3 + H$

$HOSO_2 \rightleftharpoons SO_3 + H$

$HOSO_2 + O \rightleftharpoons SO_3 + OH$

$HOSO_2 + OH \rightleftharpoons SO_3 + H_2O$

$HOSO_2 + O_2 \rightleftharpoons SO_3 + HO_2$

- SO_X formation mechanism involves equilibrium reactions
- SO_2 formation thermodynamically favoured
- Temperature ↓ : Equilibrium shifts towards SO_3 formation (iron oxide can catalyse reaction 2)

Reference: Bui, M., Fajardy, M. & Mac Dowell, N. (2017). Applied Energy, 195, 289–302.

17

Imperial College London

Coal co-firing with biomass : SO_x emissions

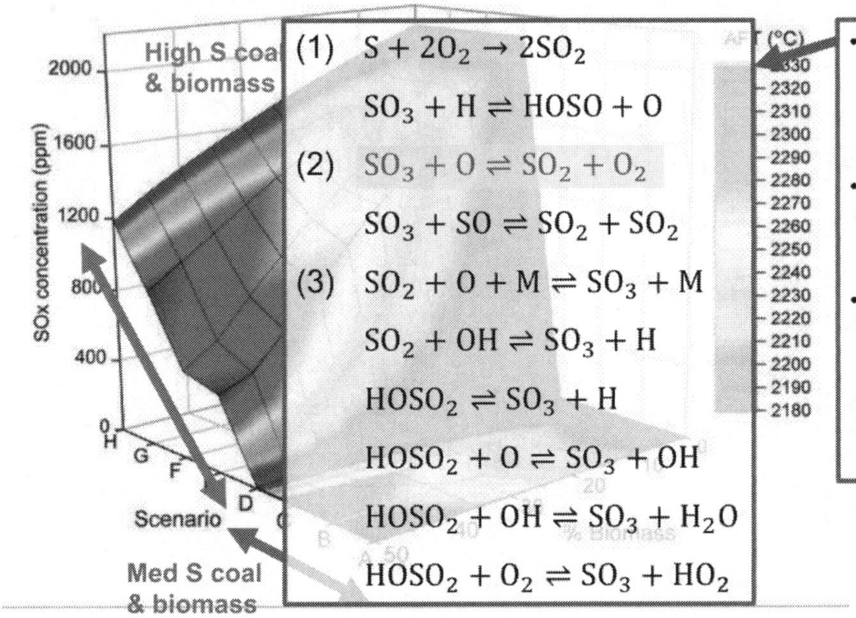

(1) $S + 2O_2 \rightarrow 2SO_2$

$SO_3 + H \rightleftharpoons HOSO + O$

(2) $SO_3 + O \rightleftharpoons SO_2 + O_2$

$SO_3 + SO \rightleftharpoons SO_2 + SO_2$

(3) $SO_2 + O + M \rightleftharpoons SO_3 + M$

$SO_2 + OH \rightleftharpoons SO_3 + H$

$HOSO_2 \rightleftharpoons SO_3 + H$

$HOSO_2 + O \rightleftharpoons SO_3 + OH$

$HOSO_2 + OH \rightleftharpoons SO_3 + H_2O$

$HOSO_2 + O_2 \rightleftharpoons SO_3 + HO_2$

- SO_X formation mechanism involves equilibrium reactions
- SO_2 formation thermodynamically favoured
- Temperature ↓ : Equilibrium shifts towards SO_3 formation (iron oxide can catalyse reaction 2)

Reference: Bui, M., Fajardy, M. & Mac Dowell, N. (2017). Applied Energy, 195, 289–302.

18

Imperial College
London

Coal co-firing with biomass : SO_x emissions

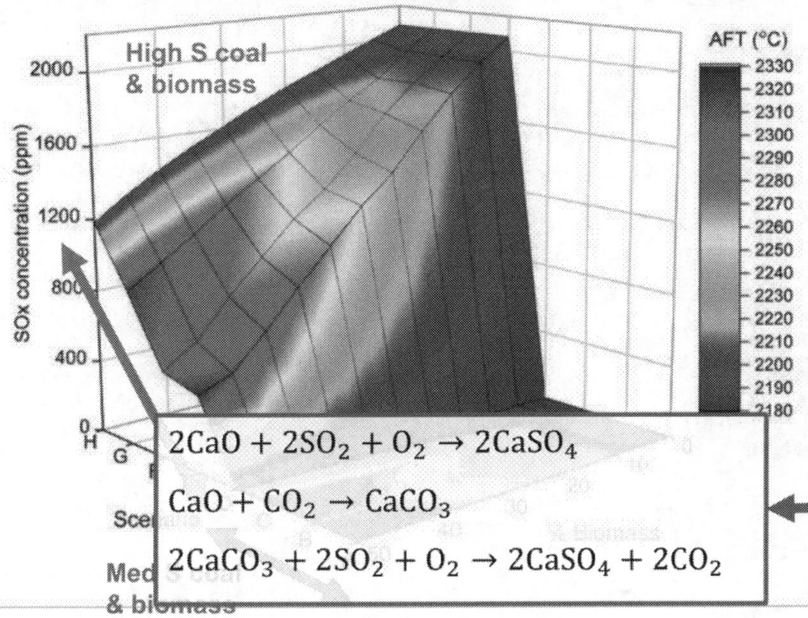

$$2CaO + 2SO_2 + O_2 \rightarrow 2CaSO_4$$

$$CaO + CO_2 \rightarrow CaCO_3$$

$$2CaCO_3 + 2SO_2 + O_2 \rightarrow 2CaSO_4 + 2CO_2$$

- SO_X formation mechanism involves equilibrium reactions
- SO_2 formation thermodynamically favoured
- Temperature ↓ : Equilibrium shifts towards SO_3 formation (iron oxide can catalyse reaction 2)
- Alkali oxides in ash capture SO_2 and SO_3

Reference: Bui, M., Fajardy, M. & Mac Dowell, N. (2017). Applied Energy, 195, 289–302.

19

Imperial College
London

High sulphur coal & biomass: SO_x emissions

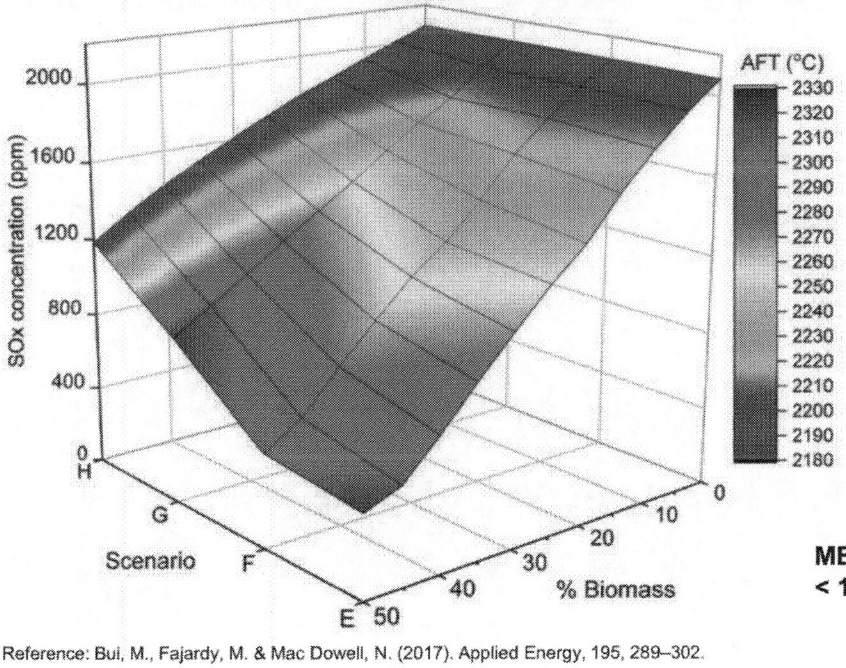

- SO_X decreases linearly with % biomass due to decreased S content

MEA tolerance to SO_X < 10 ppm

Reference: Bui, M., Fajardy, M. & Mac Dowell, N. (2017). Applied Energy, 195, 289–302.

20

Imperial College London

Medium sulphur coal & biomass: SO_x emissions

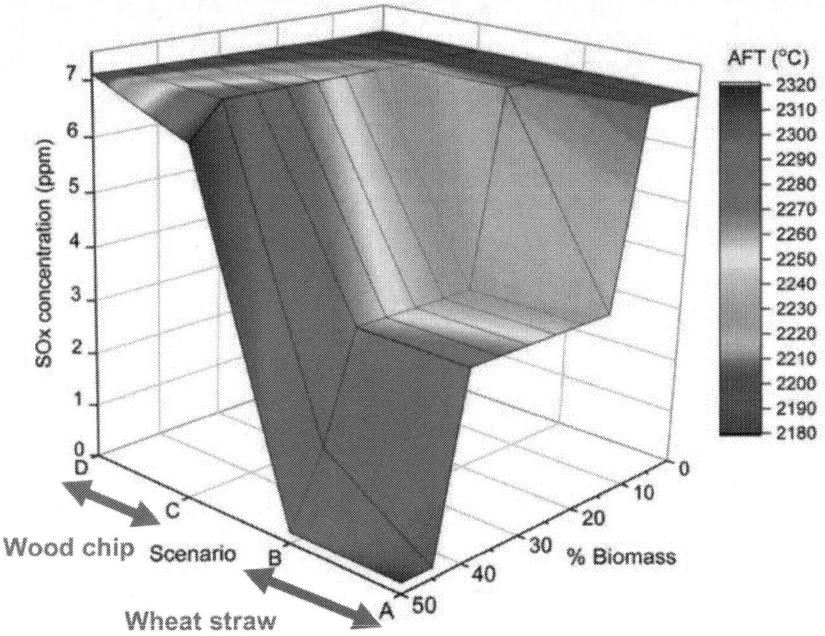

- FactSage results show that CaO, MgO, Na_2O, K_2O captures SO_X
- In FactSage, Al_2O_3, Fe_2O_3 & MnO_2 also involved in $SO_X \downarrow$
- Wheat straw has higher sulphur content than the wood chip (0.1 wt% Vs 0.05 wt%)
- Wheat straw has more ash than wood chip (5.5 dry wt% Vs 2 dry wt%)
- Alkali oxides in ash → reduce SO_X emissions

MEA tolerance to SO_X < 10 ppm

Reference: Bui, M., Fajardy, M. & Mac Dowell, N. (2017). Applied Energy, 195, 289–302.

21

Imperial College London

BECCS: Recoverable heat from the flue gas

Analysis of scenario F (high S coal and 16% moisture wheat straw)

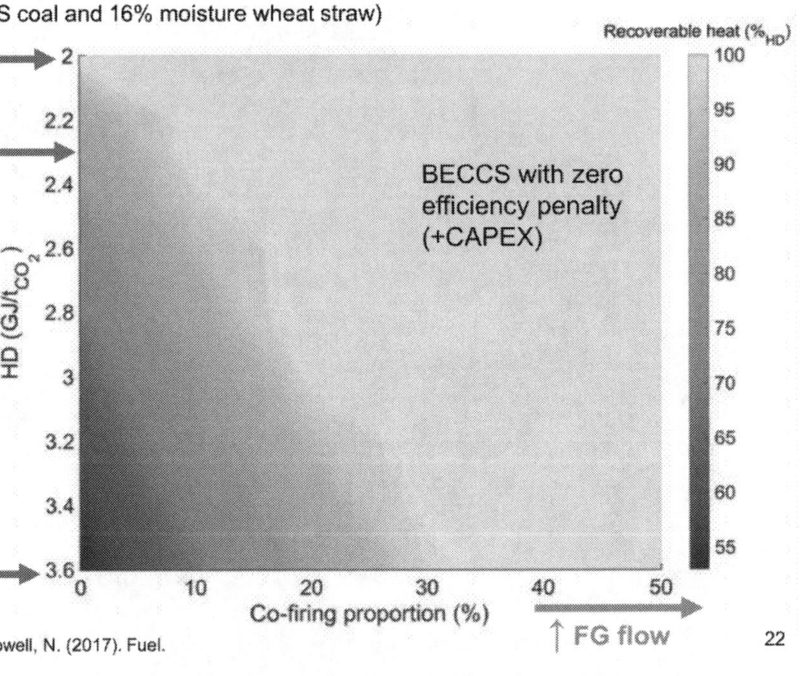

Solvent	HD (MJ/t_{CO_2})	Reg T (°C)
MEA	3600	120
Cansolv	2300	120
Biphasic	2000	80

Data for biphasic solvent systems published by Ye *et al.* (2015) and Raynal *et al.* (2011)

Reference: Bui, M., Fajardy, M. & Mac Dowell, N. (2017). Fuel.

22

Imperial College London

BECCS power plant efficiency improvement

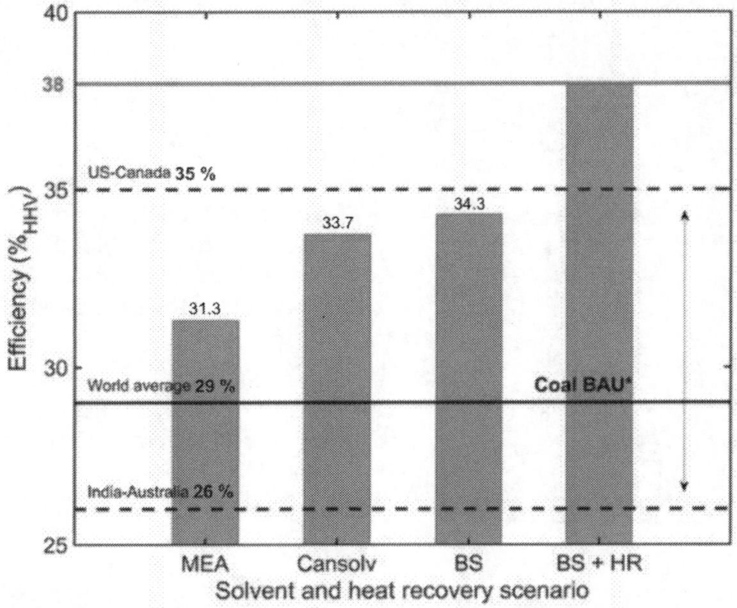

- Efficiency for a 500 MW BECCS system **co-firing 50% biomass**
- Efficiency can be significantly improved by using (i) high performance solvents, (ii) heat recovery

BS = biphasic solvent

HR = heat recovery

BAU = Business As Usual (average efficiencies around the world)

Reference: Bui, M., Fajardy, M. & Mac Dowell, N. (2017). Bio-Energy with carbon capture and storage (BECCS): Opportunities for performance improvement. Fuel.

23

Imperial College London

BECCS carbon intensity

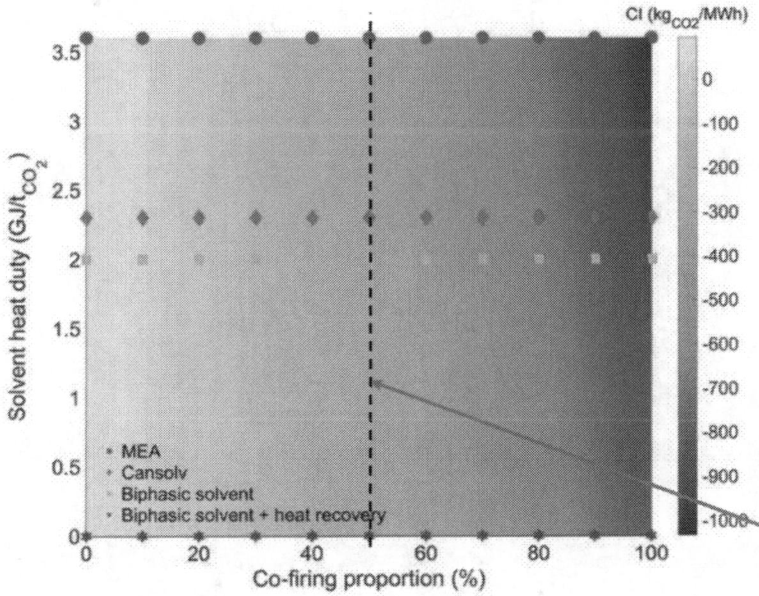

As system efficiency increases, less fuel is burned per MWh produced → less CO_2 is captured per MWh of electricity generated

At 50% biomass co-firing and 90% CO_2 capture:

System	Efficiency %HHV	Carbon intensity kg CO_2/MWh
MEA	31.3	-296
Cansolv	33.7	-275
BS	34.3	-271
BS + HR	38.0	-244

Reference: Bui, M., Fajardy, M. & Mac Dowell, N. (2017). Fuel.

24

Imperial College London

CO_2 mitigation potential of a 500 MW BECCS system

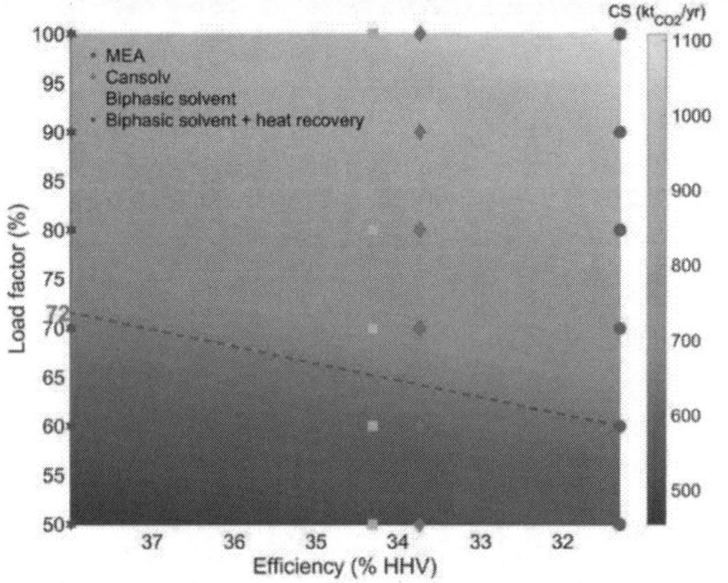

- Power plant load factor is a function of power plant efficiency.
- Improving efficiency reduces marginal costs of electricity generation, enabling operation at higher load factor (Mac Dowell & Shah, 2015; Mac Dowell & Staffell, 2016)
- The efficiency of a BECCS system will have an impact on its dispatch rate within an electricity market
  → Influences CO_2 mitigation potential

Objective: (i) more efficient at electricity generation, or (ii) more carbon negative & capture CO_2?

Reference: Bui, M., Fajardy, M. & Mac Dowell, N. (2017). Applied Energy, 195, 289–302.

Imperial College London

Effect of biomass carbon footprint (BCF) and co-firing proportion on the system carbon intensity (CI)

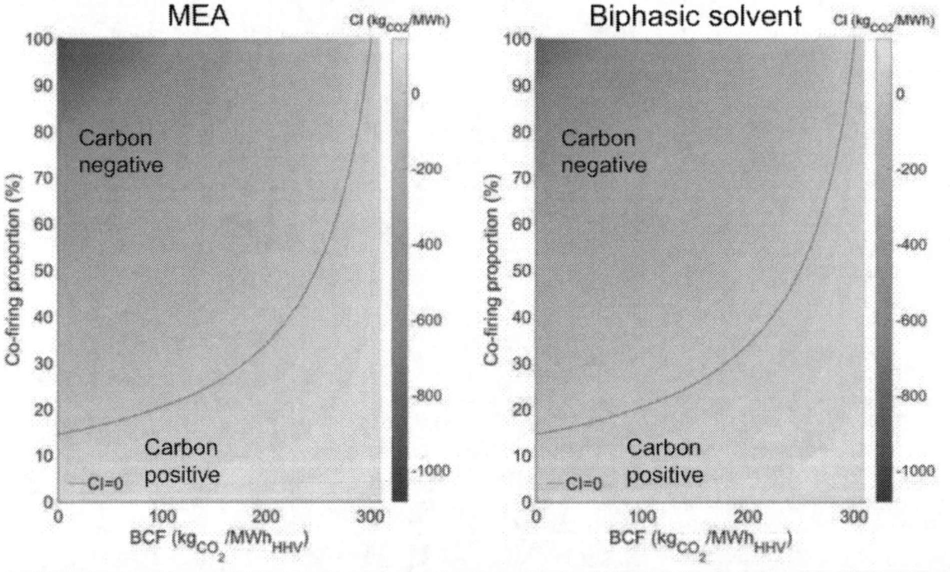

Bui, M., Fajardy, M. & Mac Dowell, N. (2017) Fuel.

Low efficiency MEA system is more carbon positive above the CI = 0 limit & more carbon negative below this limit. To ensure the BECCS system is actually carbon negative, the impact of biomass carbon footprint (BCF) must be considered.

Imperial College London

Conclusion: BECSS Performance Matrix

Performance parameter	Ash content	Sulphur content	Moisture content	HHV	Biomass co-firing %	Solvent heat duty
NO_X ↓	dependent on combustion conditions					
SO_X ↓	↑	↓	↓		↑	
AFT ↑	↓		↓		↑	
Heat recovery ↑	↓		↓	↓*	↑	↓
Efficiency ↑	↓		↓	↑↓	↓	↓
CO_2 negativity ↑	↑		↑	↑	↑	↑
Exhaust gas flow rate ↑				↓	↑	

Enhance efficiency: high performance solvents (low heat duty) and higher heat recovery (higher AFT and flue gas flow rate)
Greater carbon negativity: lower efficiency systems (burn more biomass fuel, thus more CO_2 is captured and permanently stored), low biomass carbon footprint.
Minimise pollutant emissions: Increasing biomass co-firing proportion significantly reduces SO_X emissions, fuel selection (low S, metal oxides), combustion control is required to regulate NO_X emissions.

27

Imperial College London

Conclusion

- BECCS has the potential provide negative emissions, whilst providing reliable firm power generation capacity.
- Opportunities to improve energy efficiency of BECCS include enhancing heat recovery and using high performance solvents for CO_2 capture (*e.g.*, biphasic).
- Power generation efficiency increased from $31\%_{HHV}$ (MEA system) to $38\%_{HHV}$ (biphasic system with heat recovery).
- Important to consider the complex trade-offs between carbon intensity, power plant efficiency and biomass carbon footprint.
- Inefficient BECCS → removes more CO_2 from the atmosphere.
- However, inefficient systems might have lower dispatch rate compared to high efficiency systems.

Bui, M., Fajardy, M. & Mac Dowell, N. (2017). Applied Energy, 195, 289–302.
Bui, M., Fajardy, M. & Mac Dowell, N. (2017). Energy Procedia, 114 (Supplement C), 6010-6020.
Bui, M., Fajardy, M. & Mac Dowell, N. (2017). Bio-Energy with carbon capture and storage (BECCS): Opportunities for performance improvement. Fuel.

28

493767 A Process Synthesis Approach to Natural Gas Liquefaction

Abstract:

A new approach to the design of natural gas liquefaction processes, which could significantly reduce their energy consumption, is proposed. A graphical technique based on exergy and enthalpy is used to determine optimal paths for refrigeration systems. The analysis reveals that keeping refrigeration cycles in the two-phase region, including compression and heat removal, is the most efficient way of supplying the work required to remove heat from sub-ambient processes. The approach is validated using an Aspen Plus simulation of a natural gas liquefaction process with pure refrigerant in cascade; it is shown that work efficiency higher than 90% can be achieved when the process fluid as well as the working fluids in the cascade cycles are kept in the two phase region at every stage.

499711 Recovery of Rare Earth Elements from Coal Ash with a Recycling Acid Leach Process

Abstract:

Recovery of Rare Earth Elements from Coal Ash with a Recycling Acid Leach Process

Rick Peterson1, Mike Heinrichs1, Rachid Taha1, Darwin Argumedo1, Justin Glier1, Slawek Winecki1

1Battelle Memorial Institute, Energy Resource Group, 505 King Avenue, Columbus, OH 43201

KEYWORDS: rare earth elements, coal ash, coal liquefaction

ABSTRACT

Rare earth elements (REE) have a wide range of uses in catalysis, alloying, magnets, and optics, and have become critical for the batteries, motors, and generators that enable sustainable technologies and infrastructure. Despite these critical applications, the United States has developed very little in the way of REE resources, with China producing more than 80% of the global supply of the elements1. A technology that will allow for the economic recovery of REE from coal will reduce dependence on foreign supplies and ensure US capability for renewable and defense products.

1 U.S. Geological Survey, "Mineral Commodity Summaries 2017," 19 January 2017. [Online]. Available: https://minerals.usgs.gov/minerals/pubs/mcs/2017/mcs2017.pdf.

Battelle has investigated the application of its Acid Digestion Process (ADP) to recovery of REE from coal ash produced by pulverized coal combustors (PCC), fluidized bed combustors (FBC), and Battelle's bio-based direct coal liquefaction process. The ADP was shown to recover over 90% of the nitric acid used in the process for recycle, greatly reducing chemical consumption costs. This presentation will cover results of a sampling and characterization effort for varying coal ash feedstocks, as well as a feasibility study that assessed the economics of the ADP process when considering available sources.

The sampling and characterization portion of the project analyzed coal byproducts to allow selection of a suitable feedstock for Battelle's process, and selection of key target materials for recovery to drive the process economics. Within the REE in coal, dysprosium (Dy) represents the highest value, and combined with neodymium (Nd), terbium (Tb), and praseodymium (Pr), these four elements account for over 75% of the REE value in the ash. Accordingly, Battelle's process will be tailored to focus on these REE components. Scandium (Sc) represents the highest overall value within the ash, and will be a primary byproduct. Vanadium (V), yttrium (Y), lithium (Li), and cobalt (Co) also represent a significant value, and will be evaluated as potential byproducts going forward.

Based on the analytical results, the feasibility study focused on a specific PCC fly ash as a feedstock for the ADP. The PCC plant fly ash had a higher Total REE+Y+Sc concentration than

all other operating plant ashes sampled, at 545 ppm +/- 13.4 ppm, as well as a higher Heavy REE/Light REE ratio at 0.37 +/- 0.01. Additionally, it contains significant amounts of scandium (36 ppm +/- 1.4 ppm), vanadium (279 ppm +/- 12 ppm), yttrium (104 ppm +/- 5.3 ppm), cobalt (44 ppm +/- 2.5 ppm), and lithium (~166 ppm), which can be valuable process byproducts with robust market outlets. Although the coal liquefaction ash had a greater concentration of total REE, there is some risk for implementation since Battelle's liquefaction process is not yet commercialized. The low temperature ash appeared to be more leachable, but it also demonstrated lower selectivity due to the high calcium content common with FBC systems. Furthermore, the overall capacity of FBCs is much lower than PCCs. Therefore, there would be more risk in feedstock sourcing.

Feasibility testing compared leaching efficiency and selectivity of PCC ash with samples of liquefaction ash and FBC ash to ensure and validate assumptions regarding leaching kinetics, acid loading limits, and product purities. This information was used to refine a process model built in CHEMCAD modeling software and fed into overall process economics and unit sizing used in the technoeconomic assessment.

The feasibility study focused on construction of a REE recovery plant sized to process up to 30 tonnes of ash per hour, which is approximately the ash production rate of a 2,000 MW power plant operating at full capacity.

Operating costs for both a First-of-a-Kind and an Nth-of-a-Kind plant were determined. These costs to recover rare earths from coal ash were then compared to the value of rare earths present in a variety of coal ashes around the country. Values for coal ash were obtained through determination of the concentrations of rare earth elements in coal samples adjusted to account for changes in concentration associated with the combustion process. Current market pricing information for REEs was used to determine the value associated with each of the coal sources.

The results of the feasibility study indicate that when a First-of-a-Kind plant is considered, approximately 5% of U.S. coal sources contain sufficient rare earth material to cover the costs of the recovery plant. When an Nth-of-a-Kind plant is considered, over 20% of all coal sources reviewed contain sufficient recoverable rare earths to cover recovery costs at current REE market prices. When historical high REE prices are considered, more than 25% of coal sources can be treated with a First-of-a-Kind plant, and more than 47% of sources with an Nth-of-a-Kind plant. These results are based on the current understanding and developmental status of the recovery process. It should be noted that further downstream purification costs are not included in this assessment. A probabilistic sensitivity analysis was conducted which examines how the cost to recover rare earth materials may change due to technology developments, uncertainties in capital requirements for the plant, or prices of REEs due to shifting market dynamics. Under certain sensitivity scenarios, the recovery plant can be economically viable processing an even greater range of coal sources.

This project is ongoing, and the presentation will also cover pending results of experiments designed to improve REE leaching efficiency, produce additional valuable byproducts, and upgrade the REE concentration in process products. Additionally, it will briefly address the design of a continuous bench scale ADP for REE recovery from coal ash.

This project was performed with funding from the Department of Energy's National Energy Technology Laboratory (DE-FE0027012) and the Ohio Coal Development Office (OER-CDO-D-15-10)

498633 Brief Overview of Rare Earth Research at NETL R&ic

Abstract:

Everything that is in the Earth's crust is also present in coal to some extent, and the challenge is always to use abundant domestic coal in clean and environmentally friendly ways. Trace elements present within the coal such as mercury, arsenic, selenium, cadmium, antimony, phosphorus, and chlorine, as well as sulfur and nitrogen, have long been of interest due to their roles in producing pollution emissions from combustion or gasification plants. Critical elements, such as some of the rare earths present within the coal, have recently been of great interest due to their importance in the world's economy.

Coal is an important resource, both in the United States and around the world. The United States generates approximately 30 percent of its electricity through coal combustion and, at the current overall rate of consumption, has more than a 250-year supply of coal. According to the US Energy Information Administration, approximately 1 Gt of coal has been mined annually in the United States over the past twenty-five years (1990-2014), although the 2015 total was 896 million tons, and 2016 was 739 million tons. The recent decline in domestic coal production is due to the availability of inexpensive natural gas; regulatory uncertainty with regards to carbon dioxide emissions; and the retirement of older coal-fired power plants.

In the United States, most coal is burned for power generation, but substantial quantities are also used in the manufacture of steel, chemicals, and activated carbons. Numerous industries, such as mining, power, rail transportation, manufacturing, chemical, steel, activated carbon, and fuels, are involved in the production, transportation, and use of coal.

Rare earth elements (REEs), which comprise the lanthanide elements, as well as scandium and yttrium, are present in the abundant coal and coal by-products produced domestically and worldwide. Widely used in high-technology products such as catalysts, cell phones, hard drives, hybrid engines, lasers, fluorescent lamps, batteries, magnets, medical devices, and televisions, REEs are of significant value to US national security, energy independence, economic growth, and the country's environmental future.

Most of the common inorganic lanthanide compounds, such as the phosphates found in coal, have very high melting, boiling, and thermal decomposition temperatures, allowing them to concentrate in combustion and gasification by-products. Rare earths are commercially produced from ores containing monazite (rare earth phosphate mineral) or bastnasite (rare earth carbonate-fluoride mineral), as well as from ion exchangeable clays. The lanthanides are found in coal as well as combustion by-products such as ash, coal preparation residues, gasification slags, and mining by-products. Rare earths have also been found in interesting concentrations in the strata above and below some coal seams, thereby making every process in the mining and utilization of coal a potential source of rare earth elements.

Also of interest, some coal and coal by-products have elevated concentrations of heavy rare earths (HREEs) —which are lowest in supply, rank high in criticality and price, and are projected to increase in demand— making them potentially attractive targets for REE recovery despite their overall lower concentration.

The National Energy Technology Laboratory's Research and Innovation Center (NETL-RIC) recently initiated research to support the measurement of concentration, identification of rare earth compounds, and recovery of rare earths from abundant domestic coal by-products. The NETL Rare Earth EDX Database (https://edx.netl.doe.gov/ree/) is a resource for rare earth information related to coal and by-products. Many other research organizations have also initiated efforts for the analytical characterization and recovery of rare earths from unconventional sources such as coal by-products.

Much of the recent research on coal use in the United States has focused on the capture of pollutants such as acid gases, particulates, mercury, and carbon dioxide. The possible recovery of rare earth and other critical elements from abundant coal and by-products is an exciting new research area and represents a dramatic paradigm shift for coal. Some of the early exciting results on the occurrence and recovery of lanthanides from abundant coal by-products will be discussed.

497530 Extraction of Rare Earth Elements from Fly Ash Using NaOH Hydrothermal and Ultrasound Pretreatment

Abstract:

Extraction of valuable and critical trace elements from coal and coal byproducts such as fly ash has been studied for over 100 years. Recently the volatility of rare earth element (REE) prices, and the closing of the only rare earth mine in the U.S., has renewed interest in trace element and REE recovery from abundant domestic coal and fly ash. Acid leaching, using 2M HCl to extract REE from a fly ash was studied with less than 10% total REE extraction. However, when a sodium hydroxide hydrothermal pretreatment was employed, a dramatic increase in REE extraction was observed. The sodium hydroxide pretreatment greatly increased the REE extraction to approximately 90% in the HCl acid leach. Sonication experiments using NaOH solution in combination with hydrothermal pretreatment were also examined and will be discussed.

490447 Recovery of Rare Earth Elements from North Dakota Lignite and Lignite-Related Feedstocks

Abstract:

According to the U.S. Department of Energy National Energy Technology Laboratory, rare earth elements (REE) provide significant value to our national security, energy independence, environmental future, and economic growth. REEs are utilized in a suite of high importance end-uses, such as cell phones, hybrid vehicles, magnets, computer components, catalysts and many others. The unique magnetic properties of REEs, in particular, make them crucial materials for the growing renewable energy and electric automotive markets. REEs are used in the strongest permanent magnets currently known, and are used in generators for wind turbines and in motors for hybrid/electric vehicles. These same types of magnets, as well as other critical REE-based products are used in a host of military defense applications, as well.

China, in part due to its deposits of a unique REE resource (ion-adsorbed clays) that combines high quantities of heavy and critical REE, as well as simple and low cost extraction, dominates the global supply market. In 2010, China established new quotas on exports of REEs, which resulted in huge increases in REE prices peaking in 2011 due to an expected supply shortage for critical applications. As a result, production at the California-based Mountain Pass Mine was re-started after several years of dormancy. However, after peaking in price in 2011, prices have dropped substantially to slightly above 2010 levels, challenging the profitability of non-China based production which consists mainly of hard rock carbonatite deposits, that are deficient in critical REE and heavy REE. Some researchers have noted that mining of the Mountain Pass and similar resources will neither mitigate the crisis in REE resources nor eliminate the shortage of the most critical REE, but will only result in overproduction of excessive cerium.

China accounted for about 83% of the total global REE supply in 2016, down from about 95% prior to 2010. Meanwhile, the US production was zero, with the Mountain Pass mine having declared bankruptcy and closing operations in the last quarter of 2015. The U.S. is currently 100% import reliant for REEs. Although still dominating global supply of the heavy REE, some researchers have estimated that the Chinese ion-adsorbed clays resource will only last another 15-20 years. The Chinese clays represent essentially the entire global supply of heavy REE and most of the critical REE. Further, the bulk of Chinese reserves and production is from a carbonatite-type deposit (Bayan Obo) that contains only trace amounts of heavy REE (98.7% light REE), and supplies roughly 80% of the global light REE demand. Due to its limited supply, and because the Chinese clay resource is rich in heavy and critical REE, while most other traditional resources are deficient in these less common and more valuable elements, it is imperative that new domestic sources of REEs, especially the heavy and critical REE, be identified and processes be developed to produce them. Coal and coal byproducts have recently been identified as one of these potential new resources for REEs.

As part of the US Department of Energy effort to identify alternative domestic sources of Rare Earth Elements (REE), the University of North Dakota was awarded a project to determine the feasibility of recovery of REE from North Dakota lignite coal and related feedstocks. The project team includes the University of North Dakota, Barr Engineering, and Pacific Northwest National Laboratory (PNNL) with the support of the cost share partners North American Coal Company, Great River Energy, and the North Dakota Industrial Commission/Lignite Energy Council as well as technical support from the North Dakota Geological Survey. The overall goal of this Phase I project was to develop an economically viable and environmentally benign process to concentrate REE to a goal of 2 weight percent to form a synthetic mixed REE ore. As part of the project, an extensive sampling and analysis campaign was completed to identify promising feedstocks, which was followed by laboratory-scale concentration testing and a technical and economic feasibility study. Finally, the design of a bench-scale test system was developed, which is to be constructed and tested in a subsequent Phase II project, if awarded by DOE. This paper will provide the results of the Phase I project and the plans for a subsequent Phase II project.

The originally proposed concept involved recovery of REE from the reject stream of the Great River Energy lignite drying system located at the Coal Creek Station power plant in North Dakota. However, upon further investigation in the project, it was discovered that on an ash-basis, the REE are more concentrated in certain locations within the coal seams than in the associated roof/floor sediments. Therefore, the focus for the remainder of the project was shifted towards development of methods to concentrate the REEs from these selected coal seams.

Laboratory testing of a novel concentrating method has shown very high recovery of about 80-90% of the REEs from two potential feedstocks, ranging from 580 to 2300 ppm REE content on an ash basis, with excellent selectivity/recovery of the critical and heavy REE. Details of a processing scheme to economically concentrate the REE to the goal of 2 weight percent using a simple and environmentally benign method have been established. The project team has also developed the concepts for integration of the REE recovery process within an existing coal conversion facility that will provide cost-saving synergies, multiple saleable product streams, and infrastructure to reduce costs. Based on the results of the Phase I work, the project team believes North Dakota lignite-related feedstocks to be a highly promising alternative domestic source of REE that will limit dependence on foreign supply.

489276 Froth Flotation Enrichment of Rare Earth Elements (REEs) from Different Feedstocks

Abstract:

Flotation is an aqueous, efficient, inexpensive, and well-developed commercial technology that has been widely applied in a variety of industrial processes, including mineral processing. The flotation process first generates gas bubbles within a solid-liquid or liquid-liquid suspension. The gas bubbles then attach to solid particles or immiscible liquid droplets and rise to top of the liquid surface. The floated solid particles or immiscible liquid droplets finally are skimmed off to achieve the separation. As particles/immiscible droplets are hydrophobic, they attach easily to gas bubbles and can be separated by the flotation technology.

The effectiveness of surfactants, i.e., frother, depressant, and collector, can strongly influence the flotation process, including the enrichment of rare earth elements (REEs) from coal, coal by-products or fly ashes. The properties of particle size, temperature, and pH will also affect REE enrichment. Here, we report the performance of different chemicals/process parameters for the froth flotation of different feeds in the mechanic agitated flotation cell. Different types of frothers and collectors were studied at varying flotation conditions for REE enrichment. The discussion will focus on the performance of these surfactants and the REE enrichment efficiency at these different conditions.

EERC

GEOLOGIC STORAGE OF CARBON DIOXIDE IN THE CENTRAL PLAINS OF NORTH AMERICA

2017 AIChE Annual Meeting
Minneapolis, Minnesota
November 3, 2017

Neil Wildgust
Principal CCS Scientist

Critical Challenges. Practical Solutions.

PCOR PARTNERSHIP

- Region includes:
  - Nine states.
  - Four Canadian provinces.
  - Over 3.6 million km^2.
- Several completed field projects.
- Over 4 million tons of CO_2 stored and monitored in association with CO_2 enhanced oil recovery (EOR).
- More than 120 partners!

EERC

Critical Challenges. Practical Solutions.

WHERE WE BEGAN

Phase I (2003–2005)
- Source, sink, seal characterization
- Regional vision
- 21 topical reports, multiple fact sheets
- Decision Support System (DSS)

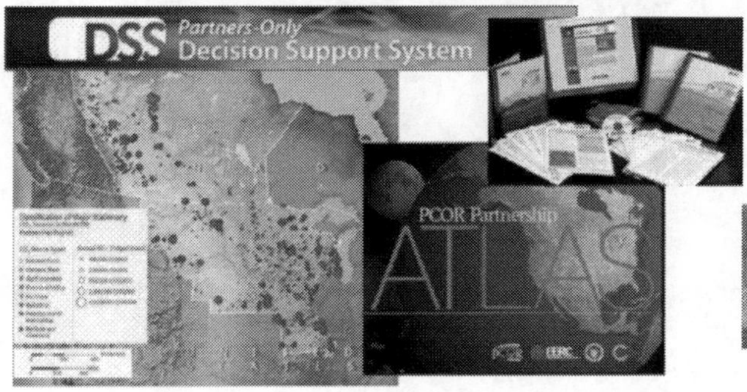

Phase II (2005–2009)
- Field validation

WHERE WE HAVE BEEN: COMPLETED PHASE III EFFORTS

Fort Nelson Carbon Capture and Storage (CCS) Feasibility Study
- CCS in a deep saline formation.

Zama Acid Gas EOR and Storage
- Operated by Apache Canada Ltd.
- Static and dynamic modeling activities completed.

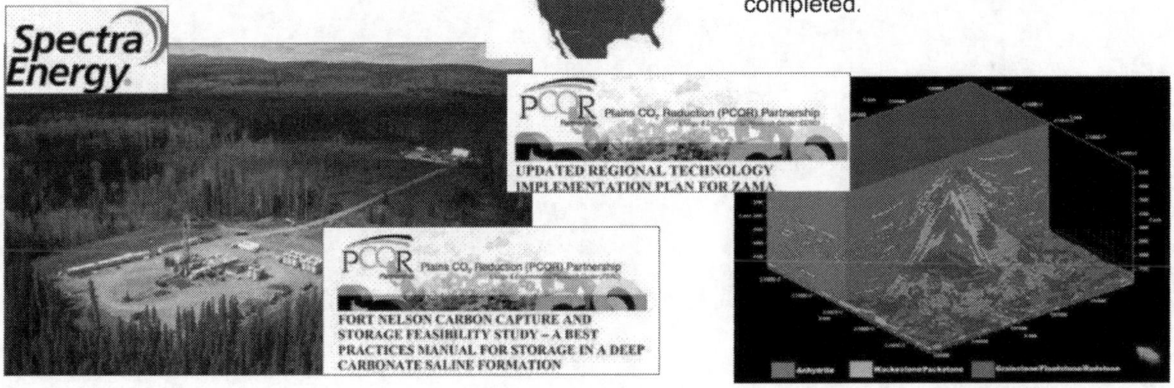

WHERE WE ARE NOW: AQUISTORE

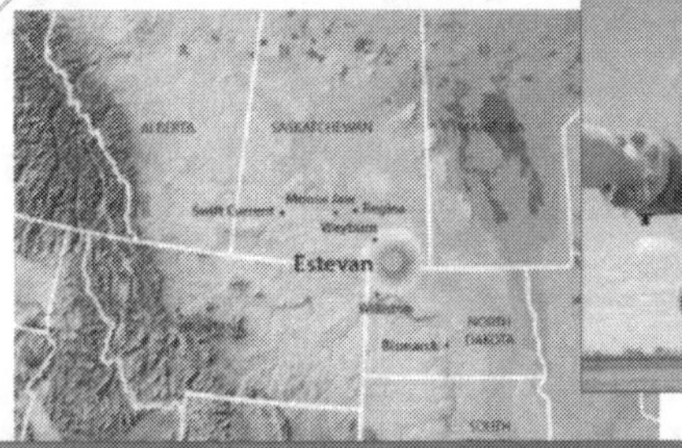

- Injection of CO_2 from the Boundary Dam power plant began in April 2015.
- Serves as buffer storage for CO_2 that is not used for EOR.

WHERE WE ARE NOW: BELL CREEK

- CO_2 storage associated with a commercial CO_2 EOR project.
- Operated by Denbury Onshore LLC.
- CO_2 is sourced from ConocoPhillips' Lost Cabin natural gas-processing plant and Exxon's Shute Creek gas-processing plant.

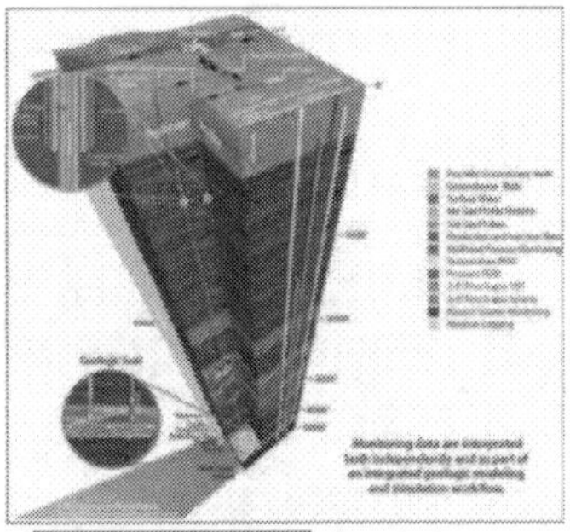

Adaptive Management Approach

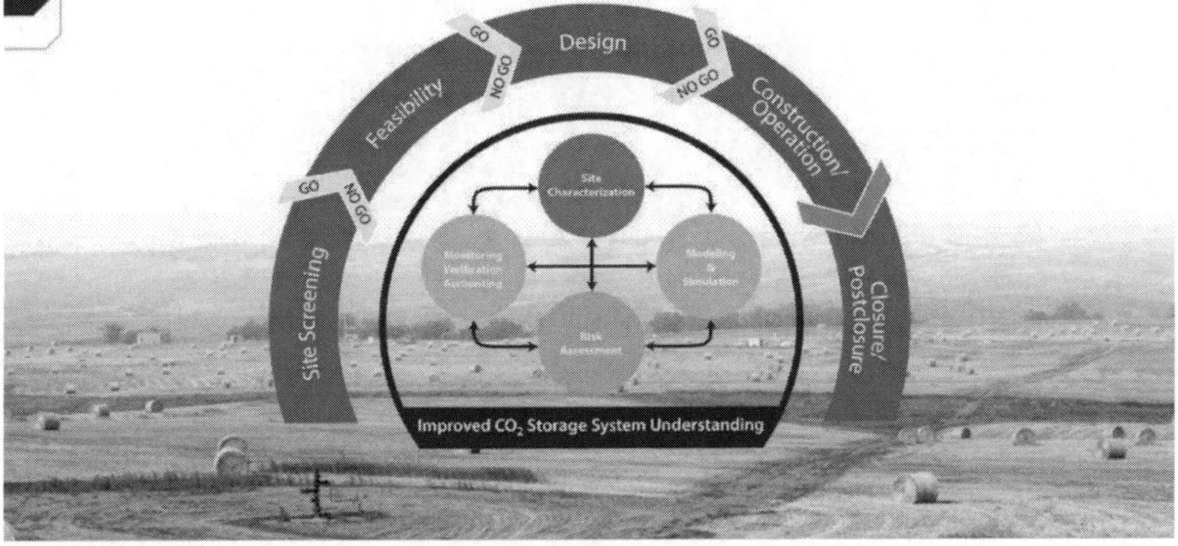

SIMULATION-GUIDED MONITORING, VERIFICATION, AND ACCOUNTING (MVA)

- Simulation was used to predict location and saturation of CO_2.
- 2-D seismic line used to confirm ability of seismic to detect CO_2 in the reservoir.
- Results supported decision to conduct large 3-D survey.

Seismic Line Overlaying Simulation

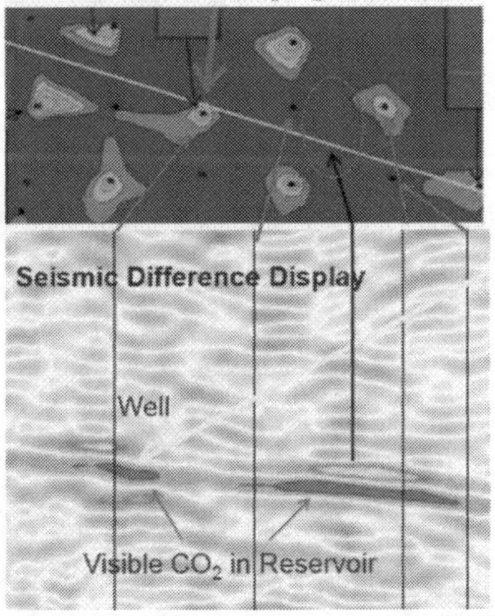

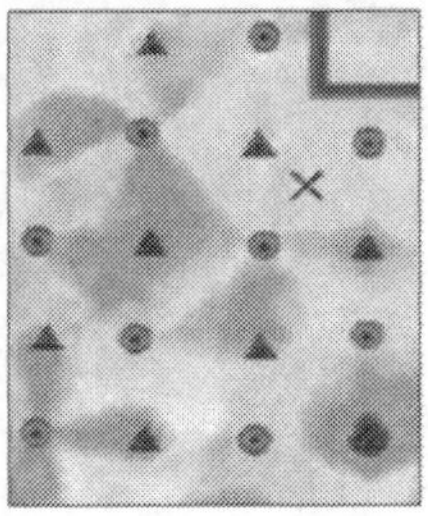

Predictive Simulation Results (CO_2 plumes)

EERC

Critical Challenges. Practical Solutions.

First Repeat 4-D Difference Interpretation

1. Utilization of Resources
2. Technical Approach
3. Rate of Progress
4. Potential Risks Considered
5. Plan to Complete

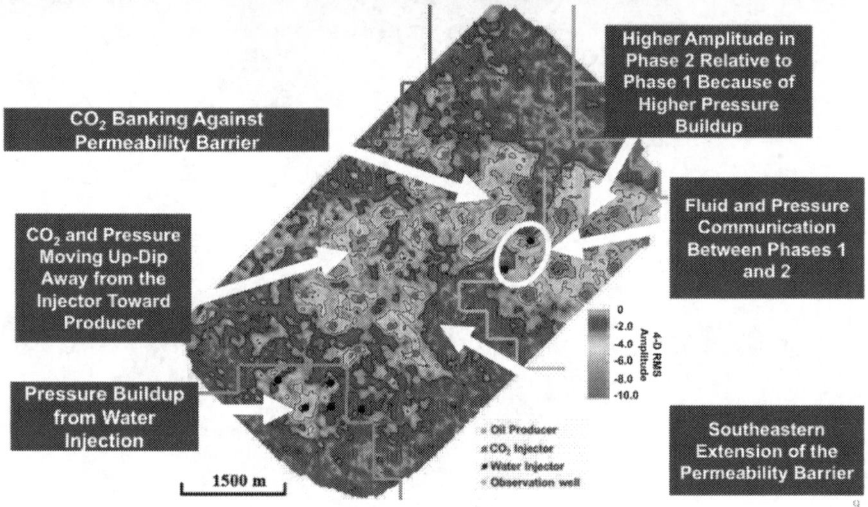

9

MVA-GUIDED SIMULATION

- Pressure and fluid communication revealed by 4-D seismic helped explain Phase 1 model history-matching issues.
- A combined Phase 1 and 2 was developed in response.

WAG, 1 HCPVI

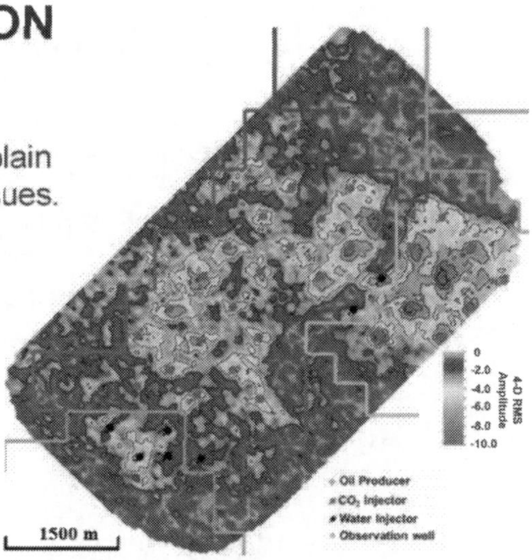

EERC

Critical Challenges. **Practical Solutions.**

LESSONS LEARNED/KEY FINDINGS

- MVA techniques are applicable to both EOR and saline storage in any region.
- The region has outstanding potential for storage opportunities.
- Adaptive management approach is readily applicable to future geologic CO_2 storage projects in this region and others.
- PCOR Partnership provided platform to test various techniques and technologies. A significant time gap can exist between data collection and analysis and integration.

11

MVA EVOLUTION

- Faster processing for quicker integration
  - Improve performance predictions
  - Inform operational decisions with actionable results
- Intelligent monitoring
- Low environmental impact
- No impact on operations
- Semiautonomous and scalable
- Viable and cost-effective long term

SASSA (SCALABLE, AUTOMATED, SEMIPERMANENT SEISMIC ARRAY)

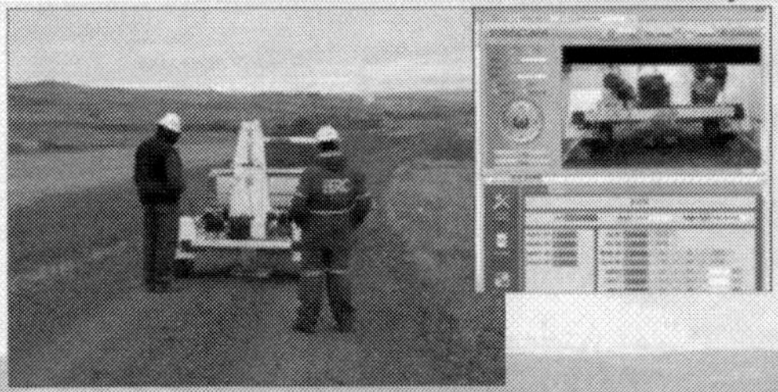

KRAUKLIS WAVE (K-WAVE)

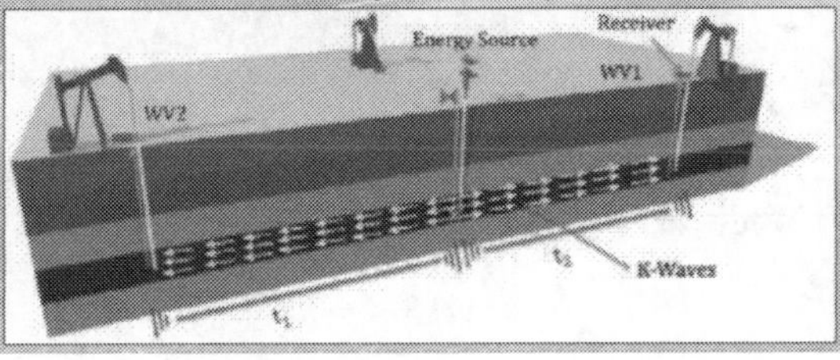

INTELLIGENT MONITORING SYSTEM (IMS)

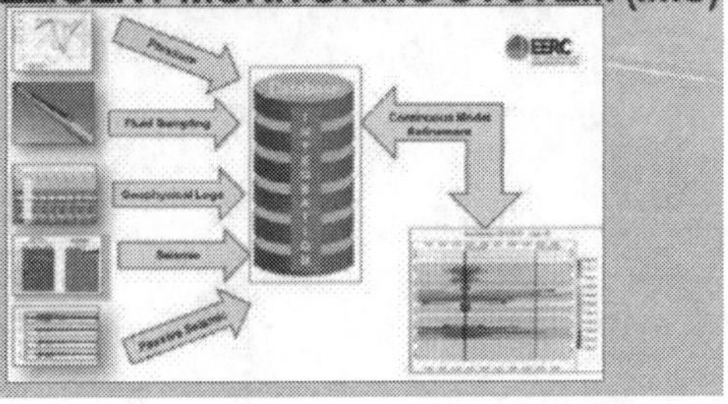

12

BRINE EXTRACTION AND STORAGE TEST (BEST)

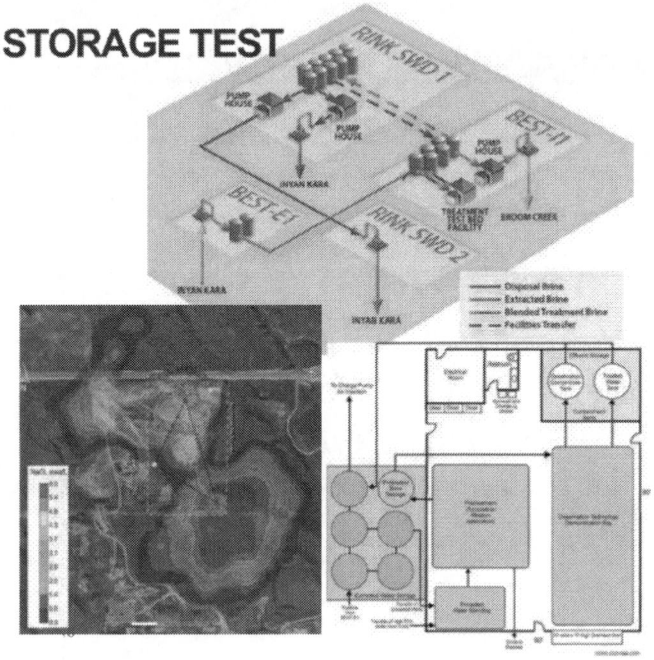

- Proxy for managing and measuring pressure and CO_2 plume movement associated with future saline CO_2 storage projects.
- **Phase I:** Development of a field implementation plan for a BEST in North Dakota (complete).
- **Phase II:** Field demonstration and validation of active reservoir management (ARM) strategies and extracted brine treatment technologies (under way).

EERC

ND INTEGRATED CARBON STORAGE COMPLEX FEASIBILITY STUDY (CARBONSAFE PHASE II)

Objective and Project Partners

- Determine the feasibility of developing a commercial-scale CO_2 storage complex able to store 50+ million metric tons of CO_2 in central North Dakota safely, permanently, and economically.

U.S. DEPARTMENT OF ENERGY | NATIONAL ENERGY TECHNOLOGY LABORATORY

ALLETE clean energy

INDUSTRIAL COMMISSION OF NORTH DAKOTA
LIGNITE RESEARCH COUNCIL

BNI energy — AN ALLETE COMPANY

BASIN ELECTRIC POWER COOPERATIVE

Prairie Public

Minnkota Power Cooperative, Inc.

Schlumberger Carbon Services

CMG

North American Coal Corporation

EERC

Critical Challenges. **Practical Solutions.**

ND CARBONSAFE PROJECT SUMMARY

- DOE's CarbonSAFE* program aims to develop integrated CO_2 storage complexes, constructed and permitted for operation in the 2025 time frame over a series of development phases.

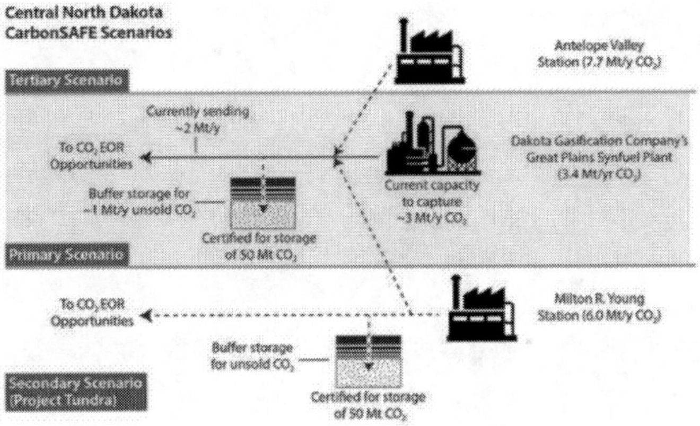

*CarbonSAFE = Carbon Storage Assurance and Facility Enterprise

EERC

Critical Challenges. **Practical Solutions.**

NEBRASKA INTEGRATED CARBON STORAGE CAPTURE PRE-FEASIBILITY STUDY (CARBONSAFE PHASE I)

- **Goal**: Determine conceptual feasibility of CO_2 capture from Nebraska Public Power District's Gerald Gentlemen Station (GGS) with geologic storage.
- **Objectives**
  - Establish CCS coordination team
  - Identify challenges to commercial-scale CCS in western Nebraska, and develop mitigation plan
  - Conduct conceptual evaluations:
    - Western Nebraska subsurface for geologic storage
    - CO_2 emissions from GGS for potential capture

U.S. DEPARTMENT OF ENERGY | NATIONAL ENERGY TECHNOLOGY LABORATORY

Nebraska Public Power District
Always there when you need us

Schlumberger Carbon Services

CMG COMPUTER MODELLING GROUP LTD

EERC

Critical Challenges. **Practical Solutions.**

CONTACT INFORMATION

Energy & Environmental Research Center
University of North Dakota
15 North 23rd Street, Stop 9018
Grand Forks, ND 58202-9018

www.undeerc.org
701.777.5193 (phone)
701.777.5181 (fax)

Neil Wildgust
Principal CCS Scientist
nwildgust@undeerc.org

EERC

17

Critical Challenges. **Practical Solutions.**

497196 Defining CO_2 Storage Options in the Upper Ohio River Valley – Advanced Characterization of Geologic Reservoirs and Caprocks Authors: Caitlin Mcneil (Presenter

Abstract:

Development of Carbon Dioxide Capture, Utilization, and Storage (CCUS) technologies represents one of the most important components in the portfolio of options for greenhouse gas reduction. The emerging international consensus under mechanisms such as the Paris Accord and the possibility of future federal or state regulations for limiting CO_2 emissions essentially rule out continued use of coal as a power source in the future without implementation of CCUS technology. The upper Ohio River Valley region including eastern Ohio has a strong history of coal production and a dependence on coal for its industrial base, which includes power generation and steel manufacturing. However, the potential reservoirs and containment zones for CO_2 storage in the area are very poorly understood due to lack of deep well data. Thus, knowledge of capacity, containment security, and injectivity is rudimentary at best and a sustained effort is needed to map the deep geology and quantify the storage potential in the region. This project attempts to start developing a CO_2 storage framework using pre-existing and new data on potential storage reservoirs and containment intervals to estimate CO_2 storage resources in eastern Ohio. The geologic data were compiled from oil and gas well logs and cores available at the Ohio Division of Geological Survey, purchase of seismic data, and through collaboration with brine-injection well operators in eastern Ohio to collect advanced well logs and operational data. The available information was used to construct cross-sections and maps of formations depths, thickness, and porosity across the study area. Based on these maps, the prospective storage resources were estimated. The porosity and limited available permeability data were used to develop preliminary numerical reservoir models in selected areas to evaluate the feasibility of sustained injection. Finally, caprocks were evaluated using geomechanical and reservoir modeling methods to determine their effectiveness for containing injected CO_2. The project findings were disseminated to the stakeholders at various venues.

The overall findings of the work show that there is significant CO_2 storage potential in eastern Ohio based on porosity estimates. This is also proven by the presence of a number of high-performance injection wells in the region. The storage resource is vertically diversified into a number of geologic layers, both carbonates and sandstones, and is laterally heterogeneous. This heterogeneous nature means that advanced characterization and sustained geologic characterization using advanced tools is needed to improve confidence in storage estimates and long-term injectivity. The assessment also indicates that the primary caprock for the Cambrian-Ordovician reservoirs in the area has appropriate characteristics to retain the injected CO_2. In this study, the focus of the caprock analysis was limited to assessing the sealing effectiveness and mechanical integrity of the two geologic units immediately overlying the reservoirs [i.e., immediately overlying the Knox Unconformity], the Black River Group and the Wells Creek Formation, even though there are hundreds to thousands of feet of additional low permeability limestone and shale (Trenton Formation, Point Pleasant/Utica) above these units. Thus, the caprock analysis is conservative because it doesn't take into consideration the entire caprock system.

This project has been conducted under grant agreement D-13-22 funded by Ohio Development Services Agency's Ohio Coal Development Office and the Midwestern Regional Carbon Sequestration Partnership (MRCSP) led by Battelle with US Department of Energy funding. The project also benefited significantly through collaboration with a number brine injection well operators, who allowed access to their wells for collection of critical geologic characterization and injection operation data. The present project complements a number of past and ongoing efforts to prove CO_2 storage and utilization options in Ohio and surrounding regions.

499289 Impact of Brine/CO_2 Exposure on the Transport and Mechanical Properties of Mt Simon Rock Samples

Abstract:

When sandstone rocks are exposed to CO_2-saturated brine their transport and mechanical properties can change due to brine/CO_2-induced chemical reactions. The present study investigates the change in the flow-through characteristics, porosity and the deformation of Mt Simon rock samples caused by exposure to brine/CO_2. The subsamples, extracted from the Mt. Simon sandstone, were first characterized for their mechanical and transport properties, and then aged in CO_2-saturated brine at pressures of 2500 psi and temperature of 50â°C for one week. Following that, tests were carried out to measure the change in transport and mechanical properties of the rock. The changes in porosity were measured by the helium porosimetry method. The alteration in pore size distribution and flow-through characteristics were determined by the flow permporometry method. The changes in the brine composition were measured by ion chromatography. In addition, a brine-saturated Mt. Simon sandstone core was aged in a specialized core holder while the deformation of the core during CO_2 invasion was detected in situ via an array of strain gauges.

498620 Assignment and Calibration of Relative Permeability By Hydrostratigraphic Units: A Novel Approach for Multiphase Flow Analyses; Case Study Example: CO_2-EOR Operations at the Farnsworth Unit, Texas

Abstract:

Among the most critical factors for geological CO_2 storage site screening, selection and operation is effective simulation of multiphase flow and transport. Relative permeability is probably the greatest source of potential uncertainty in multiphase flow simulation, second only to intrinsic permeability heterogeneity. The specific relative permeability relationship assigned greatly impacts forecasts of CO_2 trapping mechanisms, phase behavior, and long-term plume movement. A primary goal of this study is to evaluate the impacts and implications on CO_2-EOR model forecasts of different methods of assigning relative permeability relationships.

We applied a novel approach to assigning relative permeability relationships in numerical models. In previous research, most models assign relative permeability functions on the basis of geologic formation or rock type. In this study, we assign and calibrate relative permeability by hydrostratigraphic units, extending the seminal work of Maxey (1964) to multiphase flow. Ongoing work by the SWP has identified distinct regions that are believed to exhibit similar flow characteristics using the Improved Hydraulic Flow Units (IHFU) method. Core from each of these flow regions are being scanned by micro-computer-tomography (micro-CT) and relative permeability testing will be completed when the scans are finished. Specific, calibrated relative permeability relationships will be assigned to each hydrostratigraphic unit. Results of forward simulations with the newly-calibrated models will be compared to those of existing models that utilize a single relative permeability relationship for the whole model domain.

The study site is the Farnsworth Unit (FWU) in the northeast Texas Panhandle, an active CO_2-EOR operation. The target formation is the Morrow â€˜Bâ€™ Sandstone, a clastic formation composed of medium to course sands. The study was undertaken as part of the Southwest Regional Partnership on Carbon Sequestration (SWP) under the phase III project.

Development and Evaluation of an Iridium Oxide Based Chemical Sensor for Downhole CO_2 Monitoring—Part II

Sai Wang, University of North Dakota--Department of Petroleum Engineering;

Kegang Ling, University of North Dakota--Department of Petroleum Engineering;

Hongsheng Wang, University of North Dakota--Department of Petroleum Engineering;

Ning Liu, University of Louisiana at Lafayette—Department of Petroleum Engineering;

Yun Ji, University of North Dakota--Department of Chemical Engineering

Abstract

Geological carbon sequestration represents a long-term storage of CO_2, in which large-scale CO_2 is injected into the subsurface geologic formations. The in-situ CO_2 monitoring approach is highly desirable during the storage process. Compare with other monitoring technologies, a preferred method for in-situ underground CO_2 migration monitoring is to install down-hole CO_2 sensors directly in the underground. This paper presents the design and construction process of the iridium oxide based chemical sensor, including the IrO_2 working electrode electrodeposition with iridium oxide nanoparticles, Ag/AgCl reference electrode fabrication and the Severinghaus-type CO_2 sensor preparation. Then the sensor was tested under high pressure environment. Under this condition, the sensor displayed a very good CO_2 sensing performance and a linear relationship was obtained between the sensor's response potential and the Log[C_{CO2}]. In order to evaluate the CO_2 sensor performance in the carbon sequestration process, the sensor was settled in the simulated CO_2 storage condition. The results indicated that this chemical sensor could provide the CO_2 movement monitoring data continuously, timely and accurately.

Introduction

In recent decades, the CO_2 geologic storage has been considered as a feasible means for reducing anthropogenic CO_2[1, 2]. The fundamental logic of the geological storage of CO_2 is to inject the supercritical CO_2 at the storage depths, which need the CO_2 to keep on either supercritical or in solution in the water/brine present in the formation. The injected CO_2 is expected to stay in the subsurface formations for a long period. However, there are aspects of geologic storage which need to take further study, particularly in regard to the public concerns of CO_2 migration and possible leakage to surface. Since then, an in-situ CO_2 monitoring approach is highly desirable for this purpose. At present, a broad range of technologies based on geophysical, microbiological, geoelectrical and geochemical principle were exploited [3]. While due to the various disadvantages of those technologies, such as the largely time consuming or costly requirement, they were not identified as the potential methods for the application.

Because the chemical reaction between CO_2 and H_2O are pH dependent, which is transforming the CO_2 detection to the pH measurement, a preferred method of monitoring in-situ CO_2 movement and migration is to directly install the CO_2 sensors at storage depth[4]. The sensor must hold the ability to withstand harsh environmental conditions especially under high pressure circumstance. This paper describes the design and construction process of the iridium oxide based chemical sensor which fulfill the harsh environmental condition requirements. Differed from the traditional sensor, this type of the CO_2 sensor used an iridium-oxide electrode instead of a glass electrode to improve the resistant ability for the harsh downhole environments encountered. The developed CO_2 sensor includes a couple of iridium-oxide electrodes (To meet this purpose, iridium oxide nanoparticles was electrodeposited on the surface of gold substrate), a thin gas-permeable silicone membrane, a porous metal supporting material, and the bicarbonate-based electrolyte solution. With conducting several tests in different scenarios, the sensor was proved as a potential technology which can provide the CO_2 movement monitoring data continuously, timely and accurately.

Experimental

1. Sensor Preparation

1.1 Electrode Fabrication

To prepare electrodeposition solution, alkaline hydrolysis of iridium(IV) salt method was used to synthesize iridium oxide nanoparticles [5, 6]. 2.0 mM aqueous K_2IrCl_6 solution was adjusted to pH 13 with 10 wt% of aqueous NaOH to obtain a yellow solution, which then was heated at 90 °C for 20 minutes. The resulting solution was kept in an ice bath to obtain a blue solution. The cold solution was then adjusted to pH 1 by rapidly adding 3 M HNO_3 and stirred continuously for 80 minutes until the solution became deep blue. The solution was then stored in a refrigerator at 2 °C and could be adjusted to pH 1-13 by addition of 10 wt% of aqueous NaOH solution.

Then the as-prepared working electrode is electrochemically deposited with iridium oxide nanoparticles as the deposition solutions. The electrodeposition protocol was carried with cyclic voltammetry[7]. The experiments were carried out with Solartron Analytical 1470E CellTest System with Solartron Analytical 1242A frequency response analyzer. In this work, a three-electrode cell system was used where an iridium oxide layer was deposited from the deposition solution on the metal substrate working electrode by using an Ag|AgCl and Pt as reference and counter electrode, respectively. A triangular wave cyclic voltammetry method between -0.5 to 0.65 V (vs. Ag|AgCl) was applied at 200 mV/s of scan rate for 500 cycles.

Next, an electrochemical method was applied for the reference electrode preparation[8]. The bare silver wire was electroplated in 0.1-M HCl electrolyte solution to form a thin coating of silver chloride on the silver wire surface. Along this route the silver wire was used as the anode while the bare copper wire was treated as the cathode. Finally, this silver/silver chloride electrode was insulated by the plastic heat shrank tubing.

1.2 Sensor construction

For the sensor applied under high pressure conditions and simulated reservoir conditions, the construction procedure was described as follows. Firstly, the working electrode and reference electrode were fixed through a stainless steel cap and then stabilized by silicone sealant. The stabilized electrodes were dried for two days under air condition at room temperature (25 ± 2°C). Then a stain steel porous filter element was settled for the function of supporting the whole frame, which means that it can enable the sensor to work under harsh downhole environment, especially under high pressure condition. What followed is the above stabilized electrodes were inserted into the support unit where inside being full of bicarbonate-based internal electrolyte solution ($NaHCO_3$-NaCl electrolyte solution). Finally, as the above mentioned sealing method, both outlets of the gas-permeable membrane were blocked off by the silicone sealant. Meanwhile, some other tiny materials were applied in the sensor manufacture for the purpose of support and potential transfer.

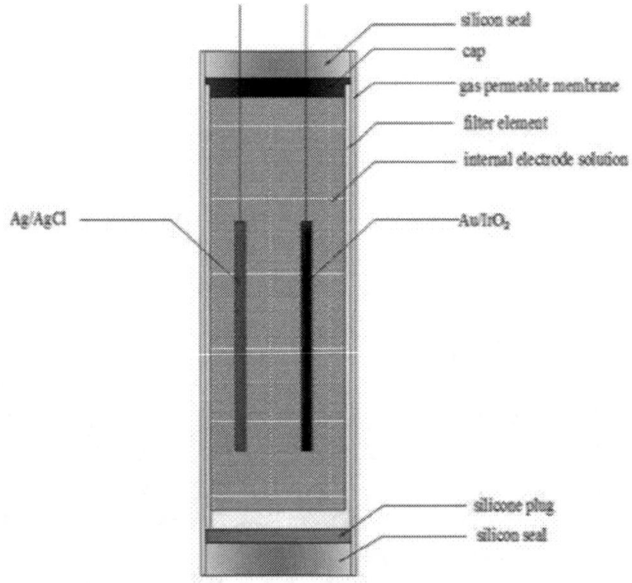

Fig. 1 Schematic of the CO_2 sensor[9]

2. *Sensor test under high pressure condition(2000psi)*

The following steps were designed to investigate the CO_2 sensor's performance. All potentiometric measurements were carried out at room temperature of 20°C±2°C. Firstly, the CO_2 sensor was settled in the stainless steel tank. Then the tank was injected with 0.2M citric acid and tightly sealed. The stainless steel tubing and the connecting wires from the sensor came out from the top of the tank. The ISCO pump was used for controlling the pump injection and the tank pressure. The connecting wires of the sensor were connected with a 34401A 61/2 digit multi-meter. Before carrying out the tests, the base potential of the CO_2 sensor was measured by maintaining the CO_2 sensor into the citric acid solution under a certain pressure. After the potential reached a stable value, the certain amount of the $NaHCO_3$ solution (0.1M) was injected in the tank and then the tank was sealed immediately. Finally, the ESCO pump began to increase the tank pressure to 2000 psi. The sensor output signal was measured by the digit multi-meter.

3. *Sensor performance under simulated reservoir conditions*

Lab-scale core flooding test was conducted to evaluate the performance of the CO_2 chemical sensor in CO_2 geological storage process. A sandstone core was applied to simulate the storage process of CO_2 in a sandstone saline aquifer. After the sensor was installed in the core, 2.0% brine was pumped into the core with flow rate of 100 ml/h until 1 pore volume was achieved. The injection pressure was around 2000 psi. Then CO_2/brine was injected into the core with the phase ratio of CO_2/brine = 1:1 and total flowrate = 100 ml/h. The response of the CO_2 sensor during this process was recorded.

Results and Discussion

1. Working Principle of the CO_2 sensor

The pH is a critical parameter need to be evaluated in the CO_2 sequestration process. Generally, the pH is defined as the degree of acidity or basicity of the solution. In its more common explanation, the pH was defined as the minus logarithm of the concentration of H^+ [10], which expressed as Eq.1.

$$\mathbf{pH = -log[H^+]} \qquad [1]$$

During the CO_2 injection process, some CO_2 molecules could dissolve into the brine aquifer and change the pH from neutral to acid dramatically. Eq.2 describes the hydrolysis of CO_2. The CO_2 molecules react with H_2O molecules to generate carbonic acid. The carbonic acid is not stable and then dissociates into bicarbonate and H^+ subsequently. As a result, the produced H^+ will decrease the pH of the surrounding circumstance. From Eq. 2, the concentration of CO_2 is proportional to the concentration of hydrogen ion.

$$\mathbf{CO_{2\,(aq)} + H_2O \leftrightarrow H_2CO_3 \leftrightarrow H^+ + HCO_3^-} \qquad [2]$$

From Eq. 2, the concentration of CO_2 is proportional to the concentration of hydrogen ion. Where the transformed equation was generated as Eq.3,

$$\mathbf{[CO_2] = k\,[H^+]} \qquad [3]$$

The pH measurement uses the electrodes to transfer the chemical activity of the H^+ into the electronic signal, which is the potential differential. Inevitably the well-known Nernstian function [11] then was substituted into the pH measurement. This equation could be described as:

$$\mathbf{E = E^0 + (2.303RT/F)\log[H^+] = E^0 + 59.15\log[H^+]} \qquad [4]$$

Where,

E^0 =Standard electrode potential	R = Universal gas constant, 8.314 472 J /K/mol
T = Absolute temperature	F = Faraday constant

By combining Eq. 3 and Eq.4, the potential differential between the electrodes can be described as Eq. 5. This generated equation reveals that there is a linear relationship between ΔE and the logarithm of the CO_2 concentration [12].

$$\Delta E = E - E^0 = 2.303RT/F * \log[CO_2] \quad [5]$$

2. ***Potential response characteristics of the CO_2 sensor at 2,000 psi***

Fig. 2 shows the potential differential curves for the CO_2 sensors to changes of the CO_2 concentration from 0.001 M to 0.006 M at room temperature and 2000 psi. At the pressure of 2000 psi, the CO_2 sensor showed very similar sensing behavior except for the initial stabilized potential response. It can be found that the potential differential is directly proportional to the CO_2 concentration. The response time of the CO_2 sensor was found to be dependent on the CO_2 concentration. The response time was 6295 seconds for the highest CO_2 concentration of 0.006 M, while a little faster response was observed at low concentration ranged from 0.001 M to 0.004 M.

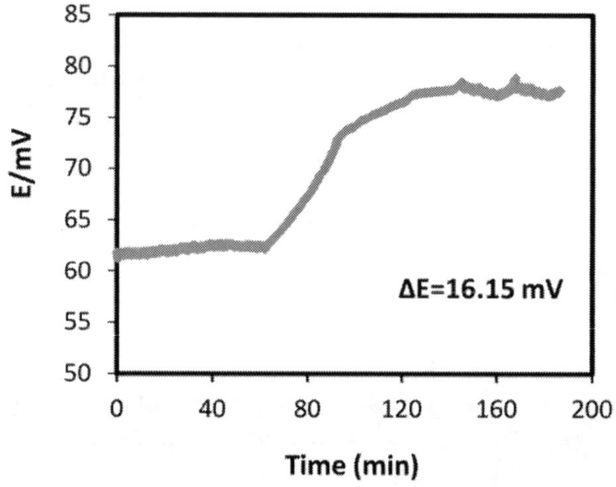

(a) 0.001 M

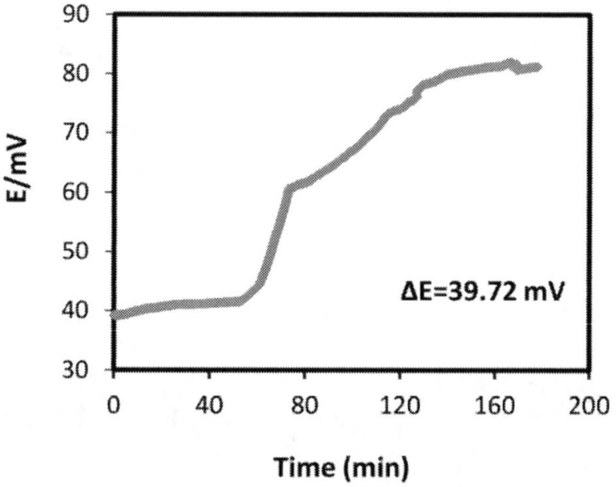

(b) 0.002 M

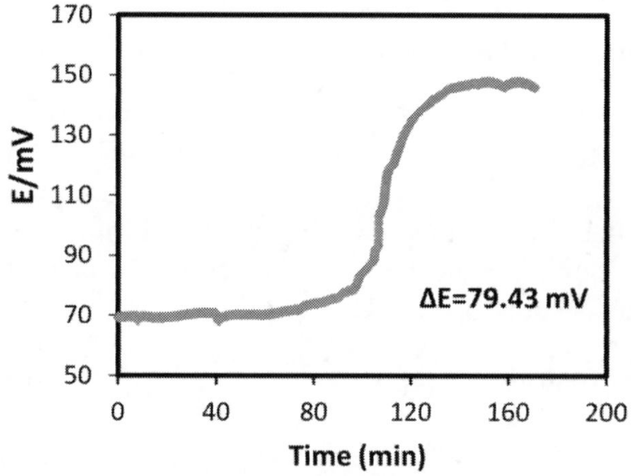

(c) 0.004 M

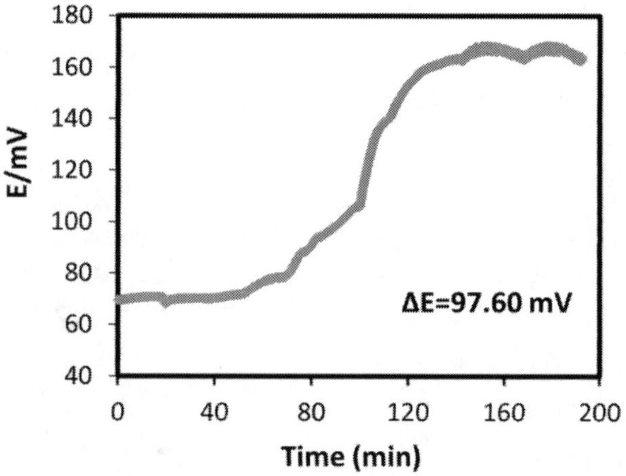

(d) 0.006 M

Fig. 2 Potential response of CO_2 sensor with the addition of $NaHCO_3$ in the 0.2 M citric acid-HCl solution at 2,000 psi, [$NaHCO_3$] is (a) 0.001M, (b) 0.002 M, (c) 0.004 M and (d) 0.006 M.

Fig. 3 illustrates the sensitivity curves of the CO_2 sensor at room temperature and 2000 psi. The linear relationship between the sensor response potential differential and the logarithm of the CO_2 concentration are quite good in the tested range of 0.001M to 0.006M. The slope of the CO_2 sensor versus Log [CO_2] is 107.5 mV/Log [CO_2], and the correction coefficient is 0.9889.

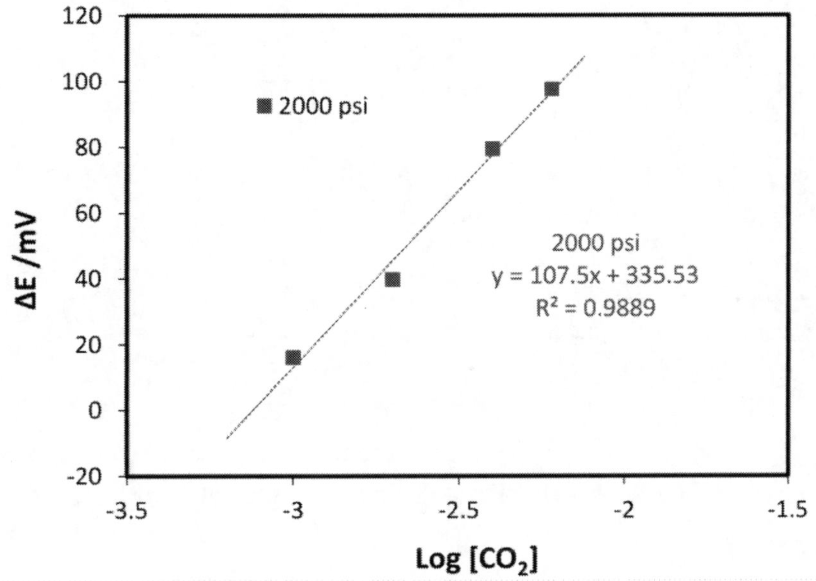

Fig. 3 Calibration curves of the CO_2 sensor tested at room temperature and 2,000 psi.

3. CO_2 sensor performance under simulated reservoir conditions

The results demonstrate that, with one pore volume of 2.0% brine was pumped into the flooding system, the output signal of the sensor was keep constantly around 42 mV. Then, when CO_2 started to inject into the core, the output potential was increased to 241 mV. At the beginning, the output potential of the sensor kept roughly constant before the CO_2 plume arrived. When the injected CO_2 arrived the sensor's location, due to the CO_2 molecule diffusion, the potential started to increase until reach the equilibrium[9, 13]. The output potential change of the sensor is 200 mV.

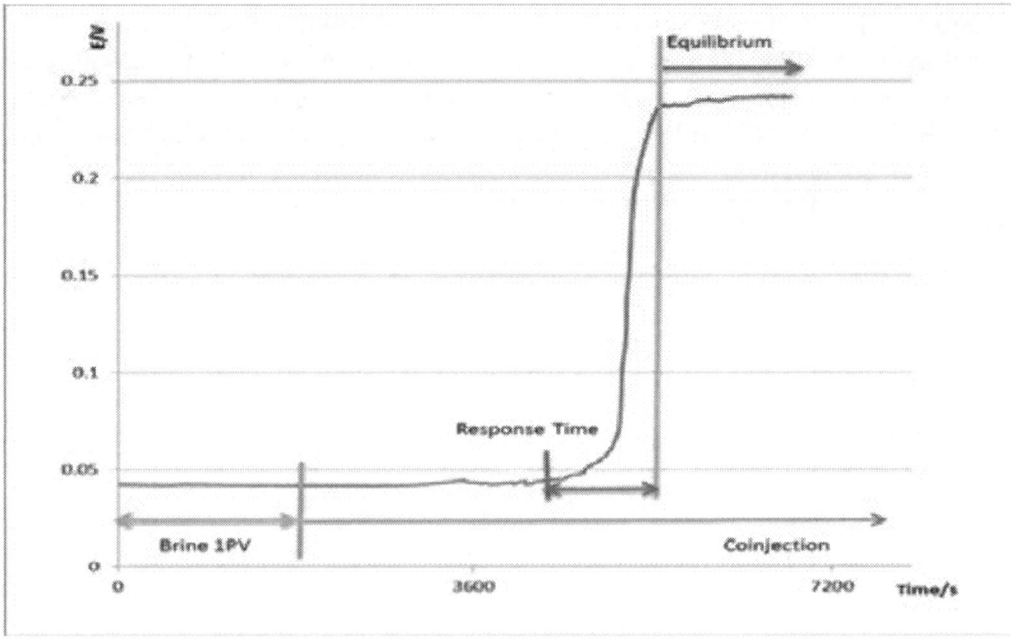

Fig. 4 Potential response of CO_2 sensor vs. Time. Brine injection rate: 100mL/h; Co-injection rate: 100 mL/h

Conclusions

An iridium oxide-based electrode was prepared by electrodeposition of the iridium oxide nanoparticles on a metal substrate with cyclic voltammetry. Then the prepared electrode was combined to develop a downhole CO_2 sensor which can provided the sensing performance under high pressure condition. With displaying Nernst performance, the sensor performed a good linear relationship between the sensor response potential and the logarithm of the CO_2 concentration. Lab-scale core flooding test demonstrated that the CO_2 sensor can measure the CO_2 movement under the reservoir condition, especially for the high pressure conditions. Also, the results indicated that the Severinghaus-type CO_2 chemical sensor may reserve the potential application in the geological carbon storage.

Acknowledgement

The authors are grateful to The Petroleum Engineering Department of University of North Dakota. This research is supported in part by the funding from Vice President for Research & Economic Development program.

Reference

[1] Allen, D.E., Strazisar, B.R., Soong, Y., Hedges, S.W., Modeling carbon dioxide sequestration in saline aquifers: significance of elevated pressures and salinities. Fuel Process. Technol. 86, 1569–1580. 2005.

[2] Wollenweber, J., Alles, S., Busch, A., Krooss, B.M., Stanjek, H., Littke, R., Experimental investigation of the CO_2 sealing efficiency of caprocks. Int. J. Green h. Gas Control 4, 231–241.2010.

[3] Giese R., Henniges J., Lüth S., Morozova D., Schmidt-Hattenberger C., Würdemann H., Zimmer M., Cosma C., Juhlin C., Monitoring at the CO_2 Sink Site: A Concept Integrating Geophysics, Geochemistry and Microbiology. Energy Procedia, 1 (1): 2251–2259, 2009.

[4] Ning L, Yang. L., Li L.X., Grigg R., McPherson B., Robert. L., Development of A CO_2 Chemical Sensor and Sensor Strategy for CO_2 Monitoring in Carbon Sequestration. Seventh Annual Carbon Capture and Sequestration Conference, 2008, Pittsburgh, PA.

[5] Zhao, Y., Vargas-Basbosa, N.M., Hernandez-Pagan, E.A., Mallouk, T.E., Anodic deposition of colloidal iridium oxide thin films from hexahydroxyiridate(IV) solutions. Small, 14, 2087-2093. 2011a.

[6] Zhao, Y., Hernandez-Pagan, E.A., Vargas-Barbosa, N.M., Dysart, J.L., Mallouk, T.E., A high yield synthesis of ligand-free iridium oxide nanoparticles with high electrocatalytic activity. Journal of Physical Chemistry Letters, 2, 402-406. 2011b.

[7] Munawar Khalil, Sai Wang, Jianjia Yu, Robert L. Lee, and Ning Liu, Electrodeposition of Iridium Oxide Nanoparticles for pH Sensing Electrodes. Journal of the Electrochemical Society, 163 (9) B485-B490 (2016).

[8] Wen-Ding H., Hung C., Sanchali D., Mu C., A Flexible pH Sensor Based on the Iridium Oxide Sensing Film. Sensors and Actuators A, 169: 1– 11, 2011.

[9] Sai W., Kegang L., Ning L., Hongsheng W., Development and Evaluation of an Iridium Oxide Based Chemical Sensor for Downhole CO_2 Monitoring--Part I, Carbon Management Technology Conference, 17-20 July 2017. Houston, Texas, USA,.

[10] Skoog D.A., West D.M., Holler F.J., Fundamentals of Analytical Chemistry, 7th ed., Philadelphia, PA: Saunders College Publishing, 1996.

[11] Orna M. V., Stock J., Electrochemistry, Past and Present. Columbus, Ohio: American Chemical Society, 1989. ISBN: 0-8412-1572-3.

[12] Sai Wang, Jingshan San, Ning Liu, A Chemical CO_2 Sensor Monitoring CO_2 Movement Under Reservoir Conditions, Carbon Management Technology Conference, 17-19 Nov,2015, Sugarland, Texas, USA

[13] Sai Wang, Jingshan San, Jianjia Yu, Robert Lee, Ning Liu, A Downhole CO_2 Sensor to Monitor CO_2 Movement In-situ for Geologic Carbon Storage, Internal Journal of Greenhouse Gas Control. 2016.

498673 Monitoring Carbon Sequestration Using Charged Wellbore Controlled Source Electromagnetics and Integrated Reservoir Models

Abstract:

Carbon capture and sequestration (CCS) can provide an effective means by which to reduce global CO2 levels. To ensure effectiveness and safety, regulatory frameworks mandate 99% storage permanence over the lifetime of a carbon sequestration project. As such, robust monitoring, verification, and accounting (MVA) technologies are needed in order for large scale CCS to be realized. Many existing MVA measurements require the costly drilling of boreholes and installation of downhole sensors. Borehole data provide minimal spatial information and their expense is prohibitive against the installation of dense monitoring grids. Surface based measurements can detect massive leaks but do not respond to changing reservoir conditions

Geophysical imaging techniques could theoretically provide this needed monitoring information if sufficient physical property change can be attributed to CO2 injection. The density of supercritical CO2 is a function of reservoir conditions, but is often similar to brine. As such gravity and 4D seismic monitoring will often fail to detect appreciable signal variation with injection activities. Seismic attributes sensitive to subsurface pressure changes can be utilized, but the expense of repeated seismic data acquisition is prohibitive for long-term monitoring. In most cases, CO2 is injected into deep saline aquifers or depleted oil and gas reservoirs. The supercritical CO2 will be much less electrically conductive than the brine which it displaces and mixes with. Therefore, electrical or electromagnetic methods, which are sensitive to electrical conductivity, are a promising techniques for monitoring CCS activities.

Many options exist for electrical survey configurations. The desired characteristics for this application are low cost—requiring minimal new infrastructure—and sufficient depth of investigation. In offshore exploration controlled source electromagnetics (CSEM) are increasingly relied upon as a de-risking tool. In this case the hydrocarbons are the resistive target within the conductive brine. We leverage this experience by applying CSEM to the monitoring of an active onshore enhanced oil recovery carbon capture and sequestration project (EOR-CCS) project by utilizing well casings as deep-reaching transmitting electrodes. Electrical current can be injected into the formation of interest by utilizing legacy wellbore casings that penetrate the injection reservoir as long deep electrodes. By monitoring electrical and magnetic fields on the surface, information about saturation changes in the subsurface can be determined through inversion and interpretation of the data. Two major advantages of the method is the ease of access on the surface and the ability to inject current to the depth of reservoir formation.

In order to improve the resolution and reliability of the quantitative interpretation, we develop a methodology for tightly integrated multiphase reservoir flow simulations with the geophysical data inversion. Electrical conductivity can be related to phase saturation using relations such as Archie's Equation, which will be calibrated and verified using auxiliary data including production and injection records. This multiphysical approach is designed to integrate structural, physical property, and reservoir simulations systematically. Structural information is retained from the reservoir models and associated characterization efforts, and incorporated into the inversion of the CSEM data as geometrical constraints. Furthermore, physical property information derived from petrophysical data, reservoir porosity, and reservoir simulation-produced fluid saturations will be incorporated into the inversion as direction constraints.

In this paper, we will present the preliminary study on the site selection, numerical simulations consisting of reservoir simulations, generation of electrical conductivity models, and electromagnetic-modeling based EM survey design, and preliminary analyses of data acquired in the baseline field survey.

488989 Development of LIBS Sensor for Sub-Surface CO_2 Leak Detection in Carbon Sequestration

Abstract:

To ensure the success of USDOEâ€™s CCS program it is important that the carbon dioxide that is injected underground for storage remains there permanently (99% permanence over 1000 years). In fact, the integrity of the storage can be ensured by monitoring for CO_2 leakage, and injection sites are the logical places to perform such monitoring. The sub-surface environment is notoriously harsh, with large potential mechanical, thermal, and chemical stresses, making long-term stability and survival a challenge to any potential in situ monitoring method. The laser induced breakdown spectroscopy (LIBS) has been demonstrated as a promising technology for chemical monitoring of harsh environments and hard to reach places. LIBS has a real-time monitoring capability and can be used for the elemental and isotopic analysis of solid, liquid, and gas samples. The flexibility of the probe design and the use of fiber-optics has made LIBS particularly suited for remote measurements. Our experiments using a stainless steel high-pressure cell demonstrated that the LIBS is a suitable technique to perform measurements in high-temperature and high-pressure (HTHP) environment, which makes it attractive for downhole measurements. The discussion will include the progress in fabricating a compact, robust, and simple LIBS sensor and distribution of many laser spark sources across a wide area for widespread subsurface leak detection.

AUTHOR INDEX

AUTHOR INDEX

AUTHOR INDEX

AUTHOR INDEX

AIChE
120 Wall Street, FL 23
New York, NY 10005-4020

ISBN 978-1-5108-5778-0

CURRAN ASSOCIATES INC.
proceedings.com